Recombination Variability and Evolution

Recombination Variability and Evolution

ALGORITHMS OF ESTIMATION AND POPULATION-GENETIC MODELS

A.B. Korol
Institute of Evolution,
University of Haifa, Israel

I.A. Preygel
Genetic Therapy, Inc.,
Gaithersburg, USA

and

S.I. Preygel
Oncor, Inc., Bethesda, USA

CHAPMAN & HALL
London · Glasgow · Weinheim · New York · Tokyo · Melbourne · Madras

Published by Chapman and Hall, 2–6 Boundary Row, London SE1 8HN, UK

Chapman and Hall, 2–6 Boundary Row, London SE1 8HN, UK

Blackie Academic and Professional, Wester Cleddens Road, Bishopbriggs, Glasgow G64 2NZ, UK

Chapman and Hall GmbH, Pappelallee, 3, 69469 Weinheim, Germany

Chapman and Hall USA, One Penn Plaza, 41st Floor, New York NY 10119, USA

Chapman and Hall Japan, ITP-Japan, Kyowa Building, 3F, 2-2-1 Hirakawacho, Chiyoda-ku, Tokyo 102, Japan

Chapman and Hall Australia, Thomas Nelson Australia, 102 Dodds Street, South Melbourne, Victoria 3205, Australia

Chapman and Hall India, R. Seshadri, 32 Second Main Road, CIT East, Madras 600 035, India

First edition 1994

Adapted from Russian language edition – *Variability of Crossingover in Higher Organisms: Methods of analysis and Population Genetic Models* – 1990, A.B. Korol, I.A. Preygel and S.I. Preygel.
Published by Shtiintsa Press, Kishinev, Moldova

Typeset in 10/12 Photina by Thomson Press (India) Ltd, New Delhi, India
Printed in Great Britain by T.J. Press (Padstow) Ltd, Padstow, Cornwall

ISBN 0 412 49410 8

A catalogue record for this book is available from the British Library

Library of Congress Catalog Card Number: 94-70270

♾ Printed on acid-free text paper, manufactured in accordance with ANSI/NISO Z39.48-1992 and ANSI/NISO Z39.48-1984 (Permanence of Paper).

Contents

Preface

It is no exaggeration to say that the recombination studies initiated Morgan school at the turn of this century determined the general lines which the development of genetics was to proceed for many years to c With time, the focus passed to the problem of mutations which affect relations of genetics with other biological disciplines, in particular with evolutionary theory and breeding. In early works on the synthetic theory of evolution, the role of recombination as a source of heritable variation was ignored (Mayr, 1980). Later, the situation changed as a result of the studies of Darlington, Dobzhansky, Huxley, Mayr, Mather and others. Nevertheless, many evolutionary conclusions were based on fairly naive notions of complete randomness of recombination variability, going back to the concept of 'bean-bag genetics'. Recombination was viewed as a purely mechanistic process ensuring reshuffling of genes in heterozygotes.

Many types of recombination are known, from recombination of whole chromosomes (the basis of Mendel's law of independent assortment of genetic factors) to crossing over, gene conversion, transformation and transduction. The study of genetics has accumulated abundant evidence on both the peculiarities and common features of these processes in various organisms, on their genetic control and on the molecular mechanisms involved. However, despite a long history and thousands of studies, recombination continues to be a puzzle with respect to its mechanisms, the diversity of genetic and evolutionary effects (or 'functions') seen and, particularly, the factors which determine its own evolution.

The problem of the evolution of recombination as part of a more general problem of the evolution of sex has become one of the 'hotspots' of population biology and is the object of intensive studies by theoreticians. Unfortunately, the progress (modest as it is) that has been made in this area is hardly matched by experimental studies and field observations. There is no doubt, however, that genuine insight into this problem can only be gained by intimate interaction between theory and experiment.

In recent years, as a result of the extensive use of molecular markers in genetics, recombination has received growing attention as a subject of applied research. The amount of work in recombination-based genetic mapping that has been done in the course of a few years exceeds that which has been carried out during all the preceding history. Modern methods of mapping have opened up entirely new opportunities for analyzing the genetic topography of quantitative traits. This has resulted in wide breeding application, and the use of this approach may be equally important in studying the genetic control of fitness traits in natural populations and in replacing 'beliefs' and a priori assumptions by rigorous analyses. It appears that joint consideration of these lines of enquiry, i.e. (a) phenomenology, mechanisms and genetic control of recombination, (b) genome mapping and marker analysis of quantitative traits and (c) the evolution of recombination, unrelated as they might seem (and really are), can shed some new light on the general problem of recombination. At any rate, we hope that the attempt to discuss these issues in an interdisciplinary way, as described in this book, will be of some interest to researchers from various fields (e.g. geneticists, evolutionists, breeders), will renovate theoretical explanation of many common features of recombination and may, as a result, provoke new experimental and field studies of the *rec* system microevolution. One of our main objectives when writing the book has been to justify the adaptationist interpretation of basic features of recombination, primarily of crossing over.

A. Korol,
University of Haifa, Haifa,
Israel

Acknowledgments

This book is a revised version of our book *Variability of Crossing Over in Higher Organisms*, which was written when the authors were based at the Institute of Genetics, Academy of Sciences of Moldova (Kishinev), and was published in Russian by Shtiintsa Press (Kishinev) in 1990. I take this opportunity to thank our friends and colleagues from the Institute and especially the collaborators of my laboratory. Some of the results represented in the book are the products of their active participation. The English revision coincided with emigration of the authors to Israel (Haifa, A.K.) and USA (Gaithersburg, Maryland, I.P. and S.P.). The work would thus have been impossible to complete without the humanitarian and professional help of my new colleagues from the Institute of Evolution, University of Haifa, Drs Y. Ronin, V. Kirzhner and Zhanna Kovalevskaya: I thank them wholeheartedly. I would like especially to thank Professor Eviatar Nevo for his profound understanding of the scientific and practical problems that I encountered, sincere personal interest and everyday encouragement.

I would also like to extend my gratitude to the patience and help of my wife, Bella, who helped me immensely in the computer text-processing work of the new English version.

G.K. Lakhman has translated this book from Russian into English. His professional skill and persistence improved the text and clarified the formulations. The authors, of course, bear the full responsibility for the textual content.

The authors are also indebted to Professors M. Soller and S.M. Gershenson, the late Dr Batia Lavie, Dr A. Beiles and to the unknown referees for reading the manuscript and for their valuable comments and suggestions. The revision was financially supported by Grant No. 3675-1-91 of the Israeli Ministry of Science, the Wolfson Family Charitable Trust and the Institut Alain de Rothschild. I gratefully acknowledge the support of these Foundations.

Introduction

In the early days of genetics many important problems were solved using higher plants as experimental organisms. Later, extensive use of *Drosophila* as a model organism resulted in rapid progress in this science, which not only enhanced the power of its techniques and the reliability of the obtained results but also broadened the range of the problems studied. However, a changeover in the leading model organism caused no disruption in the continuity of development of genetics.

Large-scale recombination studies initiated by the Morgan school, determining the general lines of enquiry for many years to come, were central to classic genetics. Recombination formed the basis of analytical methodology of the entire science of genetics (Carson, 1957). The beginning of the next (molecular) stage in the development of genetics (the 1940s to 1950s) coincided with the extensive use of fungi, bacteria and viruses in experimental work. A new changeover in model organisms has resulted in a decline in the importance of hybridological analysis based on recombination techniques, with molecular biological methods playing an ever-increasing role. The exceptional efficiency of the latter is common knowledge. Considerable progress has been made over a short period of time in trying to solve the most important issues of genetics, including advances in recombination studies, and generated reductionist overoptimism (Maddox, 1993). Recently, the opinion has gained currency that general genetics has fulfilled its task and its subject of investigation has been largely exhausted, with only biochemistry, biophysics and molecular biology being capable of providing further progress in the cognition of life.

Indeed, advances in these disciplines have revolutionized genetic studies as a result of the development of new analytical techniques. These could permit, for example, a considerable portion of the genome to be mapped and followed through generations in various organisms, including man, in whom previously this was possible but only on a very small scale. In addition, these techniques enable cloning of genes controlling the key metabolic and developmental stages of economically important traits in plants and animals as well as fitness components in natural populations. Nevertheless, many problems of paramount importance formulated during the classical period of genetics have never been solved, and, having lost their former edge, are pushed somewhat into the background following the replacement of model organisms. These processes of specialization and reduction have even affected evolutionary studies in which the systems approach is a prerequisite for success (Noordwijk, 1990).

As pointed out by Grant (1975, p. 457) 'dominance of molecular and microbial genetics.... is not a result of progressive changes within a single field, but is instead a replacement of one field by another. Scientific problems came up which called for a different breed of workers. One body of knowledge became half-forgotten, or never learnt in the case of the younger workers, while another distinct body of knowledge with the same surname 'genetics' grew up in its place. Historical continuity was partly lost in the process'. Such an appraisal of the situation in genetics of the late 1960s to mid-1970s was on the whole correct, pessimistic as it might seem. The problem lies in the inadmissibility of extrapolating to higher organisms the results from lower ones. Such an extrapolation, warranted though it may be, can result in either serious mistakes or even wrong conclusions. Nevertheless, by the late 1970s, *Drosophila* and some plants again became popular model organisms, and deservedly so.

In the last two decades, fundamentally new information on the fine structure of genetic material has been obtained on higher eucaryotes. The mosaic structure of genes has been revealed, as well as enormous differences in DNA content between related species. A special class of genetic elements, the so-called mobile dispersed genes, have been discovered in the nucleus. In view of recent findings, the eucaryotic genome is envisioned as an extremely complex non-homogeneous system in which 'islands' of stable structural gene groups are surrounded by numerous continuously varying repeated sequences (Georgiev *et al.*, 1977; Hunkapiller *et al.*, 1982; Khesin, 1984; McClintock; 1984; Golubovsky, 1985; Finnegan, 1989; Karlin and Brendel, 1993; Morell, 1993).

The importance of recombination as a crucial factor determining the balance between the stability and dynamic properties of genetic organization as well as the interplay between heredity and environment has been increasingly recognized. In recent years, research has focused on various aspects of recombination (unequal crossing over, gene conversion, chromosome rearrangements) and their impact on genome evolution, particularly its components such as multigene families (Ohta, 1986; Flavell, 1986; Slatkin, 1986; Walsh, 1987; Petes and Hill, 1988; Basten and Ohta, 1992). According to current understanding, recombination is the main factor behind the so-called concerted evolution of these most important genomic components of higher eukaryotes (Dover, 1986). One of the prerequisites for concerted evolution is a balance between the rate of mutations that disrupt the structure of gene family members and the frequency of conversions ensuring structural unity (Walsh, 1987). Conversions may also be a mechanism for mobile element transposition (Evgenjev *et al.*, 1982; Geyer *et al.*, 1988).

Directed (site-specific) recombination is a useful approach in developmental genetics (Golic, 1991; Simpson, 1993). Genetically controlled reduction of recombination based on meiotic mutants may substantially increase the stability of genomic libraries of humans and other species maintained in yeast artificial chromosomes (Brown, 1992). The problems of recombinational repatterning of foreign genetic material in transgenic organisms produced by transformation have also received growing attention in recent years (e.g. Subramani and Rubnitz, 1985; Zimmer and Gruss, 1989). One of the reasons for this interest is the effect of recombination events on the stability of transformants (Muller *et al.*, 1987). Homologous recombination is becoming a powerful tool for genetic manipulation at the cellular level (site-specific modification of genetic material), allowing the induction of disorders in target genes or, conversely, the correction of the detected disorders and the replacement of endogenous defective genes by cloned DNA sequences (Anderson, 1992).

Finally, one cannot help mentioning here such an unusual form of recombination as horizontal transfer, which allows the flow of genetic information unrestricted by reproductive barriers. While conclusive evidence on the subject is still lacking, many authors admit (Khesin, 1984; Ayala and Kiger, 1984; Krasilov, 1986), and some are strongly confident, that the process does occur in nature and plays an important role in evolution (Kordyum, 1982; Syvanen, 1985; Grandbastien, 1992; Robertson, 1993).

These findings are ever more frequently at variance with classical genetics and conventional postulations of evolutionary theory. The current situation obviously necessitates a drawing-up of an inventory of the accumulated facts and a re-examination of the older ideas. That is to say, there is a need for a new synthesis in evolutionary biology (Krasilov, 1979; Campbell, 1982; Hunkapiller *et al.*, 1982; Stebbins, 1982; Anderson, 1983; Timofeeff-Ressovsky, 1984; Pollard, 1987; Endler and McLellan, 1988). Since there are fundamental contradictions between the existing notions and theories, this synthesis should probably be preceded by extensive experimental and field testing, on a regular basis, and by the refinement of major groups of models. The next step will be to develop a 'reconciled' version of the theory of evolution and to elaborate the metatheory (Lewontin, 1974; Lyubischev, 1982; Meijen, 1984; Iwasa, 1988; Ratner, 1990).

Bridging the gap between what is known about the genome structure, on the one hand, and its functional organization, on the other, is one of the main objectives of molecular genetics. Ultimately, this knowledge should become part of the overall system of genetics and evolutionary theory and contribute towards the solution not only of new problems but also of those that have been the subject of general and evolutionary genetics for more than half a century. These are, in particular, the spatial, temporal and functional organization of the genomes of higher organisms, patterns of the genetic transformation in the course of natural and man-directed evolution and the relationship between 'germplasm architecture' (Weismann's term) and the functioning of the organisms in varying environments. Rapid progress towards the solution of the problem of the genome structural organization is now apparent. The modern molecular revolution has dramatically increased the power of genetic analysis, thus opening up opportunities for solving a wide range of problems. It is no exaggeration to say that this breakthrough will ensure a multidisciplinary synthesis in biology.

Investigation of a complex system cannot be confined to the theoretical arguments explaining the behavior of all its component parts. The synthesis must be capable of predicting the behavior of the system as a whole and perceive the mechanisms governing that behavior. Actually, such a synthesis is usually much more difficult to achieve than the analysis. Thus, population genetics has mathematical techniques for describing and analyzing the behavior of a wide variety of one- and two-locus systems. However, they cannot be used to predict the behavior of, say, a five-locus system (Lewontin, 1974). The only way out has so far been computer simulations of various particular situations or analytical studies of multilocus systems satisfying certain, sometimes very rigid and unrealistic, constraints (Kempthorne, 1988). It is common practice to make assumptions of weak selection, no linkage disequilibrium, normal distribution of the effects of mutations, lack of genotype–environment interaction (or very special forms of this interaction), and so on. As Wright (1965, p. 81) pointed out, 'The attempt to introduce exact mathematical treatment leads in all but the simplest cases to such complexity that the results are of little use in the general theory.' Despite the obvious progress in theoretical studies over the past 30 years, this appraisal remains generally true.

One can imagine a situation in which, for every individual of a population, DNA sequences of the entire genome, the whole inventory of structural and regulatory genes, etc., are known. But even with this information, it is hardly possible to predict the population composition for several generations ahead and to ascertain the probability of the population survivorship within a given community. This task appears to be a formidable one in view of the need to allow for the population size, variability, breeding system, etc. (Nunney and Campbell, 1993). Also, it is not easy to solve the reverse problem, that is to reconstruct, based on the available data, the history (even if a recent one) of a given population. This is even true for situations in which mutations are the only source of heritable variation (Levin and Castillo-Chavez, 1990).

In predicting microevolutionary changes in real systems, interactions between two classes of fundamental processes – ecological (energy and matter transformation) and genetic (genetic information transmission) – should be taken into account (Eldredge, 1986; Yablokov, 1987; Wohrmann and Jain, 1990). Formulation of the ecological genetic theory of adaptive responses is one of the problems whose solution will offer ample scope for applying genetics in managing natural ecosystems and agrocenoses and in breeding work. Obviously, these problems are intertwined with those of developmental genetics, selection theory, ecology, population and evolutionary genetics, and evolutionary theory as a whole. In all these fields, the problem of genetic variation, and, in the first place, that of recombination, is a central one.

Being a key component of the sexual system, recombination has turned out to be closely coupled with other fundamental processes such as DNA replication and repair, Mendelian segregation of homologs and assortment of non-homologous chromosomes. Some of these processes share components in genetic control and molecular mechanisms and exhibit overlapping effects on genetic variation and synergistic pattern of co-evolution of their genetic determinants. Considering that species, genotypes, chromosomes and genomic regions vary widely in detail, this makes the task of providing simple explanations, and general rules and patterns, as well as formulating reasonable predictions for experimental and field testing, a real challenge.

In our view, one of the factors limiting the rate of progress in understanding this problem is the gap between the studies of molecular mechanisms, phenomenology, genetic control and the population genetics of recombination. Important findings in one of these fields are often unknown in others. For instance, such common properties of recombination as interference of exchanges and variation in their frequency and genomic distribution with ecological conditions have almost remained unconsidered from the population genetic standpoint. Nor are the peculiarities of recombinational organization of the genome adequately taken into account in the applications of genetics to breeding, including the quantitative trait locus mapping. On the other hand, evidence from studies of genetic determination of quantitative traits, abundant data on block organization and epistatic interaction of adaptively valuable gene complexes are often ignored in present-day multilocus models.

All this implies that further progress toward a better understanding of genetic recombination as a phenomenon of general biological significance is only possible based on the joint application of various approaches and techniques at every level of life organization. It seems certain, therefore, that the future synthesis in recombination studies will be preceded by a fairly long period of analysis and accumulation of evidence on interdependence and co-evolution of processes at the level of DNA and chromosome, cell and organism, population and ecosystem. An important step in estimating the progress in understanding the evolutionary functions of recombination within the framework of such a 'vertical' approach will be the analysis of ecological–genetic population models with subsequent testing of their predictions by experiments and based on observations of natural populations.

PART ONE

Estimation of Recombination

General survey of methods for estimating recombinational variability

1

1.1 RECOMBINATION AS A SOURCE OF GENETIC VARIATION

Gene mutations constitute a primary source of genetic variation in all living forms. In microorganisms exhibiting higher reproduction rates, mutations offer possibilities for rapid genotypic adaptation under sudden environmental changes, with a population surviving as a result of individual genotypes carrying rare favorable mutations. Multicellular forms have considerably lower reproductive rates, resulting in a higher value of each individual to the population as well as in increased individual plasticity and in coordination of physiological processes (Schmalhausen, 1942). These factors necessitate an increase in both the functional and structural integrity of genetic material and a rise in the hierarchical level of elementary units of genotypic adaptation by switching from genes as units of selection to co-adapted gene blocks, thereby changing the relative roles of mutation and recombination as sources of adaptive genotypic variation (Darlington, 1939; Mayr, 1963, 1982; Schmalhausen, 1968; Severtsov, 1981; Grant, 1985; Zhuchenko and Korol, 1985).

Increased complexity ultimately means a higher number of hierarchical levels in the system, with mutations at individual loci being incorporated in the population gene pool only after having been tested in combination with many other polymorphic genes. A possibility for such testing is provided by recombination. Naturally, testing of each new gene (gene combination) occurs in the context of an already formed system exhibiting a high internal balance and adaptedness to environmental conditions (Mayr, 1970; Lewontin, 1974; Ayala and Kiger, 1984). Internal balance is also reflected in non-randomness of the recombinational variability spectrum and in the dependence of exchange frequency and distribution on environment and genotype. The process of generating primary genetic diversity by mutation is, in a sense, also optimal and internally adjusted (Tchaikovsky, 1976; Krasilov, 1979; Campbell, 1982; Borstnik, Pumpernik and Hofacker, 1987).

Estimates from various *Drosophila* species indicate that 25–40% of variation in fitness observed in natural populations is regenerated by crossing over in one generation from the gene content of a randomly drawn chromosome pair (Dobzhansky, 1959; Mayr, 1970). This led Dobzhansky to conclude that a temporary arrest of the mutation process would not result in a significant decrease in variation over a large number of generations. The same view is held by Ayala (1981), who argues that for the variation stored in a population to be accessible to selection for a number of generations recombination alone is enough without mutation.

In fact, variation generated *de novo* by the mutation process is lower by orders of magnitude. This is particularly the case in view of the large amount of genetic variability (and heterozygosity) present in natural populations discovered in the 1960s by the method of protein electrophoresis and subsequently confirmed at the DNA level (Hubby and Lewontin, 1966; Lewontin, 1974; Nevo, 1978, 1988a, 1991; Ayala, 1983; Gilbert *et al.*, 1990; Nevo *et al.*,1990; Laurie, Bridgham and Choudhary, 1991; Erlich and Arnheim, 1992; Jiang and Gibson, 1992). All other things being equal, genetic variance for a sexual population with free recombination may be several times that for a population with no recombination (Charlesworth, 1993). Notwithstanding the known theoretical difficulties, the idea of polymorphism and heterozygosity being extremely important for the adjustment of organisms to varying environments (biotic and abiotic) is the central one in modern population biology (e.g. Ayala and Kiger, 1984; Nevo, Beiles and Ben-Shlomo, 1984; Bremermann, 1987; Nevo, 1988b; Barton and Clark, 1990; Hamilton, Axelrod and Tanese, 1990; Aquadro, 1992).

The possibility of a rapid restoration of enormous genetic variability becomes of particular significance in higher eukaryotes with long life cycles and/or those experiencing the bottle-neck effect (Bryant, McCommas and Combs 1986; Bryant and Meffert, 1988; Carson, 1990). Finely adjusted systems of recombination control found in these organisms provide a trade-off between the need for maintaining high levels of fitness and preserving genetic plasticity (capacity for future variation), a compromise hardly attainable in the absence of recombination (Darlington, 1939; Grant, 1958).

One way to classify genetic variability is to subdivide it into *potential* and *free* variability. Concepts of the above types of variability were introduced by Fisher (1930) and elaborated by Mather (1943), although Chetverikov (1926) was the first to advance the idea of a large amount of heritable variability concealed in natural populations. Free variability manifests itself in genotypic differences accessible to selection. These differences can result from new mutations as well as the release of potential variability (stored in the form of heterozygosity) by recombination and segregation (Figure 1.1). Crossing is the main process

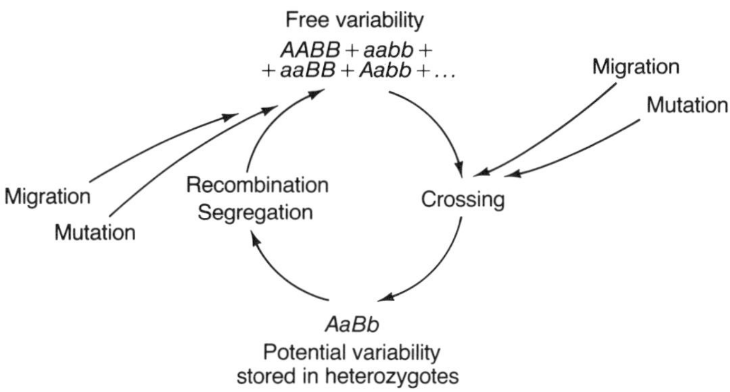

Figure 1.1 Variability conversions in populations (from Zhuchenko and Korol, 1985).

resulting in interconversions of the above two types of variability. Inbreeding leads to a '**potential → free**' transition, providing material for selection to act upon. Outbreeding produces new heterozygotes that could be considered not only as selectable genotypes but as 'reservoirs' of potential variability as well. In a population, both processes are continuously occurring, with the '**free → potential**' transition creating preconditions for the emergence of new variability in subsequent generations.

From the 'naive' point of view, the above processes constitute the essence of sexual reproduction, diploidy and recombination. Diploidy offers great possibilities for generating new alleles without a substantial reduction in current fitness. Sexual reproduction allows favorable mutations to be assembled in a single organism, with recombination producing a wide variety of combinations of newly arising alleles with those previously accumulated.

The role of recombination as a factor affecting genetic variability is controversial. It produces new allele combinations – raw material for directional or diversifying selection. Likewise, under free mating it tends to destroy any deviations from the population mean, thereby helping the stabilizing selection. Through gene duplications (by unequal exchanges), recombination creates prerequisites for increasing variability (Serebrovsky, 1938; Mather, 1953; Ohno, 1970). However, recombination is considered to be a factor maintaining homogeneity within a multigene family (Ohta, 1986). The efficiency of recombination as a source of variability depends on the breeding system, life cycle duration, ploidy level and chromosome number as well as on cross-over frequency and distribution and on their genetic and environmental variation (Chapters 4 and 6).

Although studies of factors affecting recombination have occupied a prominent place in genetics, they have always had quite different aims to pursue, namely elucidating genetic topography, mechanisms of recombination, temporal organization of meiosis, and so on, i.e. the 'structural' issues. In most cases, it is unclear what functional significance the phenomenology established in these studies has in sexual reproduction, evolution and breeding (Campbell, 1984). Furthermore, questions such as these are usually not discussed at all, and sometimes it is even considered bad taste to pose them (Hughes and Lambert, 1984). Admittedly, function is not explicable by studying structure alone (e.g. Mednikov, 1982; Rattner, 1992). Nevertheless, analysis of molecular genetic structures and of mechanisms responsible for their transformation can greatly aid, if not in the solution, then in the formulation of problems bearing on their evolutionary functions and formation (Bernstein, Hopf and Michod, 1988; Holliday, 1988). A major problem which continues to be a favorite subject of evolutionary studies is that of analyzing factors underlying the evolution of sex and recombination, i.e. reasons for maintaining meiosis despite its twofold disadvantage (Stearns, 1987a; Michod and Levin, 1988). Likewise, it is likely that a broader formulation of the problem, including the adaptationist explanation not only of the recombination level observed in nature but also its variation (environmental, genetic, intragenomic, between-sex), will provide new answers even to the main problem of sex and recombination maintenance (section 9.1).

When viewed at the level of phenotypic traits, genetic innovations produced by recombination can arbitrarily be divided into three main types:

1. Transgressions for individual traits with the range of trait values in a segregating generation exceeding the parental ranges.
2. The formation of new trait combinations of the crossed components. The resulting combinations of parental traits can be quite atypical and of high selective value even if variation in each trait does not exceed the parental range (transgression for trait combinations).
3. The appearance of entirely new traits (anomalous variation). Anomalous variation is usually a product of recombination in genetic complexes with strong non-allelic interactions; it is generally revealed in distant crosses.

Under introgressive hybridization, in particular, recombination can separate tightly linked segments, resulting in an entirely new combination with 'mutant' phenotypic expression and, frequently, with lower fitness (Anderson, 1949; Mangelsdorf, 1958; Grant, 1981, 1985).

The above forms of recombination-induced variability play an important role in the microevolution of natural populations and in breeding programs. All three types of changes occur under introgressive interspecific hybridization and provide the raw material for selection (natural and artificial). Changes of the first or second type serve as material for perfection of the adaptive norm of natural populations and for cultivar improvement.

The notion of recombination as a source of variability leads to the need to estimate the level and spectrum of recombination-induced variability (Zhuchenko, 1980, 1988; Zhuchenko and Korol, 1985). Recombination parameters (crossover frequency and distribution) should be taken into account in studying crossing over as a fundamental cytogenetic process; the genetic material topography (with respect to marker traits and sequences and quantitative traits); the genetics of speciation; evolution of the *rec* system itself; and so on. The need to estimate the variability spectrum stems from the use of recombination (both natural and induced) in breeding work as well as in population genetic studies of multilocus systems (taking account of deterministic and stochastic factors resulting from intra- and interspecific interactions and varying abiotic conditions).

1.2 SPECTRUM OF RECOMBINATIONAL VARIABILITY

Analyzing responses of *Drosophila* populations to long-term selection, Mather and Harrison (1949) observed a sudden change in the selected traits following a long period of stability (accelerated response to selection). They suggested that this reaction might be due to crossover redistribution caused by a certain environmental change. They noted that involvement in recombination of those chromosome regions in which crossing over does not normally occur would open enormous reserves of genetic variability for breeding purposes. This is one of the most urgent problems in applied genetics. Its significance becomes even greater in view of the need for a more efficient utilization of adaptive genetic resources present in wild relatives of cultivated plants and domestic animals (Zhuchenko, 1980, 1988; Nevo, 1987; Allard, 1988, 1990; Nevo and Beiles, 1989).

The difference in the degree of importance between two applications of induced recombinogenesis – altering crossing-over frequency and changing the spectrum of crossovers (let us call them problem I and problem II for convenience) – is worth noting. In the first case, changes in the progeny composition due to external effects on a heterozygous individual (or to differences in *rec* genes) are of a quantitative nature. They result from altered crossing-over rates within the same segments in which exchanges also occur under no-treatment conditions. A significant consequence of a recombinogenic treatment in this case is likely to be that of increasing the **proportion** of genotypes carrying gene combinations desirable for the breeder, with the total spectrum of gene combinations remaining identical to those in the control. In formulating problem II, a principal point of departure is the assumption that there exist genomic regions which under normal conditions are excluded from the recombination process. In view of this, it is natural to understand that 'variation in the spectrum' of genotypic diversity and changes in the population composition can be brought about by the redistribution of crossovers into such regions (Shaw, Wilkinson and Coates, 1982; Zhuchenko and Korol, 1982, 1985). These changes can be brought about by effects of genetic factors (e.g. *rec* genes, heterochromatic blocks, transposable elements or B chromosomes) or by an external effect on the parent's meiosis. The problem under consideration is also important in analyzing the pattern of variation in natural populations. It is known that stressful factors, such as inducers of mutations and recombination, can have a dramatic effect on variation. Allowing for these effects is one of the prerequisites for a better understanding of the nature of organism–environment interactions (for further discussion, see section 9.1).

The expediency of such a dual approach to recombinational variability ('frequency' and 'spectrum') can also be justified on the basis of cytogenetic evidence. Thus, some species normally have two distally located chiasmata per bivalent. At the same time, one chiasma per bivalent can be formed in a proportion of meiocytes. The location of this chiasma occasionally differs from the traditional ones, thereby promoting a release of novel genetic variability (Shaw, 1974; Fletcher, 1988). But it would be wrong to think that 'traditional' exchanges are immaterial to genetic variability formation. Thus, in the parasite *Plasmodium falciparum*, crossing over normally occurring in distal chromosome regions is assumed to be the cause of high chromosome size polymorphisms (due to deletions–duplications), with the same factor conceivably being the cause of variation at antigen loci (Corcoran *et al.*, 1988). It is not unlikely that increased recombination in the regions where most of such loci are located has an adaptive value.

It is possible for the level and spectrum of variability to be negatively correlated. There is evidence that exchange frequency and distribution within a bivalent can be controlled by different genetic systems (Lein and Lelley, 1987).

The above definition of the 'spectrum variation' concept should serve as a basis for estimation of induced changes in spectrum. Unfortunately, even in the simplest case of discontinuous variation certain difficulties are encountered, since it is impossible to differentiate operationally the situations associated with problems I and II. Indeed, if crossover form *ab* appears with a frequency of p_0 among the F_2 progeny following an exogenous treatment of the F_1 (*Ab*/*aB*) heterozygote and it is absent in the control, how should a result like this be viewed? As a change in the frequency or as a change in the spectrum of recombinants? It would seem that, by definition, the second interpretation is more justified, and this choice seems to be the only one that can be made in evaluating the results of a large experiment. Lack of recombinants in the control may actually mean that no exchanges have occurred in the segment *a*–*b* and then the second interpretation is indeed correct. However, with smaller material, this fact can readily be explained in another way: that crossing over rate rf_0 between a given pair of marker loci in the control (with the frequency of recombinants *ab*/*ab* equal to $P_0 = rf_0^2/4$ for F_2) is insufficient to permit the detection of at least one of these in a sample of a given size.

When speaking of mutation spectrum or spectrum of variation in general, a parameter is implied which characterizes diversity. Regrettably, attempts to formalize a concept as widely used as this have encountered serious difficulties. If it is the qualitative traits that we are concerned with, then it would be natural to assume that spectrum of variation is simply the number of distinct phenotypic classes (n) and Shannon's measure can be applied to characterize the diversity of the population in question as:

$$I = -\sum_{i=1}^{n} p_i \log p_i,$$

where p_i is the proportion of the ith phenotypic class. Then, to compare two variants (e.g. a recombinogenic treatment with the control, or populations in normal and stressful conditions) for the degree of diversity present in them, it is necessary to calculate p_i values for each of them and find $\Delta I = I_2 - I_1$. If $\Delta I \neq 0$, the variants can be said to be differing in the amount of diversity (variation). But how is one to decide whether or not the spectrum has been changed? For this, it is probably enough to compare the number of phenotypic classes, n, in each of the variants. Provided n has remained unchanged (and its constancy is not due to a disappearance of one of the old classes and emergence of a new one), it is natural to assume that the **spectrum** has not changed. But even in a rather trivial situation like this, a discrepancy can arise between intuitively acceptable and formal characterizations of variability.

Suppose that the control population consists of two phenotypic classes with equal numbers, i.e. $n = 2$, $p_1 = p_2 = 1/2$. Assume that a new class, in addition to those already present, is formed as a result of some effect, with the initial class frequencies changing in such a way that $p_1 = 7/9$, $p_2 = 1/9$, $p_3 = 1/9$. It is quite obvious that the spectrum, i.e. the diversity of individual types, has increased. The formal measure of diversity, however, has decreased as:

$$\Delta I = (-7/9 \log 7/9 - 2/9 \log 1/9) - (-1/2 \log 1/2 - 1/2 \log 1/2) = -0.0094.$$

A similar situation can arise in the case of continuous variation when an attempt is made to use variance (σ^2) as a measure of diversity. Consider another example. Assume that the distribution of a quantitative trait in the control backcross (BC) population shows a fairly pronounced bimodality as a result of a segregating gene block (the F_1 heterozygote structure being $++-++/--+--$) with a strong net effect on the trait, and that crossing over within the gene block is normally suppressed. Suppose further that an effective recombinogenic agent has been applied whose effect on F_1 results in a high crossing-over frequency in the region concerned, i.e. alleles $+$ and $-$ are more randomly combined within the block. Then, for a number of situations with different effects of alleles at the five loci chosen, the induced broadening of the spectrum of genotypic diversity will be accompanied by a decrease in σ^2. The inadequacy of variance as a formal measure of variation is also manifested in that, despite a decrease in σ^2, the range of variation is extended following the treatment, owing to gametes $+++++$ and $-----$.

A simple analysis reveals that even in a two-locus case an increase in crossing over rate during meiosis of an F_1 heterozygote of the type $++/--$ (from zero to a value of rf_0) results in a decreased trait variance in F_2, although genotypic diversity is certainly increased (Mather and Jinks, 1971; Korol and Zhuchenko, 1980). Variance is therefore not a sufficiently good indicator of diversity in the case of continuous variation, nor is logarithmic measure in the case of discontinuous variation. Nevertheless, for analyzing quantitative traits, the behavior of means and variances and possibly of higher order moments, in response to recombinogenic effects, can be used as an approach to studying some general characteristics of the genetic systems such as the relative frequency of *cis*- and *trans*-combinations per genome, per chromosome or per individual segment. Measuring these and other statistical parameters in marker groups of a segregating population opens up a possibility for estimating parameters of recombination within chromosome regions with strong effects on quantitative traits (Chapter 3).

As already noted, the direction of change in variance upon exposure to a recombinogenic agent is to a great extent dependent on which allele combinations prevail in heterozygotes ($++/--$ or $+-/-+$). According to Mather's polygene balance theory, *trans*-combinations play a major role in long-term storage of variability under stabilizing selection, in releasing it at moments of sudden environmental changes and in maintaining and perfecting the adaptive norm of a population (Mather, 1943; but see Zhivotovsky and Feldman, 1992). Therefore, an induced increase in recombination frequency and a redistribution of crossover sites within the chromosome can be expected to result, more often, in higher variation for traits that have not been exposed to continued directional selection. But the possibility cannot be ruled out that there exist *cis*-combinations with rather strong individual effects, particularly for traits that have undergone directional selection. Such combinations can also arise through correlated responses to selection for other traits.

From the above examples, it is clear what a difficult task it may be to choose a correct explanation of the results (a change in frequency or in spectrum) of even a simple recombination experiment. This alternative cannot be resolved at all by purely formal means. Consequently, the inference about an induced change in the spectrum of genotypic variation should be specific, related to a population of a definite size. The complexity of the problem of estimating the amount and spectrum of recombinational variability calls for an integrated approach to its solution through the application of various complementary techniques.

1.3 EVALUATING CHANGES IN VARIABILITY SPECTRUM

1.3.1 Statement of the problem

Assessment of changes in key parameters of a quantitative trait distribution in a segregating population in response to a recombinogenic treatment of an F_1

hybrid can furnish useful information on the genetic architecture of the trait and on the efficiency of treatment itself (Korol and Zhuchenko, 1980). However, such an approach is not sufficient for estimating changes in the amount and spectrum of variability. Restricted recombination, as manifested in a localized pattern of exchange distribution as well as in selective elimination of recombinant associations of alleles, reduces considerably the variability potential of the progeny (especially in interspecific crosses), thus narrowing the range of adaptive forms.

Similar restrictions result from developmental canalization. Even a comparatively small number of wild-type genes incorporated into the genome of a cultivar are likely to have a significant canalizing effect on the phenotype. The relevant estimates indicate that there is no point in looking in the F_2 generation for forms related to the cultivated type even if only one of the key traits is controlled by dominant genes with strong canalizing effects (Zhuchenko and Korol, 1985). Under these circumstances, it is worthwhile trying to obtain valuable introgressive crossovers by selfing heterozygotes but not before one to two cycles of backcrossing to the cultivated parental stock.

As a result of the above restrictions, a relatively 'standard' range of phenotypes may be observed in a segregating population (Grant, 1981; Zhuchenko *et al.*, 1982; Preygel, Garbuz and Korol, 1986). Thus, in F_2, most of the trait combinations will be constrained to fall within the limits of the so-called 'recombination spindle' (Anderson, 1949). The term has been specially coined to denote an area in the trait space within which points corresponding to F_2 individuals are clustered. Thus, in a two-trait case, the main axis of a trait-scatter ellipse for F_2 nearly always coincides in direction with the straight line connecting parental means (Figure 1.2). Moreover,

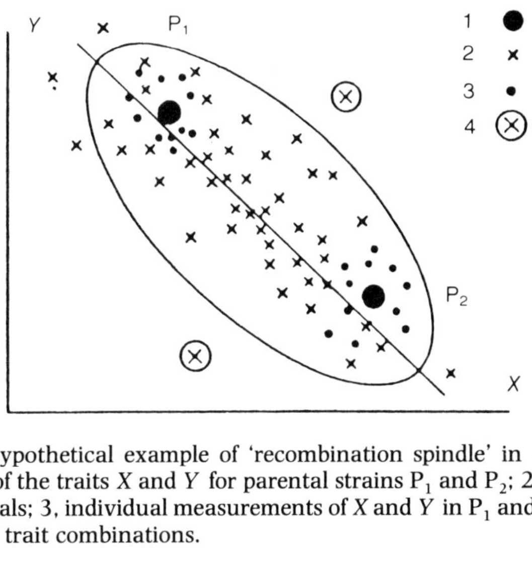

Figure 1.2 A hypothetical example of 'recombination spindle' in a bivariate case: 1, mean values of the traits *X* and *Y* for parental strains P_1 and P_2; 2, values of *X* and *Y* for F_2 individuals; 3, individual measurements of *X* and *Y* in P_1 and P_2; 4, genotypes transgressive for trait combinations.

transgression is possible for each of the traits, but the overwhelming majority of transgressive forms will tend to exhibit typical trait combinations being clustered along the main axis of the ellipse.

It is clear that the larger the difference between the crossed forms and the higher the degree of developmental canalization and the extent of restrictions on recombination, the lower is the probability of obtaining transgressions for trait combinations (two such genotypes are denoted by crossed circles in Figure 1.2). And it is with a view to obtaining genotypes of this kind that hybridization is usually performed. Therefore, the frequency of forms with unusual trait combinations may be a valuable characteristic of genotypic variability spectrum.

1.3.2 Multivariate analysis in estimating the spectrum of recombinants

The currently available methods of statistical analysis, usually useful and efficient in solving population problems, have been developed ignoring the requirements that should be met in estimating changes in the variability spectrum. This is because of the qualitative peculiarity of two classes of problems indicated in section 1.2. Indeed, formal measures of population diversity generally satisfy the following conditions (Zhivotovsky, 1984): small changes in frequencies of common morphs (phenotypes, genotypes, alleles, etc.) as well as the appearance or extinction of rare morphs have a weak effect on the diversity measure; the measure is independent of parameters specifying various morphs (fitness, breeding value, size) and the contributions of morphs to overall diversity are proportional to their frequencies.

Clearly, these conditions not only may, but also should, be violated as a result of using an intuitively acceptable criterion for variability spectrum changes resulting from crossovers moving to genomic regions previously 'excluded' from recombination. The amount and spectrum of variability in the progeny of a hybrid depend on the magnitude of effects and genomic distribution of genes and gene blocks responsible for the differences between parental genotypes, meiotic mechanisms, the fertilization process and elimination of recombinants at post-syngamic stages. All these factors can result in fairly rigid constraints on the release of potential variability (particularly in distant crosses), thereby significantly narrowing the range of new genotypes (Stephens, 1950; Grant, 1966; Rick, 1972; Sano, Chu and Oka, 1980; Shaw, Wilkinson and Coates, 1982; Zhuchenko and Korol, 1985; Ukai, 1987). Therefore, in spectrum estimations, it is desirable to take into account some 'value' measures, although the concept of 'value' will be quite different for natural and cultivar populations.

For breeding based on hybridization, the most valuable are recombinants with unique (very rare) trait combinations (the prime objective of hybridization). We have already touched upon the distribution of a set of characters within a segregating population and considered Anderson's (1949) 'recombination spindle'. The 'narrower' the spindle, the more rigorous apparently are the restrictions on combinative variability in the progeny.

A method enabling the identification of morphologically 'intermediate' and 'extreme' phenotypes with respect to a set of quantitative traits allowing for multivariate correlation has been developed by Zhivotovsky (1984), primarily for natural populations with the evolutionarily established integrated gene pool. This reduces to the following. All members of the population are ranked by their distances from the population mean using an analog of the Mahalanobis metric (which uses the correlation structure of the chosen set of traits). Then, a preset portion of individuals most distant from the population mean – a group of 'extreme' phenotypes – is taken. These extremes are subdivided into subgroups of 'small', 'large' and 'disproportionate' ones using the principal components method. Identifying groups like these may be highly desirable for improving the effectiveness of breeding programs, for, analysing genetic disorders in humans and for studying the effects of stabilizing selection at various stages of ontogenesis (Zhivotovsky, 1984; Altukhov, 1989).

Estimation of the intensity of selection in natural populations based on phenotypic traits occupies an important place in the range of evolutionary problems and is still far from being solved, though considerable progress has been made in recent years (Lande and Arnold, 1983; Manly, 1985; Endler, 1986; Mitchell-Olds and Shaw, 1987; Jain, 1990). Especially attractive in this connection seems to be the non-parametric description of fitness surfaces (Schluter, 1988), since it has been shown that the dynamics of multilocus systems that are subjected to selection may be strongly dependent on the distribution shape adopted (Turelli and Barton, 1990).

If a set of traits follows a unimodal distribution, then Zhivotovsky's method (ZM) can also be employed to detect 'rare' recombinants showing maximum deviation from the mean in the trait space. However, when these conditions are violated because of restrictions on viability of novel recombinants or segregation of 'strong' genes and/or gene blocks, data points in the trait space can form groups or clusters, for example, such as those corresponding to parental genotypes, F_1 and certain standard recombinants. Moreover, the population mean no longer corresponds to most typical, normal trait combinations (Ushakov, 1978). Therefore, the identification of unusual recombinants by the ZM in a situation like this is not possible. In the case of a cluster-structured population, the following algorithm is proposed for solving the problem stated (Zhuchenko *et al.*, 1982; Preygel, Garbuz and Korol, 1986). As a first step, the initial population is subdivided into groups using cluster analysis techniques. Incidentally, information on the parental and F_1 trait values may prove useful at this stage. Then an algorithm of the ZM type is applied to each of the resultant clusters (data points are ranked by their proximity to the cluster center based on the within-cluster correlation matrix) as well as the whole set of data (based on the general correlation matrix). The points in the trait space which are most distant from the center of their cluster are considered to be atypical.

As an example, consider the data on seedling weight and height in the F_2 progeny of an interspecific cross of tomato (Figure 1.3). Even in the bivariate case, it is apparent that the population is not unimodal and it divides into three groups (outlined by dashed lines in Figure 1.3). Table 1.1 lists some characteristics of the clusters as well as R_{cl} values for points in each cluster and for the whole set of points in two variants, including all points and culling out the 'external' ones.

One of the problems with this method is the difficulty of obtaining an objective clustering when there is no prior information on the number of clusters and their locations. In the case of more than two traits, expert testing of clustering is impossible and there is nothing for it but to give credence to algorithms of cluster analysis. Given information on parental classes (P_1 and P_2) and on F_1, a simpler alternative way of estimating the degree to which trait

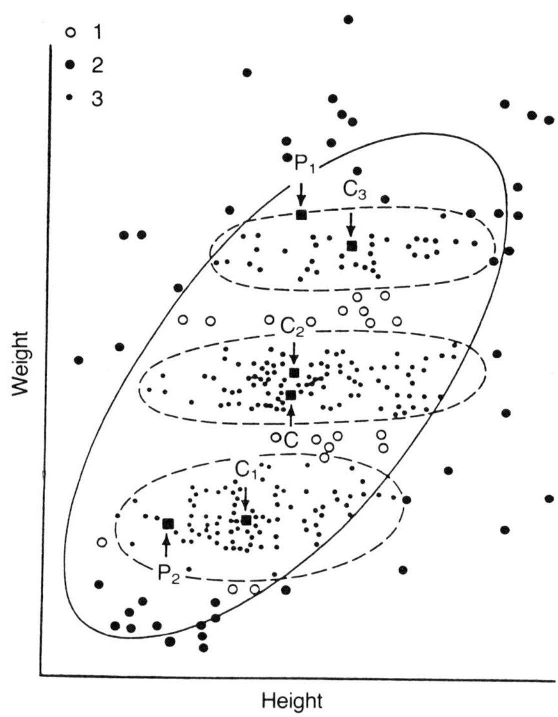

Figure 1.3 Distribution of seedling heights and weights in an F_2 (*L. esculentum* × *L. pennellii*) population: P_1 and P_2, centres for parental strains; C, the common center for F_2; C_1, C_2, C_3, the centers of the resultant clusters. Genotypes: 1, 'internal' atypical; 2, 'external' atypical; 3, the rest.

Table 1.1 Characteristics of clusters in an F_2 (*L . esculentum* × *L . pennellii*) progeny for seedling weight (X_1) and height (X_2)

Cluster number	*Number of points*	*Standard deviation* σ_{x_1}	σ_{x2}	*Correlation* $R_{x_1x_2}$	R_{cl} *when estimating by the metric of* *Mahalanobis* M_0	M_1	*Euclid* E_0	E_1
All points included								
1	122	0.34	0.53	0.46	0.75	0.70	0.49	0.49
2	125	0.22	0.58	0.09	0.97	0.79	0.93	0.79
3	65	0.48	0.67	0.20	0.83	0.89	0.69	0.77
Σ	312	0.88	0.64	0.46	0.80	0.67	0.65	0.55
With culling out								
1	110	0.27	0.45	0.36	0.68	0.59	0.34	0.32
2	120	0.21	0.51	0.23	0.95	0.73	0.83	0.69
3	44	0.17	0.50	0.27	0.75	0.56	0.49	0.48
Σ	274	0.73	0.54	0.48	0.68	0.55	0.48	0.43

Subscripts 0 and 1 refer to the general and within-cluster correlation matrices respectively (M_0 and M_1 or E_0 and E_1).

combinations of individuals of a segregating progeny are recombinant is also possible. This consists of:

1. ranking all individuals by their proximity to standard classes (P_1, P_2, F_1 and other specified areas in the trait space, for example best or 'ideal' cultivars) and assigning each to the nearest class;
2. estimating, based on clustering, distances of the genotypes from cluster centers and from the common center;
3. ranking the individuals by the degree of concordance of traits (Sano, Chu and Oka, 1980);
4. projecting multivariate data points corresponding to individual genotypes on two-dimensional planes of principal components or original traits for data visualization;
5. comparing the positions of individuals within ranked series of distances and metrics and across subsets of traits using contingency tables.

In calculating the coefficient of concordance of traits (according to Sano, Chu and Oka, 1980), transformed values of traits are used: $x_i = [X_i - (\mu + kh)]/d$, where X_i is the initial value of the ith trait of a genotype from the segregating population, $\mu = (X_{P_1} + X_{P_2})/2$, $d = (X_{P_1} - X_{P_2})/2$, X_{P_1}, X_{P_2} and X_{F_1} are mean values for parental genotypes and F_1, $h = X_{F_1} - \mu$ (if the F_1 data are not available, then h is set equal to zero) and k is the expected level of heterozygosity ($k = 1/2$ for both the BC and F_2 progeny. The values of x_i falling outside the interval $(-1, 1)$ indicate transgression and recombination, but the reverse is not true, i.e. lack of transgression does not necessarily imply the absence of recombination. Standard deviation, σ_x, of the whole set of normalized traits of an individual is proposed as a measure of its degree of recombination. If the individual resembles one of the parents in the majority of traits (values of x_i are close to 1 or to -1), then σ_x will be small. In contrast, if it is close to P_1 in some traits and to P_2 in others, then σ_x is large. Thus, σ_x is an indicator of the extent to which a given individual combines various parental alleles, i.e. of the degree of transgression for trait combinations.

Having evaluated the segregating progeny of a hybrid by means of the above techniques, one may choose the strategy for and intensity of selecting the most 'valuable' (unusual) genotypes. Note that this approach may be used in identifying genotypes that are the most deviant from (or, conversely, closest to) the specified areas in the trait space, for example from parental areas and/or F_1. We believe that these techniques could be useful in estimating spectrum of genotypic variation as affected by recombinogenic and mutagenic factors, in the progeny of interspecific hybrids and in natural populations.

An important applied problem concerning the selection of recombinants showing unusual combinations of agronomic traits based on the assessment of traits expressed at early developmental stages may also be solved with this approach. Plant breeders were using their vast experience for selecting desirable genotypes long before the formation of economically important yield component indices. Huskel (1954) mentions L. Burbank's claim that effective selection of valuable progeny is possible at the seedling stage by careful assessment of genetic correlations among traits caused by linkage and pleiotropy. In fact,. any experienced breeder makes use of correlations when performing selection. Some relationships prove to be peculiar to a particular population; others are fairly general (at the species and even the genus level). In any case, multivariate analysis is required for a more efficient practical application of these correlations. It is worth noting that multivariate techniques have long been successfully used in systematics and generally in evolutionary biology.

1.3.3 Methods of estimation in artificial environments

One of the main reasons for extending the spectrum of variation in selected populations is the low adaptability of cultivars. It is generally believed that introgression into cultivar genomes of adaptively significant gene blocks from wild relatives or land races with a view to improving the resistance to adverse environmental factors is the principal way to solve this problem (Zhuchenko, 1980, 1988; Nevo, 1987, 1991; Allard, 1988; Soller and Beckmann, 1988). In view of this, the direct method of estimating the variation spectrum in various ecological conditions is of great practical significance (Nevo, 1987; Nevo and Beiles, 1989).

When high resistance is due to dominant or intermediate gene expression in the heterozygote, the incorporation of resistance into a cultivar can be ensured through successive backcrossing and progeny testing in the appropriate environments. Moreover, the proportion of donor genes, not associated with the selected trait, decreases exponentially ($1/2^n$, where n is the number of successive backcrosses). The elimination of undesirable genes from chromosome regions carrying resistance genes proceeds much more slowly: in the absence of negative selection, the corresponding value is about $1/n$ (Bartlett and Haldane, 1935; Fisher, 1949). Actually, this process can proceed still more slowly due to crossing-over suppression in segments transferred into the recipient genome from a wild species (Rhyne, 1962; Rick, 1972; Korol and Bocharnikova, 1988; Doebley and Stec, 1991).

What kind of environment should be used for estimating variation spectrum and selecting highly resistant forms? When the conditions are too severe, intense selection rapidly eliminates the bulk of interesting gene combinations which could otherwise become raw material for the formation of valuable recombinants in the next generation. This is especially important in the case of gene pools of different species being integrated in a single population undergoing recombination. The selection intensity should not be preset for such cases but rather be modified from generation to generation, the extent of modification at each step being dependent on the results obtained at the preceding ones.

Interesting problems arise when there is a need for combining in a single genotype resistances to several adverse factors. When doing so, which environmental regime should be adopted and should intense selection be applied to a whole set of factors or alternated as 'intense × weak' and 'weak × intense'? Further, which scheme will be optimal, the one that produces forms resistant to a whole set of limiting conditions or the one that accumulates, step by step, genes for resistance to different factors? Clearly, answers to these questions depend on the genetic architecture of adaptive systems in the donor species as well as on the features of the system controlling recombination. But the questions themselves are so difficult that, even given a complete description, it is not easy to provide clear answers, let alone to propose an optimum solution. One general approach to the problem is combining experimental studies with computer simulations (Marani, 1975; Preygel, Preygel and Korol, 1991). In addition, it should be borne in mind that stressful conditions can not only contribute to the exhaustion of genetic variability but, conversely, can also promote its preservation, converting a generally less useful type into that of high adaptive (and breeding) value (Acharya and Jana, 1982).

The above approaches based on multivariate analysis (under normal and/or stressful conditions) may be used in population genetics for ascertaining targets of selection, i.e. for estimating fitness components (Zhivotovsky, 1984; Altukhov, 1989). For both breeding and population biology problems, a substantial improvement in the resolving capacity of genotype identification techniques can be provided by passing from phenotypic descriptions to the corresponding genes and gene blocks. For quantitative traits, the basis for such a transition is formed by marker-based genetic analysis (Chapter 3). Marker analysis is also a powerful tool, especially in combination with cytological analysis, for studying the effects of various factors (genetic and environmental) on the frequency and genomic distribution of recombination events.

1.4 MARKER AND CYTOLOGICAL ANALYSIS OF RECOMBINATION

In comparing *Drosophila* genetic maps based on between-marker recombination distances with polytene chromosome maps, regions with strongly inhibited exchanges have been found (e.g. Roberts, 1965). In tomato, comparisons between cytological and recombination maps yield similar results (Rick, 1971). Marker analysis in *Drosophila* has revealed significant differences in crossover distribution along the chromosome length for single, double and triple exchanges (Charles, 1938). Using multiple-marked strains, an induced crossover redistribution can also be detected. Evidence in the literature indicates that such a redistribution does occur in a number of cases, e.g. in meiotic mutants as well as upon exposure of heterozygotes to exogenous factors (reviewed by Zhuchenko and Korol, 1985).

The use of multimarker lines is of considerable interest in other aspects as well. Multiple exchanges have been known to be relatively rare even in the absence of interference. These vary bètween 1 and 3 for the majority of higher organisms, with the range of 5–6 occurring rather seldom, although values up to 12 have been registered (see Holliday, 1977; Ukai, 1988). For example, it has been demonstrated, in *Drosophila*, that the frequency of tetrads with four or more crossovers is extremely low (usually much less than 1%) (Charles, 1938). The chiasma number per bivalent reaches 8 in *Fritillaria* and 12 in the M chromosomes of *Vicia faba*.

The development of molecular markers, e.g. RFLPs (restriction fragment length polymorphisms), has opened up fundamentally new possibilities for marker analysis (Tanksley, 1983; Beckmann and Soller, 1983, 1986a; Lander and Botstein, 1986, 1989; Tanksley *et al.*, 1989; O'Connell *et al.*, 1989). New molecular techniques allow for analysis of a nearly unlimited number of sites throughout the genome, forming the basis of modern worldwide efforts on mapping the human genome and genomes of several model organisms (mice, *Drosophila, Caenorhabditis*, yeast, *Escherichia coli* and *Arabidopsis*) (Olson, 1993; see also the special issue of *Trends in Biotechnology*, volume 10, No. 1). Detailed genetic maps based on molecular markers have already been constructed for tomato (Tanksley *et al.*, 1992), barley (Graner *et al.*, 1991), maize (Helentjaris *et al.*, 1986), rice (McCouch *et al.*, 1988), lettuce (Landry *et al.*, 1987), potato (Gebhardt *et al.*, 1991), peas (Ellis *et al.*, 1992), sugar beet (Pillen *et al.*, 1992), bean (Nodari *et al.*, 1993), pine (Carlson *et al.*, 1991), *Arabidopsis* (Reiter *et al.*, 1992), man (Donis-Keller *et al.*, 1987; Weissenbach *et al.*, 1992) and other organisms.

The advantages offered by RFLP markers are obvious: they exhibit co-dominant inheritance, usually have no phenotypic side-effects and are expressed regardless of environmental conditions, i.e. allow the genotyping of each individual with a high resolution. Algorithms for visualizing the results of RFLP analysis have been developed, thus permitting 'graphical genotype representation' of each individual of a segregating generation and of any chromosome or segment derived from either of the parents as well as the segments in the heterozygous condition. The per chromosome number of such segments depends on the number of restriction sites examined, and it may be very large (Young and Tanksley, 1989; Gebhardt *et al.*, 1991). A procedure like this provides, in particular, an insight into the genomic distribution of crossovers based on the results from a single generation (see also Avner *et al.*, 1988).

Further application of molecular marker-based mapping will obviously result in basically new solutions of a wide range of genetic problems. These may, for example, include:

- mapping and analyzing the genetic nature of the recombination hotspots (Rasooly and Robbins, 1991);
- revealing the genetic determination of human disease and genetic counseling (Newmark, 1984; Gusella *et al.*, 1984; Lander and Botstein, 1986; Gershon *et al.*, 1990);
- elucidating the genetic architecture of economically important organisms and ensuring reliable selection of genotypes in breeding programs (Soller and Beckmann, 1988; Tanksley *et al.*, 1989; Williams *et al.*, 1991; Edwards *et al.*, 1992);
- controlling between-species transfer of genes and gene blocks (Rogowsky *et al.*, 1991; DeVincente and Tanksley, 1993);
- studying population genetic and breeding problems associated with measuring heterozygosity, polymorphism and adaptation (Mitton and Grant, 1984; Chakraborty, 1987; Nevo, 1987, 1988b; Altukhov, 1989; Shattuck-Eidens *et al.*, 1990; Tomekpe and Lumaret, 1991; Zhang *et al.*, 1992);
- identifying parental inbred lines of the hybrids (Soller and Beckmann, 1983; Welsh *et al.*, 1991);
- elucidating species origin and systematics, as well as the genome evolution and the genetic basis of interspecific differences (Miller and Tanksley, 1990; Doebley and Stec, 1991, 1993; Gebhardt *et al.*, 1991; Wang, Second and Tanksley, 1992).

Another class of molecular markers includes the polymorphism for tandem repeat number – minisatellites and microsatellites. The length of an individual minisatellite repeat depends on its genomic location and ranges, e.g. in man from 14 to 70 nucleotides (Jeffreys, Wilson and Thein, 1985). The simplest microsatellite probes are repeated bases, di-, tri- and tetranucleotides, etc.

Genetically, this means that markers with large number of alleles corresponding to variation in tandem repeat number (VNTR) can be used in the analysis. These, along with other markers, have been successfully utilized for fingerprinting and comparing individuals, populations and species (Jeffreys, Wilson and Thein, 1985; Nakamura, Leppert and O'Connell, 1987; Gilbert *et al.*, 1990; Kashi *et al.*, 1990; Amos, Barrett and Dover, 1991; Burke *et al.*, 1991; Ellegren *et al.*, 1992). It has also been suggested that heteroduplexes – regions with mismatched bases – constitute a promising class of molecular genetic markers (Roberts *et al.*, 1989). Like mini- and microsatellites, they seem to be much more polymorphic than RFLPs.

A new and probably the most useful class of polymorphic molecular markers based on polymerase chain reaction (PCR) should be considered (Williams *et al.*, 1990; Welsh and McClelland, 1990; Arnheim, White and Rainey, 1990; Erlich, Gelfand and Sninsky, 1991; Navidi, Arnheim and Waterman, 1992). Random amplified polymorphic DNA sequences (RAPD) based on short primers (e.g. 10 base pairs) and arbitrary-primed (AP) DNA markers can be developed to target specific genomic regions. This new approach has been shown to be very effective in gene mapping, revealing genotype and species differences, searching for single-pair changes in the DNA of tested genotypes, detecting specific RNA and DNA sequences, characterizing collections of inbred lines, identifying the parental lines of hybrids, and so on. Microsatellites (e.g. dinucleotide repeats) seem to be very promising probes for PCR analysis especially in map construction (Love *et al.*, 1990; Weissenbach *et al.*, 1992). Another class of markers based on PCR analysis with a level of variation similar to that of microsatellite repeats, single-strand DNA conformation polymorphism (SSCP), has been suggested by Beier, Dushkin and Sussman (1992). Half of the primer pairs they tested were polymorphic between the compared inbred lines of mouse and up to 90% between mouse species.

The most important advantage of PCR markers over the standard RFLPs is their much higher resolution and sensitivity. In particular, the PCR technique enables analysis of DNA from individual cells, e.g. sperm (Li, Cui and Arnheim, 1990; Goradia and Lange, 1990) or meiocytes, as well as on degraded ancient tissue extracts (Golenberg *et al.*, 1990; DeSalle *et al.*, 1992; Cano *et al.*, 1993). Another merit of the PCR method is that the differences revealed occur within the amplified segments themselves (Williams *et al.*, 1991). By contrast, RFLPs may be due to insertions and deletions in unknown sites outside the hybridizing region of DNA. In using PCR fingerprints for marker analysis, of special value may be the primers based on sequences of intron–exon junctions (Weining and Langridge, 1991). This may help avoid targeting heterochromatic regions, thereby increasing the power of analysis. A very attractive approach has been suggested by Giovannoni *et al.* (1991) allowing the isolation of RAPD markers specific for any segment flanked by chosen RFLP loci and polymorphic in the given mapping population.

The cytological approach based on analysis of chiasma distribution at diplotene–diakinesis can be an accurate technique for evaluating the effects of genetic and external factors on recombination spectrum in a variety of organisms. For example, G.H. Jones (1967) isolated a meiotic mutant of rye with random crossover distribution along the chromosome arms; rye normally exhibits terminal localization of chiasmata.

Estimation of recombination from cytological data has numerous advantages. These include the possibility of defining the integral characteristics of the chromosomes and of the nucleus, independent test of the differential elimination of gametes and zygotes and the applicability of the analysis to a wide range of species, including those for which genetic maps are either unavailable or incomplete. All these features of the cytological method make it a good adjunct to marker analysis. Until quite recently, the two methods were also complementary in their accuracy. Despite their high overall accuracy (given a sufficient sample size), cytological estimates of recombination show low local precision. By contrast, marker analysis permits a fairly accurate estimation of recombination within any segment. However, simultaneous estimation over the entire genome has until recently been almost impossible, even in *D. melanogaster*, a genetically well-studied organism with a small chromosome number.

The situation improved in recent years, with respect not only to marker but also cytological analysis of recombination. Now, application of new techniques for staining meiotic chromosomes (e.g. bromodeoxyuridine + Giemsa) allows a more precise localization of crossover points (Jones and Tease, 1981). This has led to a fundamental conclusion that the widespread localized pattern of chiasma chromosome distribution is not an artifact but rather a reflection of non-random crossover distribution (Jones, 1987). Unfortunately, the procedure is technically complex and has so far been used in only a few species. Of considerable importance, however, is the analysis of distribution of the so-called recombination nodules as locations of these structures are believed to coincide with crossover sites (Carpenter, 1975, 1988; von Wettstein, Rasmussen and Holm, 1984; Croft and Jones, 1989; Zickler *et al.*, 1992). Analysis of the recombination nodule distribution indicates that non-random exchanges as manifested in crossover interference are also characteristic of nodule locations, but only at the late pachytene state (in early pachytene, the nodules are randomly located) (Rasmussen and Holm, 1984). In the last decade, some progress has also been made in the use of meiocyte spreading techniques in this field (Gillies, 1981, 1985; Stack and Soulliere, 1984; Albini and Jones, 1988; Bourd and Klimenko, 1990).

A useful adjunct to recombination-based mapping is cytological localization using somatic-cell genetic techniques (Williams *et al.*, 1979; Serov *et al.*, 1987; Gillois, 1991). This approach has allowed numerous human (and other mammalian species) loci to be assigned to particular chromosomes and segments. This has been achieved largely by the differential elimination of chromosomes of one of the species in a hybrid cell line at mitosis (for human mapping the commonly used combinations are man × mouse or man × Chinese hamster) and inferring the locations of human structural genes by comparing cytological and biochemical data. Special collections of murine monosomic cell strains with substituted human chromosomes as well as strains with various human segment insertions are currently being established. This material will help to shorten not only the chromosomal but occasionally regional localizations of genes or other DNA fragments (Lange and Boehnke, 1992).

Recombination-based mapping as a method of studying the genetic material topography has been a major analytical approach in genetics for nearly 80 years. However, the technique has little to offer in terms of measuring actual physical distances between genes. As already noted, this limitation can be partially overcome by using a combined marker and cytological analysis. New possibilities have opened up in this area with the advent of the pulsed-field electrophoresis allowing physical mapping of cloned (marker) genes and other DNA sequences based on separation of relatively large DNA fragments (Meagher, McLean and Arnold, 1988; Ganal, Young and Tanksley 1989; Lai and Birren, 1990; Gillois, 1991). Application of this technique has revealed enormous variation in the frequency of recombination events per unit length (per kilobase) of DNA sequences both between species and within single genomes (Meagher, McLean and Arnold, 1988) thereby supporting the classical cytogenetic findings (for further discussion, see a review in Chapter 4). A useful subsidiary tool in physical mapping of genes and in establishing the relationships between physical and recombination distances is the *in situ* hybridization of specific labeled DNA sequences (e.g. single-copy cDNAs) in preparations of meiotic cells (see, for example, Gustafson, Butler and McIntyre, 1990). Another possibility is to employ *in situ* hybridization with interphase nuclei, which gives a mapping resolution in humans of approximately one marker for every 100 kbp (van den Engh, Sachs and Trask, 1992).

An interesting approach has been proposed for physical mapping of nucleotide sequences based on DNA isolated from single cells (Dear and Cook, 1989). The authors suggest that the DNA of individual haploid cells (sperm) should be fragmented with the aid of gamma irradiation and the resultant material partitioned at random into a number of equal aliquots. The two DNA sequences (markers) tested are then hybridized with each of the aliquots and the above procedure repeated on a large number of cells. The tighter the linkage between the marker sequences tested, the higher the probability that they will yield a positive result upon hybridization with DNA from only one of the aliquots and, conversely, unlinked markers will tend to be found more often in different aliquots. This technique can sharply increase the power of genetic analysis, allowing the recombinational measure of chromosomal distance between sites to be supplemented with a variable which represents the physical distance more adequately. Recently, a promising method of mapping human genes has been developed based on yeast artificial chromosome (YAC) technique. It includes: identification and cloning of the cDNA of interest into a special yeast fragmentation vector; transference of the obtained plasmid into the YAC strain containing the target human gene; mapping the cloned sequence along the YAC employing the yeast recombination system (Das Gupta *et al.*, 1993).

2 Marker analysis

2.1 FORMULATION OF GENETIC RECOMBINATION

2.1.1 Basic parameters

Map distance, or the average number of crossovers in a given segment, is the generally accepted measure of recombination events in a region of the genome. The unit of measure is such that on average one exchange per meiotic cell occurs within a segment of unit length resulting in 50% of recombinant gametes. Map distance is of additive nature, i.e. for any subdivision of a segment into subintervals the segment length is the sum of lengths of its subintervals. It determines a scale of uniform density with respect to reciprocal recombination allowing comparison in terms of crossover rates between different linkage groups, different species, etc. (cytogenetic chromosome length would obviously be unsuitable for such an analysis).

A serious limitation of map distance as a measure is that it cannot be estimated directly, since an odd number of exchanges in a segment marked by two loci can lead to the appearance, in the progeny, of the same recombinant genotypes as those resulting from one exchange, whereas an even number results in no observable recombinants. Recombination frequencies (rf) estimated from the observed proportions of recombinant and non-recombinant individuals coincide with map distances only for short segments when multiple exchanges can be ignored and therefore is not a sufficiently good estimator of the frequency of crossover points between the marker loci. In view of this, transformation of experimentally estimated rf values into map distances (x) appears to be necessary. Relations of the form $rf = rf(x)$ are referred to as mapping functions (Sturt, 1976; Felsenstein, 1979; Liberman and Karlin, 1984; Foss *et al.*, 1993). By mapping function some authors understand a reverse transformation, i.e. the dependence of map distance on rf (Owen, 1950; Bailey, 1961; Morton and MacLean, 1984; Ott, 1985). Difficulties encountered in constructing mapping functions are due to the necessity of allowing for multiple exchanges and for the influence of crossovers upon one another (interference of exchanges).

Interference is one of the crucial factors determining the spatial pattern of crossover distribution and the frequency of multiple exchanges, which explains its important role in regulating recombinational variability. The existence of interference was demonstrated both genetically and cytologically (Muller, 1916, Haldane, 1931; Sybenga, 1975; Jones, 1987). It is ubiquitous in eucaryotes. Hitherto, only three species have been found (*Shizosaccharomyces pombe, Aspergillus nidulans* and *Ascobolus immersus*) in which crossing over (presumably) proceeds without interference (Egel-Mitani, Olson and Egel, 1982; Hastings, 1988).

2.1.2 Interference – its measurement and basic properties

Consider two adjacent marked chromosome segments m_1–m_2 and m_2–m_3 whose crossing over frequencies are rf_1 and rf_2 respectively. If exchanges in adjacent segments occur independently of each other, the probability of a double crossing over within the $m_1 - m_3$ segment can readily be determined as the product of probabilities, $rf_{12} = rf_1 rf_2$. Under the same assumption, recombination frequency rf_3 between loci m_1 and m_3 is given by

$$rf_3 = rf_1(1 - rf_2) + rf_2(1 - rf_1) = rf_1 + rf_2 - 2rf_{12}. \qquad (2.1)$$

From experimental evidence obtained on numerous organisms, it follows, however, that the observed frequencies of double crossovers are, as a rule, considerably lower than the expected values, i.e. an exchange in one segment affects the probability of exchanges occurring in adjacent segments. This phenomenon is termed 'genetic interference' (Muller, 1916). The degree of interference is measured by the coefficient of coincidence (c):

$$c = rf_{12}\,(\text{observed})/rf_{12}(\text{expected}) = rf_{12}\,(\text{observed})/(rf_1\, rf_2).$$

Allowing for interference,

$$rf_3 = rf_1 + rf_2 - 2crf_1\, rf_2. \qquad (2.2)$$

The values of rf_1, rf_2 and rf_3 can readily be obtained experimentally. Therefore, one can easily find c from equation 2.2. Of course, this is not the best way of estimating interference. More reliable estimates of c can be obtained based on the analysis of co-segregation of three or more loci rather than of marginal (for locus pairs) segregations (Lathrop *et al.*, 1984; Thompson, 1984).

If occurrence of crossing over in one segment reduces the probability of exchange in an adjacent segment ($c < 1$), then we have positive interference; when one exchange increases the frequency of another, we speak of negative interference ($c > 1$). A large amount of negative interference is a phenomenon associated with intragenic recombination, and its study resulted in considerable progress in our understanding of mechanisms of genetic recombination in general (Kushev, 1974; Whitehouse, 1982).

Two types of interference have been distinguished (e.g. Kushev, 1974). One is **chiasma interference**, under which an exchange occurring in a segment affects the probability of another one taking place in an adjacent segment. Clearly, chiasma interference determines the pattern of distribution of distances between successive exchanges along the chromosome length. In analyzing chiasma distribution patterns, Haldane (1931) observed a deviation from the Poisson distribution (significantly decreased variance) and interpreted this as a manifestation of positive chiasma interference. The other type is **chromatid interference**, which determines the pattern of chromatid configurations in successive pairs of chiasmata in a bivalent, i.e. the proportions of two-, three- and four-strand double exchanges. In the absence of chromatid interference, the ratio should be 1:2:1. Any deviation toward one of the types can be regarded as chromatid interference; evidence on chromatid interference is, however, scarce.

Consider briefly the main findings and general patterns emerging from studies of interference in higher organisms.

1. The amount of interference within a segment depends on its distance from the centromere. In the near-centromeric region, interference between exchanges in the adjacent segments of the same chromosome arm is maximum ($c \approx 0$), while it is practically non-existent ($c \approx 1$) across the centromere.
2. Two parameters have been introduced to describe the effect of the centromere on chiasma distribution: differential distance (d) and interference distance (i) (Mather, 1936a). The variable d measures the average distance between the centromere and the first chiasma and is specific for each particular bivalent. Parameter i, the distance between the neighboring chiasmata of the same arm, is (presumably) constant for all the chromosomes of the complement. According to Mather, chiasmata are formed successively along the length of the bivalent, starting from the centromere. However, for this hypothesis to fit experimental data from some species, it should be assumed that chromosome ends and other points can act as initiation sites (Henderson, 1963; Gillies, 1985; Chadov, Chadova and Hotskina, 1985). Findings reported in maize and *Drosophila* translocation heterozygotes disagree with Mather's hypothesis (Burnham, Stout and Weinheimer, 1972; Hawley, 1980).
3. According to Rendel's (1957) estimates, coefficient of coincidence increases in an approximately linear manner with increasing crossover distance between segments, provided crossing over rate variations reflect those in physical length. However, when rf variability results from the effects of specific internal or external factors (e.g. age or inversion in another region), the coefficient of coincidence increases less rapidly than one could expect under the linear relationship between c and rf. A similar discrepancy has been reported in maize (Goldman and Coe, 1968). Based on the data on translocation heterozygotes affecting marked segments and, on the whole, reducing crossing over, the parameter c has been found to be tending to

exceed unity (especially in samples with lower than average *rf* values). Among the significant deviations of coefficient of coincidence from the value of $c = 1$ corresponding to no interference, the values of $c > 1$ occurred three times as frequently as those of $c < 1$ for segments free of any rearrangements. Studies of the effects of temperature on crossing over and interference in *Drosophila* have revealed that changes in *rf* and *c* are negatively correlated: interference is decreased rather than increased upon an induced reduction in *rf* and, conversely, it becomes higher with increasing *rf* (Grell and Day, 1974; Grell, 1978). Similar results have been obtained in recombination studies of meiotic mutants (see Moens, 1969; Carpenter and Sandler, 1974; Lindsley and Sandler, 1977; Szauter, 1984).

4. The above results from maize indicate that the frequency of double exchanges can considerably exceed the level expected under independent crossing over in adjacent segments. This deviation can be regarded as negative interference for relatively long distances. It is not only for structural heterozygotes that such results (i.e. $c > 1$) have been obtained. Thus, negative interference in chromosome 1 has been reported in barley (Sogaard, 1974). In *Drosophila*, an excess of multiple exchanges characteristic of high negative interference has been found in a short segment spanning the centromere of chromosome 3, with some classes of reciprocal crossing over products differing significantly in numbers (Sinclair, 1975). Similar results have also been obtained in *Drosophila* by other workers (Morgan, Bridges and Sturtevant, 1925; Green, 1975; Dennell and Keppy, 1979; Korol *et al.*, 1984). Since the mechanism of the observed negative interference is currently unknown, one should take into account the results of formal analysis showing that values of $c > 1$ can also result from data pooling (because of heterogeneity of individual values of rf_i) (Sall and Bengtsson, 1989).

Using specially constructed strains with rearrangements in region *ri–p* on chromosome 3 of *Drosophila*, Dennell and Keppy (1979) obtained evidence ruling out the possibility of gene conversion as an explanation of the observed negative interference. They suggested that negative chromosome interference is characteristic of all chromosome regions with a very low average number of crossovers per unit physical length (the *ri–p* region in question is just among these: it accounts for 15% of the whole length of chromosome 3 chromatin and for as little as 1% of the genetic map length). In an analogous situation with recombination in a proximal region of the X chromosome, interference is, however, positive ($c \ll 1$) (Lakc, 1986). The mechanism of high frequency of double exchanges in proximal regions of autosomes is currently unknown.

2.1.3 Mapping functions

Mapping function is a measure of the dependence of the level of recombination (*rf*) between loci on the corresponding map distance allowing for the phenomenon of crossover interference (Sturt, 1976; Felsenstein, 1979; Risch and Lange, 1979; Karlin 1984; Liberman and Karlin, 1984; Foss *et al.*, 1993). The first function of this kind was developed by Haldane (1919) for the case of no interference ($c = 1$).

Let crossover points be randomly distributed along the chromosome length. Suppose also that the average number of exchanges is small. Then, the probability of k exchanges having occurred between two loci separated by the distance x can be calculated based on the Poisson distribution as:

$$P_k(x) = \frac{e^{-x}x^k}{k!}, \quad k = 0, 1, 2, \ldots \tag{2.3}$$

It is only at odd numbers of exchanges that gametes recombinant for a given pair of loci are formed. Therefore, recombination frequency (rf) can be calculated as the sum of the probabilities of odd exchanges (Haldane, 1919)

$$rf(x) = \sum_{k=0}^{\infty} P_{2k+1}(x) = \frac{(1 - e^{-2x})}{2}. \tag{2.4}$$

At short distances, $rf(x) \approx x$; the value of rf becomes increasingly different from that of x with increasing x. When x is rather large, rf approaches the level of 0.5 (50%), which corresponds to independent segregation. Note that this kind of approach defines recombination in purely genetic terms, i.e. based on gametic types produced by a given pair of parental chromosomes. Therefore, there is no need for consideration of all possible four-strand configurations in the absence of chromatid interference (Fisher, 1948a).

There are two main approaches to constructing mapping functions in the presence of crossover interference: (a) based on setting up a differential equation; and (b) formulating the process of crossover formation and distribution along the chromosome length (from some initial point) based on probability density functions of lengths of intervals between successive crossovers. The former case deals with the crossover interference in a segment of length x and in a short adjacent segment, dx. For intervals x, dx and $x + dx$, the formula for combining recombination fractions (equation 2.2) can be used in the following form (Bailey, 1961, 1967):

$$rf(x + dx) = rf(x) + dx - 2c\,(rf)\,rf dx,$$

where $rf(x)$ is crossing over rate within a map segment of length x and $c(x) = c[rf(x)]$ is the marginal coefficient of coincidence for segment x with a very short adjacent interval. Hence

$$\frac{drf(x)}{dx} = 1 - 2rf(x)\,c\,[rf(x)]. \tag{2.5}$$

Differential equation (2.5) is an important tool for generating mapping functions under various assumptions of the nature of interference (i.e. about how coefficient of coincidence c depends on rf or x). For $c = 0$, we obtain $rf = x$, i.e. an approximation for tightly linked loci (Morgan's function). Solving for $c[rf(x)] = 1$ results in Haldane's function (equation 2.4). Kosambi (1944) used the relation $c[rf(x)] = 2rf$ satisfying the condition of complete interference ($c = 0$) for short genetic distances ($rf \approx 0$) and of no interference when x is large (i.e. $c \approx 1$ when $rf \approx 0.5$). Solving equation 2.5 gives Kosambi's mapping function:

$$rf = 0.5 \tanh(2x) \quad \text{or} \quad x = 0.25 \ln\left[\frac{(1+2rf)}{(1-2rf)}\right].$$

Kosambi's function is a rather good approximation for *Drosophila*, rice, house mice and other organisms (Bailey, 1967). The formula for combining recombination fractions in adjacent intervals, corresponding to Kosambi's function, takes the following form:

$$rf_3 = \frac{(rf_1 + rf_2)}{(1 + 4rf_1 rf_2)}. \tag{2.6}$$

Various modifications of equation 2.5 have been proposed (reviewed by Liberman and Karlin, 1984), for example:

$$\frac{drf}{dx} = 1 - p \cdot rf \rightarrow rf = \frac{(1 - e^{-px})}{p};$$
$$\frac{drf}{dx} = 1 - (2rf)^n,$$

where p and n are constant parameters. In the latter case, solving for $n = 1$ gives Haldane's function, for $n = 2$ Kosambi's function and for $n = \infty$ Morgan's function ($rf = x$), based on which Sturtevant constructed the first chromosome map for *Drosophila*. For $n = 4$, Carter and Falconer (1951) derived the relationship between rf and x as

$$x = \frac{[\tanh^{-1}(2rf) + \tan^{-1}(2rf)]}{4}.$$

Neither an explicit form of the reverse relation $rf(x)$ nor the formula for combining recombination fractions of the type of equation 2.6 has been obtained in this case. However, the rf value corresponding to any distance x can readily be found numerically based on iterative procedure (see Ott, 1985):

$$rf_{i+1} = 0.5 \tanh[4x - \tan^{-1}(2rf_i)], \text{ with } rf_0 = x.$$

Mapping functions of this kind are a good approximation to recombination only in median segments of long chromosome arms. However, they do not take account of multiple strand exchanges and the distance of the segment from the centromere, nor do they describe the cases (rare as they are) of $rf > 50\%$.

Felsenstein (1979) extended Kosambi's idea to include the assumption that the marginal coefficient of coincidence depends on a certain parameter, k, such that $c\,(rf, k) = k - 2\,(k-1)\,rf$, and, therefore, is a good approximation for both short and long segments. For $rf \approx 0$, the amount of interference depends on k; for $rf = 0.5$, there is no interference and $c = 1$. This reduces equation 2.5 to

$$\frac{drf}{dx} = 1 - 2krf + 4\,(k-1)\,rf^2.$$

Solving this, subject to the initial condition $rf(0) = 0$, gives

$$x = \frac{1}{2\,(k-2)} \ln \frac{1-2rf}{1-2\,(k-1)\,rf}, \quad rf = \frac{1}{2}\,\frac{1-e^{-2(2-k)x}}{1-(k-1)\,e^{-2(2-k)x}}.$$

For $k = 1$, we obtain Haldane's function, and for $k = 0$ Kosambi's one. Permissible values of k fall within the $0 \leqslant k \leqslant 2$ interval, with $1 < k \leqslant 2$ corresponding to negative interference. Using negative k values, one can simulate the map expansion effect apparent at short distances (Kushev, 1974). The formula for combining recombination rates takes a relatively simple form:

$$rf_3 = \frac{(rf_1 + rf_2 - 2krf_1\,rf_2)}{[1 - 4\,(k-1)\,rf_1 rf_2]}.$$

For two adjacent intervals, the coefficient of coincidence is calculated based on recombination fractions within these intervals:

$$c_{12\cdot 23} = \frac{(rf_1 + rf_2 - rf_3)}{2rf_1 rf_2} = 1 + (k-1)\,(1 - 2rf_3),$$

from which the value of k can be found.

Rao *et al.* (1977) used a different method for generalizing mapping functions:

$$w\,(\theta, p) = [p\,(2p-1)\,(1-4p)\ln(1-2\theta) + 16p\,(p-1)\,(2p-1)\tan^{-1}(2\theta) + 4p\,(1-p)\,(4p+1)\tanh^{-1}(2\theta) + 6p\,(1-p)\,(1-2p)\,(1-4p)\,\theta]/6.$$

At various values of the mapping parameter p, $w\,(\theta, p)$ ensures approximation to one of the above types. For $p = 1$, we obtain Haldane's function, for $p = 0.25$ that of Carter and Falconer (1951), and finally, for $p = 0$, Morgan's function. The actual value of p can be estimated from experimental data. Different chromosome regions (e.g. near-centromeric, median or telomeric) may show different values of p (Morton, MacLean and Lew, 1985). Note that this method was pioneered by Haldane (1919), as a combination of functions corresponding to $p = 1$ and $p = 0$.

Mapping functions derived from the differential equation 2.5 are a good approximation to recombination, but only in median segments of long chromosome arms. They do not take account of multiple strand exchanges and the distance of the segment from the centromere, nor do they describe the cases (rare as they are) of *rf* > 50%. All mapping functions of this class have one grave weakness: frequencies of the recombinant classes cannot be calculated (based on map distances) for cases of more than three loci.

A second class of mapping functions is based on probability density functions of interval lengths between crossover points along the chromosome arm (Owen, 1950). Owen's metrical theory and its further modifications are detailed in Bailey (1961) and other works (Cobbs, 1978; Stam, 1979; Foss *et al.*, 1993). Various modifications of this approach yield maximum *rf* values ranging from 52.2% to 60% and permit the description of four-strand crossing over (explicitly allowing for both types of interference–chromosome and chromatid). One of the functions of this type was derived by Owen (1951): $rf = (1 - e^{-2x}\cos 2x)/2$. Algorithms for analysis based on this approach are rather complex, with most of the models being obviously based on Mather's interpretation of chiasma formation as a process proceeding from the centromere (Bailey, 1961). At the same time, numerous data indicate that Mather's model is an oversimplified one (Henderson, 1963; John and Lewis, 1965; Burnham *et al.*, 1972).

An advantage of using metrical mapping functions is the possibility of simultaneously analyzing data from many loci, although this is not assured automatically for all mapping functions of this type (Liberman and Karlin, 1984). Functions that always allow the frequencies of recombinant classes expected under multimarker segregation to be computed can be constructed by formulating the scheme 'exchange number–exchange map positions' (Karlin and Liberman, 1979; Karlin, 1984; Liberman and Karlin, 1984; Goldar and Fain, 1988; Goldar, Fain and Kimberling, 1989). In this scheme, the probability distribution is specified for the number of exchanges (0, 1, 2, etc.) and, subject to the condition that k crossovers have occurred, their locations are determined as independent events based on the specified distribution $F(x)$ describing crossover positions within the chromosome region in question. Note that the scheme is convenient not only for deriving mapping functions, but also for computer simulations of recombination.

In view of the rapid development of new molecular marker analysis techniques, the adequacy of a mapping function in a multimarker case seems to be a very important consideration in selecting an appropriate formulated description of the recombination process.

2.2 ESTIMATING RECOMBINATION FROM EXPERIMENTAL DATA

In terms of estimating exchange distribution within the chromosome, of greatest interest is a combined analysis of multilocus data. This is subject to a number of limitations of both experimental and mathematical nature. To carry out a multilocus analysis, multimarker strains are needed, but their production involves considerable inputs, even with such model organisms as *Drosophila* and *Arabidopsis*. The use of multilocus analysis of RFLPs and other molecular markers opens up entirely new possibilities for genetic research. Even here, however, the number of identifiable genotypes in the segregating progeny (in one generation) has so far been limited from tens to a few hundreds, the method being a highly laborious one (Helentjaris *et al.*, 1986). To allow for multiple exchanges and to estimate (with reasonable accuracy) the parameters describing chromosomal distribution of crossovers, the number of segregating progeny (e.g. F_2) of the heterozygote should be relatively high. But even having obtained the necessary data, the experimenter still faces a serious problem in their interpretation. This is particularly so in the presence of differential viability effects as well as when it is necessary to study a multilocus system of linked genes based on complex pedigrees (Lathrop *et al.*, 1984; Ott, 1985; O'Connell *et al.*, 1989) and, in many cases, even in controlled crosses, e.g. F_2. The most serious difficulties have been encountered in analyzing F_2 data in which marker segregation is distorted. This problem may be of special interest for researchers working with self-pollinating species in their search for factors inducing crossover redistribution and, simultaneously, segregation distortion, in studying hybrid dysgenesis (causing changes in both *rf* and segregation), in estimating recombination in interspecific crosses or in meiotic drive systems, etc.

Let the results of a recombination experiment be represented as the observed k numbers of phenotypic or genotypic classes $n_i\,(i=\overline{1,k})$. Unless otherwise specified, a situation is assumed corresponding to controlled crosses. In the simplest case of complete dominance $k=2^r$, where r is the number of loci examined; under intermediate expression and in the absence of epistasis for all the markers, we obtain, for F_2, $k=3^r$, or $2^r \leqslant k \leqslant 3^r$ in the general diallelic case. For further analysis, a model describing the experimental situation should be adopted. According to such a model, the expected class frequencies depend on s unknown parameters: $f_i = f_i(\theta_1, \ldots, \theta_s)$, $i=\overline{1,k}$. Parameters $\theta_i\,(i=\overline{1,s})$ may represent crossing over frequencies in different segments under different conditions, gamete and zygote viability coefficients, misclassification coefficients, the degree of genetic interference, etc. Upon the completion of an experiment, the researcher should obtain numerical estimates of unknown parameters based on the formulated model and observed phenotypic (or genotypic) classes. These estimates should have certain optimum properties such as, for example, to provide the best approximation of the theoretical segregation obtained to the actually observed one and/or minimum standard errors of the parameters estimated. In solving problems of this type, the maximum likelihood (ML) method introduced into statistics by R. Fisher, especially in connection with the problem of estimating genetic linkage, has gained the widest currency (Fisher, 1948b; Allard, 1956; Bailey, 1961; Kendall and Stuart, 1967).

A set of numbers $\hat{\theta} = \{\hat{\theta}_1, \ldots, \hat{\theta}_s\}$ providing a maximum of the likelihood function or of the log-likelihood function, which is the same thing, are considered to be ML estimates of unknown parameters $\theta = \{\theta_1, \ldots, \theta_s\}$ (Bailey, 1961; Kendall and Stuart, 1967):

$$\ln L = \ln L(\theta_1, \ldots, \theta_s) = \sum_{i=1}^{k} n_i \ln f_i(\theta_i, \ldots, \theta_s) + \text{const.}$$

The maximum can be found by solving the following set of equations:

$$S_i = \frac{\partial \ln L(\theta)}{\partial \theta_i} = 0, \quad i = \overline{1, s}.$$

If analytical solution is impossible, an iterative procedure for numerical solution is constructed which has the form

$$\theta^{(m+1)} = \theta^{(m)} + SM^{-1}, \tag{2.7}$$

where S is a vector of partial derivatives S_j at point $\theta = \theta^{(m)}$ and M is the matrix:

$$M = -\left.\frac{\partial^2 \ln L(\theta)}{\partial \theta_i \partial \theta_j}\right| s, \quad i, j = 1$$

computed at the point $\theta = \theta^{(m)}$. Variances and covariances are obtained by inverting the **information matrix** $I(\theta)$, which is defined as the mathematical expectation of M and can be estimated as M at the point $\theta = \hat{\theta}$ corresponding to the max$L(\theta)$.

In the simplest case of segregation for two recessive genes a and b in the absence of deviation of monohybrid segregation ratios from the expected values, theoretical frequencies for four phenotypic classes AB, Ab, aB and ab will be, for example in F_2, as follows (Fisher, 1948b):

$$f_1(\theta) = \frac{(2+\theta)}{4}, \quad f_2(\theta) = f_3(\theta) = \frac{(1-\theta)}{4}, \quad f_4(\theta) = \frac{\theta}{4}, \tag{2.8}$$

where $\theta = (1-p)^2$ and p^2 for coupling phase ($F_1 = AB/ab$) and repulsion phase ($F_1 = Ab/aB$), respectively, and p is the rate of recombination between loci A/a and B/b. In order to obtain an estimate of parameter θ, the maximum of the log-likelihood function should be calculated:

$$\ln L(\theta) = n_1 \ln(2+\theta) + (n_2 + n_3)\ln(1-\theta) + n_4 \ln(\theta) + \text{const.}$$

Setting the derivative with respect to θ equal to zero

$$S(\theta) = \frac{d \ln L(\theta)}{d\theta} = \frac{n_1}{2+\theta} - \frac{n_2 + n_3}{1-\theta} + \frac{n_4}{\theta} = 0,$$

we obtain the estimate of θ as the solution of the quadratic equation $N\theta^2 - (n_1 - 2n_2 - 2n_3 - n_4)\theta - 2n_4 = 0$, where $N = n_1 + n_2 + n_3 + n_4$. The variance of this estimate can be found as

$$\sigma_\theta^2 = -E\left(\frac{d^2 \ln L}{d\theta^2}\right)^{-1} = \frac{2(2+\theta)(1-\theta)\theta}{[N(1+2\theta)]}.$$

The amount of information, I, on parameter θ contained in experimental data is calculated as $-E(d^2\ln L/d\theta^2)$ (Fisher, 1948b). Thus, each observation from F_2 contains the amount of information equal to $I/N = (1+2\theta)/[2\theta(1-\theta) \times (2+\theta)]$. Calculating the value of I/N enables one to compare different mating schemes and data types with respect to their statistical efficiency for estimating recombination (Fisher, 1948b; Allard, 1956; Bailey, 1961; Serebrovsky, 1970; Ott, 1985). The estimates generated by the ML method have been shown to be effective: they yield minimum ${\sigma^2}_\theta$ (thereby ensuring maximum extraction of information on the desired parameter contained in the data) (Kendall and Stuart, 1967).

Effective estimates may also be obtained by the method of products and minimum X^2 method and other approaches (Immer, 1930; Fisher, 1948b; Bailey, 1961; Serebrovsky, 1970).

In the example just considered, out of the three degrees of freedom (the number of classes minus one), one has been used for estimating the parameter and the remaining two are used in testing the model based on the χ^2 test:

$$X^2 = \sum_{i=1}^{4} (n_{i_o} - n_{i_e})^2 / n_{i_e},$$

where index o refers to observed and e to expected class numbers. For any θ, the value of the X^2 is the sum of three positive components, two of which are caused by deviations from the expected monohybrid ratios and the third depends on the difference between the given value θ and the value $\hat{\theta}_{ML}$, which is the solution of ML equation (Fisher, 1948b). This is important in interpreting situations where the computed value of X^2 significantly exceeds the tabulated one.

Sophisticated molecular techniques currently used in genetic mapping enable simultaneous analysis of tens of markers (polymorphic sites of restriction, VNTR - and RAPD - markers, etc.). To this end, it is desirable, in controlled crosses (backcross or F_2) of animals and plants, to use as parental stocks forms sufficiently divergent at the level of DNA sequences (e.g. different species). This approach has been termed interspecific backcross or intercross mapping (Avner *et al.*, 1988; McCouch *et al.*, 1988; Paterson *et al.*, 1988, 1991; Gebhardt *et al.*, 1989; Siracusa *et al.*, 1990). With a sufficiently large number of markers per chromosome, it is possible, for each genotype of the interspecific backcross progeny, to determine the positions of crossing-over points in the meiocyte which gave rise to the individual in question (see also Young and Tanksley, 1989). With this approach it is necessary to take into account the possible influence of interspecific hybridization on recombination (Rhyne, 1962; Rick, 1972; Korol and Bocharnikova, 1988; Seldin, Howard and D'Eustachio, 1989; Gebhardt *et al.*, 1991; Doebley and Stec, 1991; see also pp. 183–6). When only molecular markers with unknown chromosomal locations are used, cytogenetic techniques, e.g. trisomic or translocation analysis as well as *in situ* hybridization, can be employed to determine the correspondence between chromosomes and the resulting linkage groups (McCouch *et al.*, 1988; Weber and Helentjaris, 1989; Gustafson, Butler and McIntyre, 1990).

With controlled crosses, there also exists another interesting possibility for estimating linkage; this is offered by the so-called **recombinant inbred lines** RILs) (Burr *et al.*, 1988; Burr and Burr, 1991). RILs result from inbreeding of the F_2 progeny derived from crosses between two inbred parental lines. It has been assumed that it takes 6–7 generations for RILs to attain the required level of homozygosity. As with F_2 or backcross progeny, linkage estimation for RILs is based on the ML method, with the expected frequencies of recombinant genotypes being computed from the corresponding model (Haldane and Waddington, 1931). Given that N is the total number of inbred lines produced, j is the number of recombinant lines for the chosen pair of marker loci and θ is recombination frequency in the marked segment, then the log-likelihood function is computed as follows:

$$\ln L(\theta) = j\ln\left[\frac{4\theta}{(1+6\theta)}\right] + (N-j)\ln\left[\frac{1+2\theta}{(1+6\theta)}\right] + \text{const.}$$

Hence (Silver, 1985; Smouse, 1988):

$$\hat{\theta} = j/(4N-6j), \quad \sigma^2_\theta = \frac{\theta(1+2\theta)(1+6\theta)}{4N}.$$

RILs are considered now as an especially useful instrument to study in model mammals (primarily mouse) the genetic basis of human diseases, including multigenic diseases (Justice, Jenkins and Copeland, 1992).

The ML method also permits the simultaneous estimation of θ and σ^2 from F_2 (or backcross) and RIL data (see Smouse, 1988). Another important problem in practical mapping is that of estimating *rf* based on the progeny resulting from several (k) generations of selfing. In this case, the analytical expression for θ and σ^2_θ is virtually impossible to derive. Nevertheless, an iterative procedure for the numerical calculation of ML estimates can easily be constructed for any value of k (Snape, 1988). For situations of this type, algorithms can also be developed for estimating recombination in adjacent regions, allowing for exchange interference and distorted segregation.

The method based on RILs has a number of advantages over the analysis of a single segregating generation (Burr *et al.*, 1988). One of these is the possibility of estimating small values of *rf* (corresponding to map distances of about 5 centimorgans and less) with smaller confidence intervals. This advantage is particularly important in constructing genetic maps based on molecular markers. Homozygous original lines and RILs can easily be maintained for a number of generations and tested for new DNA markers with the possibility of culturing them under different ecological conditions. By drawing on the information previously accumulated in the database, mapping of new markers becomes relatively easy and the entire process of map construction is reduced to a cumulative procedure (Burr *et al.*, 1988).

A disadvantage of using RILs is, in our opinion, a possible strong bias in estimates under selective elimination of genotypes which can distort the ratio of recombinant to non-recombinant genotypes in a set of RILs. Another important constraint results from the implicit assumption about the constancy of recombination parameters during inbreeding. Crossing and inbreeding can produce dramatic changes in recombination, which are usually opposite in sign in obligate outbreeders and self-pollinators (reviewed by Zhuchenko and Korol, 1985). Thus, the class frequencies in the resultant population will depend on the *rf* values averaged over generations. This last remark also applies to the production of homozygous recombinants (double haploids) through anther culture (see, for example, Graner *et al.*, 1991; Zivy *et al.*, 1992).

Detection and estimation of linkage among the pedigrees of arbitrary structure is carried out by the so-called lod score method. Lod score is a decimal logarithm of the ratio $P(\theta)/P(0.5)$ (see Ott, 1985):

$$Z(\theta) = \lg \frac{P(\theta)}{P(0.5)},$$

where $P(\theta)$ is the probability of sampling for a specific family under a certain fixed level of recombination, θ, and $P(0.5)$–under free recombination. The total value of lod for the entire sample is the sum of values for individual families (Sturt, 1975; Ott, 1985). Based on maximization of $Z(\theta)$, or of $P(\theta)$ (which is equivalent in the case of one parameter), the ML estimate and confidence interval for θ can be obtained. It is recommended that θ should be calculated after pooling the results from various sources, as well as data from various families, by summing lod functions. This procedure, unlike that of averaging the estimates themselves, yields an unbiased overall estimate. For a detailed treatment of complicated problems pertaining to genetic analysis of linkage (testing of hypotheses and estimation of parameters) from complex pedigrees the interested reader is referred to Ott (1985). While applying the lod score method to an unknown situation one usually has to assume that the loci are in linkage equilibrium, which may not be the case (Smith, 1986).

In the above calculations reflecting the main principles of estimation procedure, segregations correspond to the 'ideal' type. However, differential viability of genotypes, incomplete penetrance of markers, meiotic drive and other factors may result in distorted monohybrid ratios and, consequently, in biased *rf* estimates. Also, sex differences in crossing-over level and other departures from the 'ideal' (standard) patterns of inheritance are possible. The problems of estimating recombination parameters in situations like these are discussed in the next section. Here we would like to outline the potentialities of linkage analysis under multilocus segregation.

The expected frequencies of each of the 2^r gametic classes in a heterozygote for r linked loci can be calculated by specifying the values of $2^{r-1}-1$ parameters and assuming equal frequencies for reciprocal types (Lewontin, 1964). Thus, for $r = 5$, the number of parameters describing the probabilities of different types of exchanges is 15, namely: probabilities S_1, S_2, S_3 and S_4 of only single exchanges occurring in four segments; probabilities $D_{12}, D_{13}, D_{14}, D_{23}, D_{24}$ and D_{34} of double exchanges; probabilities $T_{123}, T_{124}, T_{134}$ and T_{234} of triple exchanges; and probability Q_{1234} of quadruple exchanges. At the same time, the number of recombination frequencies between all possible pairs of markers is $r(r-1)/2$, which is substantially lower than $2^{r-1}-1$. This implies that pairwise recombination analysis based on all of the dihybrid segregations entails considerable loss of information (Bailey, 1961; Morton and MacLean, 1984). Thus, the power of three-locus analysis based on family data in humans may be nearly four times as high as in the case of computation based on combinations of two-locus tests (Lathrop *et al.*, 1984; Thompson, 1984; Smith, 1986).

The maximum likelihood method yields estimates of parameters (recombination frequencies, coefficients of coincidence), provided an appropriate model has been chosen in which the expected frequencies of phenotypic classes are expressed as functions of unknown parameters. In principle, there is no need to explicitly formulate relations of the $f_i = f_i(\theta_1, \ldots, \theta_s)$ type in order to obtain numerical estimates. It is sufficient to specify an algorithm for computing the expected frequencies (f_i) from a range of permissible values of a set of parameters ($\theta_1, \ldots, \theta_s$) provided the latter is fixed. It is this kind of approach that we have used in implementing a computer-assisted system of estimating recombination in controlled crosses (Preygel, 1986). The standard version provides for the following types of analysis: estimation of data heterogeneity (based on segregation of all r loci as well as on mono-, di- and trihybrid segregations); test of fit of the observed monohybrid segregation ratios to the expected ones; calculation of *rf* values and their standard errors; point and interval estimation of coefficients of coincidence for any pair of adjacent and non-adjacent chromosome segments, etc. The general approach remains, in principle, the same for non-standard situations in which formulation of special models allowing for a particular type of deviation from the 'ideal' pattern of inheritance is needed for parameter estimation (sections 2.3 and 2.4).

The major problem with multilocus analysis is the necessity of estimating a large number of parameters – 3^r-1 in the general case, 2^r-1 under complete dominance, $2^{r-1}-1$ under complete dominance – and assuming equal frequencies of reciprocal classes. At $r = 5$, their number for the last case is 15, at $r = 6$ it is 31, and so on. At the same time, it is possible to consider segregation at seven or eight or many more loci (e.g. molecular markers) simultaneously, i.e. the number of parameters, even under the ideal behavior of markers, approaches 100, thus excluding the possibility of directly applying the ML method as an estimation procedure to the whole body of data. If that is the case, further analysis is only possible provided some techniques for dimensionality reduction are used (leading to a loss of generality). One of the solutions is to formulate some simplifying assumptions (e.g. of interference of multiple exchanges) or to specify coefficients of coincidence according to some metric, and so on. But, even with this approach, it is reasonable to confine oneself to a lower dimension basic model, say a five-locus one (e.g. Smith, 1986). Then, segregations with $r > 5$ can be analyzed by recurrent application of the basic model. Of course, in doing so, a fraction of the information may be lost. But this limitation can be viewed as an insignificant one since exchanges of the orders higher than 3 are relatively rare for the majority of higher eucaryotes.

Furthermore, in view of the above remark, we think that it will be advantageous to choose a four- rather than a five-locus model as a basic one for problems involving assessment of the effects of various factors on recombination. For an ideal segregation this means the estimation of seven parameters. An increase in dimensionality associated with the need to allow for various disturbances is, in this case, not large enough to make the problem of estimation an intractable one.

Consider briefly a possible way of reducing, through simplifying assumptions, the problem of dimensionality when estimating interference. For example, in the case of a four-locus basic model, one can assume that for bivalents with three exchanges the probability of a third crossing over taking place, when two already exist, depends solely on the exchange within the segment nearest to the given one, i.e. $p(3|12) = p(3|2)$ (the Markow assumption) (Morton and MacLean, 1984). The probability of a triple exchange being $\theta_{123} = p(3|12)\,\theta_{12}$, we obtain $p(3|12) = \theta_{123}/\theta_{12}$. On the other hand, $\theta_{23} = p(3|2)\,r_2$ or $p(3|2) = \theta_{23}/r_2$. Then, the assumption $p(3|12) = p(3|2)$ gives $\theta_{123}/\theta_{12} = \theta_{23}/r_2$ or $\theta_{123} = \theta_{12}\theta_{23}/r_2$. The latter relation permits the coefficient of coincidence c_{123} to be calculated instead of introducing it as an independent parameter of the model. Indeed, by definition (p. 31):

$$c_{123} = \frac{\theta_{123}}{(rf_1 rf_2 rf_3)} = \frac{\theta_{12}\theta_{23}}{(rf_1 rf_2 \cdot rf_2 rf_3)} = c_{12}c_{23}\,.$$

However, in spite of all the attractiveness of such an analysis, the question of how realistic the Markow assumption is remains open.

A further reduction in the number of parameters is possible through the use of mapping functions (Morton and MacLean, 1984; Morton, MacLean and Lew, 1985; Ott, 1985). Assume first that the lengths of the three segments are known (x_1, x_2, x_3). Let the mapping function be of the form $rf = f(x, t)$, where t is a parameter specifying the degree of interference in a given segment; then we calculate $rf_i = f(x_i, t)$, $i = \overline{1, 3}$. In order to determine the expected segregations (i.e. to calculate the frequencies of all 16 haplotypes), it is necessary to have $2^{4-1}-1=7$ parameters. Frequencies of double exchanges θ_{12} and θ_{23} in adjacent segments are computed as $\theta_{i,i+1}=rf_i+rf_{i+1}-rf_{i,i+1}$, where $rf_i=f(x_i,t)$, $rf_{i+1}=f(x_{i+1},t)$, $rf_{i,i+1}=f(x_i+x_{i+1},t)$. The remaining two parameters, θ_{123} and θ_{13}, are calculated as follows. For non-adjacent segments 1 and 3, we obtain, subject to the condition $\theta_{1(2+3)}=\theta_{12}+\theta_{13}-2\theta_{123}$, $\theta_{13}=\theta_{1(2+3)}-\theta_{12}+2\theta_{123}$, where $\theta_{1(2+3)}=(rf_1+rf_{2+3}-rf_{1+2+3})/2$, $\theta_{123}=\theta_{12}\theta_{23}/rf_2$. Thus, from three parameters, x_1, x_2 and x_3, we estimate, using a mapping function, seven parameters allowing the expected frequencies of 16 gamete classes and those of 16 phenotypic classes to be calculated (Morton and MacLean, 1984).

Consequently, it is also possible to solve a reverse problem, i.e. to calculate map distances from the observed numbers of phenotypic classes based on the mapping function chosen, and to test the adequacy of the model. Incidentally, the problem is reduced to estimating as few as $r-1$ parameters instead of the original $2^{r-1}-1$. This is an important advantage. Such a considerable reduction in dimensionality allows the model to be easily modified to include the cases of various distortions of monohybrid ratios, to provide for the possibility of analyzing not only test crosses but also intercrosses. Another advantage is the possibility of testing the applicability of various mapping functions, studying the effects of various factors (genetic and environmental) on subtle features of crossing-over distribution, including the pattern of interference of exchanges (double and multiple).

The multilocus recombination analysis strategy has been used in processing family data (e.g. in humans), in gene ordering and in constructing genetic maps based on these data (Lathrop *et al.*, 1984; Ott, 1985; Lander and Green, 1987; Buetow and Chakravarti, 1987; Kammerer and MacCluer, 1988, Edwards, 1991). In doing so, the general procedure of study is that of successive approximation to an optimum solution by first allowing for pairwise lod scores and then identifying groups of linked genes, ordering the groups and the genes within groups (O'Connell *et al.*, 1989). In some cases, multilocus analysis enhances the probability of detecting linkage by several orders of magnitude (Thompson, 1984). Family data pooling problems encountered under the multilocus approach also become urgent in genetic mapping using data from controlled crosses. The main reason for this is the difficulty of selecting parental pairs ensuring heterozygosity for the whole set of markers in question. Thus, Beavis and Grant (1991) constructed an RFLP map based on joint analysis of partially overlapping segregations in four F_2 families of maize. Clearly, to achieve a good efficiency of map construction procedure one should provide a high proportion of informative meioses (Ott, 1985). Especially helpful for that can be preliminary selection of multiallelic polymorphic markers (e.g. microsatellites) (Weissenbach *et al.*, 1992). Multilocus analysis also permits sex differences in *rf* to be allowed for and separate maps for each sex to be constructed (Donis-Keller *et al.*, 1987). Clearly, genetic recombination analysis of a scope like this providing segregation at tens or even hundreds of loci and ensuring a sharp increase in the information gain is only possible with the aid of fast computers and is carried out using specially developed packages, such as MAPMAKER, CRI-MAP, LINKAGE, GENLINK and others (reviewed by Weaver *et al.*, 1992).

One should not think that the multilocus approach to estimating *rf* values is worth using for statistical reasons alone. Multilocus analysis enables one to take account of the recombinational behavior of the chromosome as an integrated system (a principle established by cytogenetics) which cannot be achieved through methods based on the use of successive, three-point overlapping marker groups (see, for example, Blank *et al.*, 1988). Thus, at a different level of analysis, there is a return to older ideas of mapping underlying the metrical theory of crossing over (Owen, 1950; Bailey, 1961, 1967). However, in nearly all of the realized algorithms the integrity features are not yet taken into account; in particular, the assumption of no exchange interference considerably reduces the accuracy of results. Therefore, the problem of allowing for interference in processing multilocus data appears to be a particularly urgent one in view of the increasing importance of mapping techniques in genetic analysis, primarily because of the use of molecular markers (enzyme loci, RFLP and PCR) (Morton, 1988; Goldar, Fain and Kimberling, 1989).

2.3 SOME EXAMPLES OF CROSSING-OVER FREQUENCY ESTIMATION

The maximum likelihood method of analyzing linkage between marker genes offers the possibility of estimating parameters whenever the results of genetic analysis can be represented as frequencies of phenotypic (genotypic) classes. In order to apply the estimation procedure, it is necessary that the genetic hypothesis explaining the observed segregations (normal or abnormal) be formulated as expected class frequencies which depend explicitly on certain parameters, such as crossing-over rate, viability of genotypes, etc. It seems to be worthwhile to consider several examples of applying this method of estimate crossing-over rates in some specific situations.

Studies of natural and induced variation in genetic recombination involve solving a number of problems, and the direct application of classical methods in these cases is fraught with pitfalls. These may be due to factors such as segregation distortion, departure from random syngamy, incomplete penetrance of marker(s) and data heterogeneity. A simple approach is to try to avoid using markers that show distorted segregation ratios. Consistent realization of this approach implies that solving such problems as the relationship between recombination and meiotic drive, recombination induction by stress conditions and linkage analysis under distant hybridization has to be abandoned.

The widespread occurrence of segregation distortion in distant crosses established during the classical period of genetics based on morphological markers has been confirmed by recent evidence based on molecular (e.g. RFLP) markers (Paterson *et al.*, 1988; Seldin, Howard and D'Eustachio, 1989; Gebhardt *et al.*, 1991; Doebley and Stec, 1991; Eshed *et al.*, 1992). Thus, difficulties resulting from segregation distortion are, in many cases, fundamental. If there is no abandoning the problem completely, then certain trade-offs should be made, and in particular abnormalities resulting from incomplete penetrance or differential viability allowed for in the data. The best method to eliminate viability effects, the 'balance' one, is based on a joint maximum likelihood analysis of data from two test crosses: *cis*, $AB/ab \times ab/ab$; and *trans*, $Ab/aB \times ab/ab$ (Bailey, 1967; Hastings, 1972). The ML method is also applicable to more complex situations when differential viability is combined with incomplete penetrance in the same or in another locus (Bailey, 1961; Laing 1972; Clark, 1981).

We have developed for IBM PC and PDP-11 computers a system of algorithms for the genetic analysis of recombination data (Preygel, 1986). The system offers the possibility of processing data with monohybrid segregation distortions owing to various 'standard' causes (differential viability of gametes and zygotes, incomplete penetrance), is suitable for a joint consideration of results from the test crosses and the F_2 generation (the latter case allowing for situations then $rf_{♀} = rf_{♂}$ or $rf_{♀} \neq rf_{♂}$, including the case when crossing over is completely suppressed in one of the sexes). It also enables estimation of segregation heterogeneity, identification of deviant families, and so on. Parameter estimation is performed numerically by the maximum likelihood (ML) method. The algorithms also allow for treatment of situations in which the analytical form of expected frequencies for each of the phenotypic classes is not preset. This ensures a flexibility in the usage of the system and allows recombination to be rapidly estimated in diverse situations. For example, *rf* values during micro- and macrosporogenesis could be easily calculated based on a combined analysis of $P_1 \times F_1$ and $F_1 \times F_1$ data (Vizir and Korol, 1990). Other options of the system: estimate *rf* when identifying progeny from a cross of an $ABCD/abcd \times ABcd/abcd$ type; study the situations when individual markers are homozygous lethals or when complete dominance is observed at some loci and intermediate expression at others, etc. Provided the number of degrees of freedom is greater than the number of parameters estimated, there is a possibility of testing statistically if the hypothesis fits experimental data (Bailey, 1961).

2.3.1 Estimating *rf* between two loci when one of the homozygotes is lethal

The rate of crossing over can be determined based on the progeny produced by test crossing F_1 ($AaBb$) $\times$ $aabb$ or by selfing $F_1 \times F_1$. Consider the latter case in more detail, in a situation in which homozygotes BB are lethal. In the case of complete dominance for the phenotypic expression at both loci, the number of classes is 4 and, allowing for the elimination of *BB*, the expected proportions of phenotypic classes can be calculated as follows:

$$[2-2p(1-p)]AB:[1-(1-p)^2]Ab:2p(1-p)aB:(1-p)^2ab,$$

where p stands for crossing-over rate. Then, based on the observed numbers (n_1, n_2, n_3, n_4), the log-likelihood function is obtained as

$$\ln L(p) = n_1 \ln[2-2p(1-p)] + n_2 \ln[1-(1-p)^2] + n_3 \ln[2p(1-p)] + n_4 \ln(1-p)^2 + \text{const.}$$

This attains a maximum at a certain point $\hat{p}$ which is assumed to be an optimum estimate of parameter p. In order to find p, it is necessary to solve the equation $d\ln L(p)/dp = 0$, or to find $\max \ln L(p)$ numerically (section 2.2). It should be noted that in the situation under consideration one of the possibilities of accurately estimating p without a computer (i.e. based on 'manual' calculations) is simply to apply an exhaustive search of various p_i values by calculating the corresponding expected phenotype distributions and comparing these with the observed ones. It is this course that some authors have taken (see, for example, Butler, 1965, although the expected frequencies have been calculated incorrectly in this work); however, the disadvantages of the approach are not only the low accuracy and efficiency, but also the difficulties encountered in estimating standard errors of the estimates. Certainly, one may completely ignore the segregation distortions present and estimate p as in the case with normal segregation of the markers. But, clearly, this is fraught with the danger of obtaining a biased estimate.

Consider an example. Butler (1965, 1973) reported the results of testing linkage between the *Wo* (*woolly*) marker and other loci on chromosome 2 of tomato. The *Wo* gene is phenotypically dominant over the normal allele Wo^+ and homozygotes *WoWo* are lethal. Therefore, the $Wo:Wo^+$ ratio in F_2 is approximately 2:1. Ignoring the segregation distortion due to lethality of *WoWo* may result in a highly significant bias of the *rf* estimate. The bias is especially large in the case of tight linkage in repulsion phase (Table 2.1).

2.3.2 Allowing for sex differences in *rf*

When estimating crossing-over rate from the F_2 data, an assumption of equal exchange frequencies in female and male meiosis is usually made. Then, $rf = 1 - \sqrt{\theta}$, or $rf = \sqrt{\theta}$ [where $\theta = (1-rf)^2$ if $F_1 = AB/ab$, or $\theta = rf^2$ if $F_1 = Ab/aB$]. For some animal and plant species $rf_♀ \neq rf_♂$ (section 9.4). What is actually estimated in such a case is the average value, $rf = 1 - \sqrt{[(1-rf_♀) \times (1-rf_♂)]}$ or $rf = \sqrt{(rf_♀ rf_♂)}$. In maize, in particular, $rf_♀ < rf_♂$ in the majority of the segments (Robertson, 1984). A similar situation exists in *Arabidopsis* as well, in which some of the differences are as high as 300–400% (Figure 2.1). With differences between $rf_♀$ and $rf_♂$ as large as these, there is no point in obtaining

Table 2.1 The effects of the tomato *WoWo* homozygote lethality on the accuracy of *rf* estimates (from data of Butler, 1965, 1973)

Pair of markers			*Dihybrid segregation*				*rf estimate (%)*	
A(a)	*B(b)*	*Phase*	*AB*	*Ab*	*aB*	*ab*	*MP*	*ML*
Wo	*d*	c	7816	2027	2815	2198	34.9 ± 0.5	33.0 ± 0.5
		r	6542	1817	3097	238	31.8 ± 0.8	28.7 ± 0.6
Wo	*dil*	c	844	59	71	402	8.7 ± 0.8	7.5 ± 0.7
		r	1111	146	517	44	40.0 ± 1.9	19.7 ± 1.3
Wo	*Me*	c	1525	67	56	685	4.9 ± 0.5	4.1 ± 0.4
		r	1966	226	916	12(8)	22.4 ± 1.7	11.6 ± 0.8

ML and MP estimates have been obtained respectively by the maximum likelihood method taking account of the lethality of *WoWo* and by the product method (e.g. Immer, 1930); c and r are the linkage phases of coupling and repulsion respectively. The bias of the MP estimates is especially large in the *r*-phase under tight linkage.

an average estimate, rather it is worthwhile to estimate *rf* separately for micro- and macrosporocytes. This is achieved by reciprocal crosses of F_1 to the recessive tester *ab*: $F_1(♀) \times ab(♂)$ and $ab(♀) \times F_1(♂)$. However, such a procedure proves to be ineffective in some cases, as for example when the *ab* genotypes exhibit low fruiting capacity or when there is unilateral incompatibility (which is rather frequently observed in distant hybrids). In these circumstances, in order to estimate sex differences it would be desirable to use data from one of the test crosses and the F_2 progeny. An approach such as this is especially useful for obligate selfers in which production of F_2 progeny is much less labor-intensive than backcrossing.

Let an F_1 hybrid be used as a pollen parent in the test cross. Then, assuming normal segregation of markers, we obtain the following proportions (*AB*:*Ab*:*aB*:*ab*):

$$\begin{array}{lllll} \text{for } F_2: & 2+\theta: & 1-\theta: & 1-\theta: & \theta \\ \text{for } BC_1: & 1-rf_{♂}: & rf_{♂}: & rf_{♂}: & 1-rf_{♂} \end{array}$$

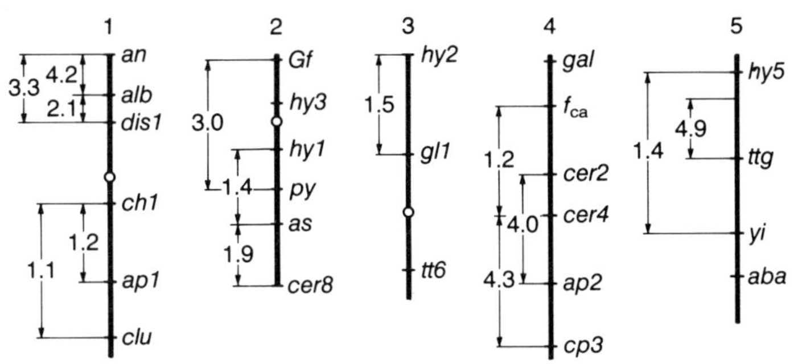

Figure 2.1 Sex differences in recombination rate in *Arabidopsis*; the numbers indicate the ratio $rf_{♂}/rf_{♀}$ (based on the data from Vizir and Korol, 1990).

where $\theta = (1 - rf_♀)(1 - rf_♂)$. For each of the progenies, a likelihood function is generated (L_{F_2} and L_{BC}) and the estimates of $rf_♀$ and $rf_♂$ and the variances and covariance are computed for the general function $L = L_{F_2} L_{BC}$ using the ML procedure. Note that when $\hat{rf}_♀$ and $\hat{rf}_♂$ are to be compared the need for calculation of covariance or $\hat{rf}_♀$ and $\hat{rf}_♂$ arises. The following equation can be used to test the significance of difference: $u = (\hat{rf}_♀ - \hat{rf}_♂)/\sqrt{(m^2_{rf♀} + m^2_{rf♂} - 2\text{cov}_{rf♀\, rf♂})}$. Some examples based on our data on *Arabidopsis* are given in Table 2.2. This approach allows simultaneous confidence regions to be determined for the $\hat{rf}_♀$ and $\hat{rf}_♂$ estimates (see Ott, 1985; Preygel, 1986; Ginsburg and Mullokandov, 1987; Conneally *et al.*, 1989).

Another approach, proposed to allow for sex differences in humans, is based on the assumption that the ratio of map distances $s = x_♀/x_♂$ is constant throughout the genome. If that is the case, then the relationship between $rf_♀$ and $rf_♂$ depends on the mapping function. In particular, with no interference:

$$rf_♀ = \frac{[1 - (1 - 2rf_♂)^s]}{2}$$

With Kosambi interference:

$$rf_♀ = \frac{1}{2}\frac{(1 + 2rf_♂)^s - (1 - 2rf_♂)^s}{(1 + 2rf_♂)^s + (1 - 2rf_♂)^s}.$$

For $s = 2$ this gives $rf_♀ = 2rf_♂/(1 + 4rf^2_♂)$ (see also Ott, 1985).

Surprising as it might seem at first sight, there exist situations in which it is possible to estimate separately $rf_♂$ and $rf_♀$ based on the analysis of progeny from the $F_2 = F_1 \times F_1$ cross only. For this to be the case, it is necessary that the heterozygous F_1 individuals involved in the cross differ in at least one allele at one or both loci in question, i.e. crosses of any of the following types, A_1B_1/A_2B_2(♀) × A_1B_1/A_2B_3(♂), or A_1B_1/A_2B_2(♀) × A_1B_1/A_3B_3(♂),...,A_1B_1/A_2B_2(♀) × A_3B_3/A_4B_4(♂) will be suitable (Lawrence, Cornish and Hyward, 1979; Mather and Lawrence, 1979). Moreover, homo- and heterozygotes for both loci should be distinguishable among the progeny. Satisfying this condi-

Table 2.2 Examples of estimating sex differences in *rf* based on a combined analysis of the F_2 and backcross data in *Arabidopsis* (from Vizir and Korol, 1990)

Segment	*Population*	*Segregation*					*rf* (%)
as–cer8	BC	333	50	54	341	♂	24.84 ± 2.76
	F_2	172	24	19	38	♀	13.37 ± 1.06
ap2–cer2	BC	192	15	7	147	♂	24.48 ± 1.93
	F_2	805	81	78	177	♀	6.09 ± 0.62
lu–ttg	BC	303	152	108	355	♂	28.32 ± 1.05
	F_2	621	83	70	162	♀	5.73 ± 2.67

In all the three cases the difference $rf_♂ - rf_♀$ is highly significant ($P \ll 0.001$).

tion are, in particular, the VNTR markers. These correspond to variability in the number of tandem repeats and have found a wide application in recent years (Jeffreys, Wilson and Thein, 1985; Nakamura, Leppert and O'Connell, 1987; Georges *et al.*, 1990). Sufficient allele diversity of the tested system of marker loci is an important prerequisite for determining the origin of a given allele or segment of an offspring when both parents are heterozygotes (White and Lalouel, 1988).

Another possibility for differential estimation of $rf_{♂}$ and $rf_{♀}$ based on the F_2 data is associated with dosage effect. In plant endosperm cells all four states *AAA*, *AAa*, *Aaa* and *aaa* of an enzyme locus may be distinguished by electrophoretic analysis. This implies that among the progeny the *Aa* heterozygous genotypes resulting from syngamy of $a(♀) \times A(♂)$ and $A(♀) \times a(♂)$ can be recognized. This fact offers the opportunity to estimate simultaneously $rf_{♀}$ and $rf_{♂}$ and it is sufficient that either of the loci under consideration shows the indicated effect.

2.4 ALLOWING FOR DATA HETEROGENEITY IN ESTIMATING LINKAGE

Owing to the peculiar nature of recombination data, use of adequate measures of variation of *rf* is a prerequisite for the proper analysis. It is common practice to employ binomial distribution as a basis for calculating the variance σ^2_{rf}. We thereby expect that differences in *rf* between replications are purely statistical, i.e. due to sampling, and the true value of *rf* is the same in all the replications. But it is this basic assumption that may be violated for reasons that are biological rather than statistical.

If segregating progenies (e.g. F_2) of different replications have been obtained from different heterozygotes (albeit presumably of one genotype), it is almost inevitable that an environmental component will affect crossing-over frequency. This component is superimposed on statistical variation as a result of finite sample size. Therefore, not only the sample estimate of *rf* but also the 'true' value of *rf* should be assumed to be randomly distributed and one should speak of a distribution $\psi(rf)$ of recombination values instead of that of *rf* itself. Genetic variation, sex, age differences and variation in the selective elimination of recombinants may, along with environmental variation, be important sources of heterogeneity of *rf* values (see, for example, Conneally and Rivas, 1980; Laurie, Palmer and Hulton, 1982; Ott, 1985; Zhuchenko and Korol, 1985; Beavis and Grant, 1991; Tanzi *et al.*, 1992).

In the presence of those sources of recombination rate variation over which the experimenter has no control, the confidence interval of $\widehat{rf}$ obtained from the binomial distribution variance $\widehat{rf}(1 - \widehat{rf})/N$ for test crosses, or the variance calculated from the ML covariance matrix, appears to be underestimated. In particular, in conducting replicated experiments, a much higher than expected portion of observations fall outside the confidence interval (Butler, 1977). It appears more reasonable, therefore, to calculate the variance of $\widehat{rf}$ directly from the between-replication deviations of the rf estimates as:

$$\sigma^2_{rf} = \sum_{i=1}^{k} \frac{(\widehat{rf}_i - \overline{rf})^2}{(k-1)}.$$

According to Butler's results, standard errors obtained in this way are, in some cases, higher than those calculated from the binomial variances. Thus, for the prediction of probable outcomes of subsequent experiments (as implied by the standard error estimate) to be reliable, it is necessary to take account of both the stochastic component of variation and the biological components (the so-called 'extrabinomial' variability). Ignoring such heterogeneity of recombination data can lead to erroneous conclusions about the significance of differences between the variants. This heterogeneity is also of primary importance in studies of genetic variability of recombination, e.g. in natural populations.

Butler's method of allowing for differences in the 'true' values of rf between replications has two serious drawbacks: (1) possible differences in sampling errors of $\widehat{rf}_i$ are not allowed for; and (2) the question of estimating the standard error for the grand mean over all replications is not discussed. We shall, therefore, consider this problem in more detail here. Assume that in addition to stochastic variation about 'true' values of rf_i for each replication, the 'true' values themselves exhibit some variation whose exact form can only be determined in an ideal experiment with samples of infinite size. Let each of the k replications have its own 'true' crossing-over rate between the markers in question, $rf_i = (i = \overline{1, k})$. It is necessary to test $\widehat{rf}_i$ values for homogeneity, to calculate the mean value of rf and the variance of the overall distribution, σ^2_{rf}, based on a sampling experiment.

Consider first a simple situation with binomial samples of equal size which corresponds to a test cross. According to the scheme of binomial testing, if we regard an event a 'success' when it results in the appearance of a recombinant for the region in question, then estimates of the mean and variance of the mean number of successes for the whole population will be: $\overline{X} = N\overline{rf}$, $\sigma^2 = N\overline{rf}(1 - \overline{rf}) + N(N-1)\sigma^2_{rf}$, where $\overline{rf}$ is the mean of sample values $\widehat{rf}_i$ and σ^2_{rf} is their variance (Kendall and Stuart, 1963). Hence, the variance of estimated population mean is $\sigma^2_{\overline{X}} = \sigma^2/k = [N\overline{rf}(1 - \overline{rf}) + N(N-1)\sigma^2_{rf}]/k$. For the proportion of 'successes' in the overall distribution, we obtain $\sigma^2_{\overline{rf}} = \overline{rf}(1 - \overline{rf})/(Nk) + [(N-1)/N](\sigma^2_{rf}/k)$. The augend of this sum takes account of the variability due to the finite sample size and the addend reflects the differences between replications due to the effects of internal and external factors over which the experimenter has no control. Similar arguments apply to the situation when data differ in accuracy (Korol, Preygel and Preygel, 1990).

The theory of the maximum likelihood method (Fisher, 1946; Kramer and Burnham, 1947; Allard, 1956; Bailey, 1961) provides a means for estimating the heterogeneity of rf values. Consider a situation in which sample sizes in each of the k replications are fairly large, so that it is possible to calculate individual $\hat{rf_i}$ values. By the ML method, the variance of estimate is the reciprocal of information (p. 40): $\sigma^2_{rf} = -E(d^2\ln L/drf^2)^{-1}$. The mean value of rf, as averaged over the entire experiment, is computed from individual group estimates:

$$\overline{rf} = \sum_{i=1}^{k} \hat{rf_i} I_i \Big/ \sum_{i=1}^{k} I_i = \left[\sum_{i=1}^{k} \hat{rf_i}/\sigma^2_{rf_i}\right] \Big/ \left[\sum_{i=1}^{k} 1/\sigma^2_{rf_i}\right].$$

To test for homogeneity of rf_i values, the χ^2 test is used:

$$X^2_{k-1} = \sum_{i=1}^{k} (\hat{rf_i} - \overline{rf})^2 I_i = \sum_{i=1}^{k} (\hat{rf_i} - \overline{rf})^2/\sigma^2_{rf_i}.$$

Consider now a more general case in which size of each group is arbitrary. In particular, some (or all) of these may be so small as to make calculations of individual $\hat{rf_i}$ values pointless. Nevertheless, applying the approach proposed by Fisher (1946), one can test the hypothesis of recombination rate homogeneity. First, the ML value of rf is calculated based on the overall distribution, i.e. the equation $S(rf) = d\ln L/drf = 0$ is solved (section 2.2). Then, $S_i(\hat{rf})$ and $I_i(\hat{rf})$ are calculated, which allows X^2 to be computed as

$$X^2_{k-1} = \sum_{i=1}^{k} S^2_i(\hat{rf})/I_i.$$

It is worth noting that the signs of individual values of S_i indicate the direction of deviation of rf_i from $\hat{rf}$ (even though $\hat{rf_i}$ values themselves are not calculated because of limited sizes of the groups). The above approach is also applicable to a multivariate case when several parameters are to be estimated simultaneously (see Bailey, 1961).

A similar idea has been proposed for estimating family heterogeneity in the case of multilocus mapping based on $\ln L$ (and not on their derivatives) (Lander and Green, 1987; Beavis and Grant, 1991). As a test of heterogeneity, the statistic

$$\chi^2_{k-1} = 2\ln 10\left[\sum_{i=1}^{k} \ln L_i(\hat{\theta}) - \ln L(\hat{\theta})\right]$$

is calculated, where $L_i(\theta)$ and $L(p)$ are the likelihood functions for the ith family and for the whole body of data, taking into account the complete set of recombination parameters $\theta = \{\theta_1, \ldots, \theta_{m-1}\}$ (for m-locus mapping).

Finally, still another approach to estimating family heterogeneity without directly calculating individual rf_i values is possible. It is based on standard analysis of the $k \times n$ contingency tables, where n is the number of phenotypic classes and k is the number of families compared (see, for example, Blank *et al.*, 1988).

In view of the necessity to allow for the heterogeneity of *rf* values, one may ask the question: what is the optimum unit of sampling for estimating recombination parameters from the observed numbers of phenotypic classes in segregating generations? Clearly, such a unit must meet the following criteria: crossing over in the meiotic cells and fertilization and embryo development should proceed under the most uniform conditions possible (Zhuchenko and Korol, 1985). Of course, the choice of an optimum unit satisfying these criteria depends, to a large extent, on biological features of the organism under study.

For a more rigorous analysis, an explicit specification of the model allowing for extrabinomial variation, estimation of the model parameters and testing of each hypotheses is needed. There are several models of this type: beta-binomial, probit (with random effects) and various non-parametric models (Haseman and Kupper, 1979; Brooks, 1984; Preisler, 1988; Risch, 1988). Consider a beta-binomial model, as applied to our problem, assuming that the proportion of recombinants p $(0 < p < 0.5)$ follows the β-distribution with the density function

$$f(p) = \frac{p^{\alpha - 1}(1 - p)^{\beta - 1}}{B(\alpha, \beta)},$$

where $B(\alpha, \beta)$ is the beta-function with parameters $\alpha, \beta > 0$. Then, for sample size n (e.g. for the progeny of a single fruit, plant or a single brood of *Drosophila* as a unit of analysis), the probability of obtaining x recombinants follows not the binomial distribution $P(x) = C^x{}_n p^x(1 - p)^{n-x}$, but the beta-binomial one:

$$P_\beta(x) = \frac{C^x{}_n B(\alpha + x, n + \beta - x)}{B(\alpha, \beta)},$$

where $C^x{}_n = n!/[x!(n - x)!]$. Parameters are estimated numerically by the ML method both for the same (Griffiths, 1973) and for varying sample size n (D.A. Williams, 1975). In carrying out the analysis it is assumed that within a variant the data represent several segregating families corresponding to the chosen units of analysis. For better convergence, transformed parameters are usually used: $\mu = \alpha/(\alpha + \beta)$, $\theta = 1/(\alpha + \beta)$.

The estimates obtained for several variants and the corresponding values of the log-likelihood function allow the tests of a number of hypotheses concerning variant homogeneity, presence of extrabinomial variability and equality (or inequality) of sets of parameters (μ, θ) in different variants (D.A. Williams, 1975; Paul, Liang and Self, 1989). Thus, beta-binomial models can reveal subtler differences between variants than those detectable under the standard analysis based on a binomial model (Aeschbacher *et al.*, 1977).

Another approach to the estimation of individual variation in recombination directly from experimental data without calculating individual *rf* values has been suggested by Smith (1983). The idea is to use the ML method in estimating the moments of distribution. In doing so, there is no need to specify the probability density function.

2.5 MODELS OF QUASI-LINKAGE

2.5.1 The phenomenon of quasi-linkage

Earlier we discussed approaches to estimating parameters and testing hypotheses concerned with gene segregation under linked inheritance. Similar problems may also arise in studying systems of unlinked loci. It is generally assumed that the behavior of loci of non-homologous chromosomes follows the law of independent assortment. Test of independence is an important component of genetic analysis and is used to assign the loci to this or that linkage group. Of course, independent segregation is also characteristic of traits controlled by genes located rather far apart on the same chromosome. A departure from random assortment of genes, as revealed by statistical tests during dihybrid segregation analysis, indicates possible linkage between genes. As a result of translocation, genes of the chromosomes involved in reciprocal exchange of segments prove to be linked, but this case of non-random assortment of genes on different chromosomes is easy to detect by cytological and genetic analysis.

Michie (1953) and Wallace (1953) observed a departure from random segregation of markers on non-homologous chromosomes in crosses between different strains of house mouse (*Mus musculus*), but it was Gates who first detected this phenomenon in backcrosses between *M. musculus* and *M. bacterianus* (Gates, 1926; quoted by Bailey, 1961). Michie assumed that the departure was due to the mutual attraction of the centromeres of segregating chromosomes of the same species, i.e. to the tendency (but no more) of chromosomes of different species to migrate to opposite poles (the affinity hypothesis). As a result, gametes with parental combinations of alleles at loci of non-homologous chromosomes are formed with a higher frequency than are the recombinant gametes. Such a departure from independent assortment of unlinked genes has been termed 'quasi-linkage'.

In working with multimarker strains of *M. musculus*, specific features of quasi-linkage have been established which can best be explained by the affinity model (Wallace, 1957, 1958; Mike, 1977). Reviewing the data in the literature, Wallace (1960a) postulated possible chromosome affinity in cotton. Malinowsky (1927) explained, based on the chromosome affinity hypothesis, a number of cases of abnormal segregation in hybrids of lettuce, peas, tobacco, beans and wheat. He was probably the first to introduce the term 'affinity'. A model based on the assumption of non-random segregation of centromeres in meiosis was used to explain linkage data from *Saccharomyces*, *Neurospora* and *Ascobolus* (see Surzycki and Paszewski, 1964). It has been suggested that complete affinity of centromeres may be responsible for the rarely observed fertility in the female mule (interspecific cross between horse and donkey) (Chandley, 1981). Mikhailov (1991) has revealed significant deviations of *rf* from the expected level of 50% for markers on chromosomes 6 and 9 in interstrain hybrids of maize. A pronounced effect of quasi-linkage has been reported in an interspecific hybrid of *Coix* (Sapre and Deshpande, 1987; also in this work, see a review of a number of cases of quasi-linkage established in other organisms). There is evidence for non-random segregation of non-homologs during the first meiotic division in man (Driscoll, Palmer and Melman, 1979). Non-random assortment of maternal and paternal chromosome complements in meiosis has been established in a dipterous insect, *Sciara coprophila* (Kubai, 1987).

As an extreme example of 100% affinity at the genomic level, one may consider the evolutionarily established mechanism for preserving the intactness of the chromosome complement in a unisexual fish, *Poecilia* (Schultz, 1973). Crosses between *P. monocha* and *P. lucida* yield only female individuals in the F_1. In these hybrids, a unipolar spindle is formed in the course of the last mitosis prior to meiosis, with which the whole of the maternal chromosome complement is associated while the paternal complement is lost. In this case, the *P. lucida* gene pool is effectively utilized by *P. monocha* for its own reproduction at diplophase while being completely isolated from the latter's gene pool with respect to recombination. This mechanism, referred to as **hybridogenesis**, is also found in the water frog *Rana esculenta* (for a review, see Vrijenhoek, 1990).

Quasi-linkage effect has also been established in experiments with multimarker strains, various semicultivated varieties and wild species of tomato (see Fogle and Currence, 1950; Butler, 1952; Wallace, 1960b; Zhuchenko *et al.*, 1977). There is extensive evidence for both quasi-linkage itself and its modification in tomato (Zhuchenko and Korol, 1985; Korol, Preygel and Nyutin, 1989). For pairs of loci *c-lut* (chromosomes 6 and 9) and *lut-a* (chromosomes 9 and 11), significant deviations of *rf* from the level expected under independent segregation have been revealed (30–40% instead of 50%) with the effect having being recurrently reproduced in replicate F_2 samples identified under different conditions. Besides, there is good agreement between F_2 populations and test crosses. In general, deviations from independence are observed for many pairs of morphological marker loci (Table 2.3).

Permanent quasi-linkage has also been established in our experiments for another pair of loci, *wt–j* (chromosomes 5 and 11): the value of rf_{wt-j} has ranged from 24.0 to 37.9%, remaining significantly lower ($P < 0.001$) than the 50% level (Korol, 1976; Zhuchenko *et al.*, 1977). Estimates differing from the 50% level were also obtained for this pair of loci elsewhere: $rf = 32\ (\pm 2.2)\%$ for F_2 and $30\ (\pm 3.6)\%$ for the test cross (Fogle and Currence, 1950) and $rf = 38\%$ (Butler, 1952). This prompted Rick and Butler (1956) to conclude that *wt* and *j* were linked (linkage group V). However, in tomato, genetic maps based on trisomic tests and linkage of markers have revealed that loci *wt* and *j* belong to chromosomes 5 and 11 respectively, indicating that they are not linked (see O'Brien, 1990). Consequently, significant deviations of rf_{wt-j} from 50% observed by different groups (Fogle and Currence, 1950; Butler,

Table 2.3 Quasilinkage between marker loci of different tomato chromosomes in estimating *rf* from F_2 segregation data

Type of cross	*Marker (chromosome)*	F_1 *genotype*	*Number of* F_2 plants	*rf* (%)	χ^2 *test of independence*
61 × 17†	*c*(6), *a*(11)	*ca*/+ +	2173	59.3 ± 1.6	34.5***
328 × 529	*c*(6), *lut*(9)	*clut*/+ +	1522	37.7 ± 1.8	45.4***
328 × 529	*a*(11), *lut*(9)	*alut*/+ +	1522	36.9 ± 1.7	50.2***
61 × 394	*ps*(2), *c*(6)	*psc*/+ + +	426	32.1 ± 4.3	11.6***
61 × 385	*wt*(5), *c*(6)	+ *c*/*wt* +	682	41.0 ± 2.7	11.4***
61 × 385	*wt*(5), *j*(11)	+ +/*wtj*	682	32.8 ± 2.4	53.4***
61 × 406	*c*(6), *dl*(8)	*c* +/+ *dl*	310	59.2 ± 4.0	5.1*
61 × 406	*c*(6), *gq*(5)	*c* +/+ *gq*	310	62.5 ± 4.8	6.7**
393 × 504	*d*(2), *m2*(6)	*d* +/+ *m2*	3924	43.1 ± 1.5	21.7***
393 × 504	*aw*(2), *m2*(6)	*aw* +/+ *m2*	3924	41.6 ± 1.5	30.4***

Deviations from independence are significant at *$P < 0.05$, **$P < 0.01$ and ***$P < 0.001$.
†Numbers of marker lines are those registered in the catalog of the Institute of Genetics in the Academy of Sciences of Moldova (Kishinev).

1952; Korol, 1976; Zhuchenko *et al.*, 1977) suggest quasi-linkage between these loci.

Numerous data on quasi-linkage have been obtained for markers *aw*, *d* (chromosome 2) and *m2*, *c* (chromosome 6) (Zhuchenko and Korol, 1985; Korol, Preygel and Nyutin, 1989). In most cases, rf_{aw-m2} and rf_{d-m2} are significantly lower than the level expected from independent assortment ($rf_{aw-m2} \approx$ 25–40%; $rf_{d-m2} \approx$ 30–45%), although occasionally an increase up to 60–65% is observed for both pairs. Deviations of rf_{aw-m2} and rf_{d-m2} from the 50% level are positively correlated ($R = 0.7$–0.9), which is natural since *aw* and *d* are rather closely linked ($rf \approx 10\%$). The results showed good reproducibility regardless of the time and method of experimentation, differential environmental conditions for F_1 hybrids and F_2 progenies.

2.5.2 Formal analysis of possible mechanisms of quasi-linkage

For a correct assessment of the theoretical and the applied significance of quasi-linkage it is necessary to understand the underlying mechanism. Of most interest in this connection is the hypothesis of affinity of non-homologous chromosomes (Wallace, 1953, 1957, 1959, 1960a,b). It postulates possible differences between centromeres within each chromosome pair, i.e. a certain 'centrotype' (α or β) is assigned to the centromeres. The centromeres of two non-homologous chromosomes, say *i* and *j*, can attract (identical centrotypes – α_i and α_j) or repel each other (opposite centrotypes – α_i, β_j or β_i, α_j). The corresponding situations under heterozygosity for markers on chromosomes *i* and *j* is described in Table 2.4.

Table 2.4 Quasi-linkage between markers of non-homologous chromosomes resulting from affinity of centromeres (adapted from Wallace, 1959)

Genotype and centrotype of F_1	*Phase of*		*rf between markers (%)*	*Description of the situation*
	Markers	*Centromeres*		
$\frac{m_i\alpha_i}{M_i\beta_i}\ \frac{m_j\alpha_j}{M_j\beta_j}$	c	c	< 50	Convergent coupling
$\frac{m_i\alpha_i}{M_i\beta_i}\ \frac{m_j\beta_j}{M_j\alpha_j}$	c	r	> 50	Divergent coupling
$\frac{m_i\alpha_i}{M_i\beta_i}\ \frac{M_j\alpha_j}{m_j\beta_j}$	r	c	< 50	Convergent repulsion
$\frac{m_i\alpha_i}{M_i\beta_i}\ \frac{M_j\beta_j}{m_j\alpha_j}$	r	r	> 50	Divergent repulsion
$\frac{m_i\alpha_i}{M_i\beta_i}\ \frac{m_j\alpha_j}{M_j\alpha_j}$	c	—	50	Independence
$\frac{m_i\alpha_i}{M_i\beta_i}\ \frac{M_j\alpha_j}{m_j\alpha_j}$	r	—	50	Independence

c (coupling) and r (repulsion) denote the situations in which F_1 results from crosses $m_im_j \times M_iM_j$ and $m_iM_j \times M_im_j$ respectively; for the centrotypes c is $\alpha_i\alpha_j \times \beta_i\beta_j$ and r is $\alpha_i\beta_j \times \beta_i\alpha_j$ (i and j are the indices of chromosomes).

The above formal scheme does not suggest anything about specific mechanisms of affinity. Here, peculiarities of relative positions of bivalents with respect to each other, to equatorial plane and to nuclear membrane are likely to play a role. Some authors observed interchromosomal filaments connecting whole groups of chromosomes. This has been assumed to be responsible for the tendency of centromeres of common origin to co-segregate, resulting in 'false linkage' (Terzi, 1974; Sharma, 1978; Semenov and Smirnov, 1981). Suggestions have been made about high ordering (non-randomness) in locations of chromosomes within nuclei of both somatic and meiotic cells (Kozhevnikov, 1947; Khesin, 1984; Bennett, 1982; Prokofieva-Belgovskaya, 1986). Sved (1966) assumed possible involvement of telomeric affinity in quasi-linkage. To account for hybrid dysgenesis in *Drosophila*, he proposed a modification of the model in which the nuclear membrane plays a significant role (Sved, 1976). It has been postulated that the maternal chromosome complement is transmitted to the F_1 hybrid in such a way as to ensure its association with the nuclear membrane, while chromosomes of the paternal complement may or may not exhibit the conformation determining their proper organization within the hybrid nucleus. In the latter case, a whole range of abnormalities known as hybrid dysgenesis (deviation of monogenic segregations from expected ratios, F_1 female sterility, male recombination) arise. Simultaneously, there may be a departure from independent segregation of non-homologous chromosomes.

In mammals, chromosomal imprinting, i.e. structural and molecular changes occurring during gametogenesis and apparently playing an important role in embryo development (Cattanach, 1986; Barlow, 1993), may, in principle, also be a source of asymmetry in segregation behavior of male and female chromosome complements. Unfortunately, it is not known whether these changes persist until meiosis. It is of interest to note that chromosomal imprinting has so far been most thoroughly studied in the mouse, an organism exhibiting quasi-linkage.

In the few existing models of affinity (centromeric or telomeric), it has usually been assumed that it is non-specific heterochromatin association that is responsible for interactions between non-homologs (Driscoll, Palmer and Melman, 1979). It cannot be ruled out that the similarity of highly repetitive sequences of satellite DNA in non-homologous chromosomes may be the molecular basis of attraction (Flamm, 1972). One way to test this hypothesis is to evaluate the effects of near-centromeric heterochromatin deletions of different size on the degree of quasi-linkage. In *Drosophila*, this approach has been used to elucidate the role of satellite DNA in chromosome pairing and recombination (Yamamoto and Miklos, 1978; Yamamoto, 1979a, b). If these hypotheses are valid, one can assume independent assortment of non-homologs to result from special selection inactivating the affinity mechanism. That is, the fundamental Mendelian rules of genetic material segregation should be regarded, under this approach, as a product of evolution (presumably adaptive) in the past (Darlington, 1939, 1979, 1981; Malinowsky, 1941; Lloyd, 1984; Zhuchenko and Korol, 1985; Feldman and Otto, 1991).

Quasi-linkage can also be explained by invoking mechanisms of differential viability. Consider the M_iM_j/m_im_j heterozygote resulting from a cross between M_iM_j/M_iM_j and m_im_j/m_im_j. Suppose that parental combinations of whole chromosomes (M_iM_j and m_im_j) and recombinant combinations (M_im_j and m_iM_j) have differential selective values. Experimental evidence of this type has been obtained in *Drosophila* (Dobzhansky, Spassky and Anderson, 1965). The effect has also been demonstrated for different segments of the same *Drosophila* chromosome (Gibson and Thoday, 1962; Templeton, Sing and Brokaw, 1976), and its extreme manifestation is recombinational (synthetic) lethals (Dobzhansky, 1946; Wallace *et al.*, 1953).

Selective differences between parental and recombinant combinations of intact chromosomes are reflected in the differential viability of zygotes, embryos and adult individuals. The effect can also be expressed at the gamete stage in the form of differential fertilizing capacity of spermatozoa and in differential rates of pollen tube growth, etc. The idea of such an explanation is based on the concept of **genetic architecture** (Harland, 1936) according to which a certain evolutionarily established architecture of dominance mechanism and of modifier systems is peculiar to each species. Even though a whole range of genotypes are possible in segregating generations, it is only balanced genotypes whose architecture closely resembles that of parental types that are viable. This preserves the integrity of the parental genome and is one of the causes of restricted recombination, leading to reduction in the values of *rf* in experiments. Recombinant gametes and genotypes can, in principle, exhibit higher viabilities (competitive ability) than parental types. This means that differential selective values of different combinations of non-homologous chromosomes are, theoretically, a likely cause of *rf* exceeding the 50% level, although in reality cases like these seem to be very scarce.

In order to analyze experimental data on quasi-linkage, we have developed a special system of formal models, based on the above-mentioned hypotheses (Korol, 1987; Korol, Preygel and Nyutin, 1989). Despite the considerable combinatorial complexity of the models and the need to estimate a relatively large number of parameters, a good algorithm convergence has been achieved. We consider here the situation with a four-marker case by taking the *AwDm2c/awdM2C* heterozygote as an example and analyzing the behavior of its progeny.

Non-random assortment of non-homologous chromosomes at meiosis due to affinity of centromeres or other points affinity (Wallace, 1958)
Here, three cases of relative positions of affinity points (α, β) and of markers are considered (Figure 2.2). In addition to the situation shown in Figure 2.2, also corresponding to type (a) are cases in which 'loci' $\alpha_1\beta_1$ and $\alpha_2\beta_2$ are located distally with respect to marked segments *aw–d* and *m2–c*, or one is located distally and the other proximally. Initially, let both of them be located proximally (as in Figure 2.2a). Let also $rf_{aw\text{–}d} = p$ and $rf_{m2\text{–}c} = q$. The recombination fraction between loci *aw* and *m2* depends on recombination between three pairs of loci:

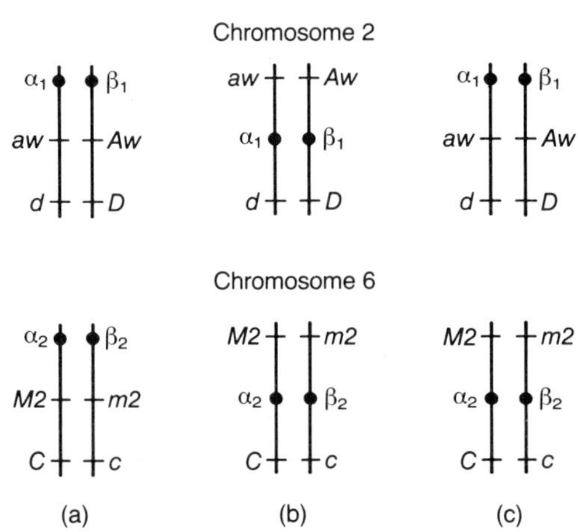

Figure 2.2 Relative positions of markers and affinity points. (a) Both points of affinity are located outside marked segments. (b) Both points are located within the segments. (c) One point is outside and the other is within the segment.

$aw{-}(\alpha\beta)_1$, $(\alpha\beta)_1{-}(\alpha\beta)_2$, and $(\alpha\beta)_2{-}m2$. The value of $rf_{aw-m2} = \pi$ will differ from 50% provided $rf_{aw-(\alpha\beta)_1}$ and $rf_{m2-(\alpha\beta)_2}$ are not too large and combinations $\alpha_1\alpha_2$ are formed with higher or lower probability than $\alpha_1\beta_2$ and $\beta_1\alpha_2$ are. In view of the introduced designations and assuming the absence of gametic selection, expressions for the doubled frequencies of gametes are:

	AwD	*Awd*	*awD*	*awd*
M2C	$(1-p)(1-\pi)(1-q)$	$p(1-\pi)(1-q)$	$p\pi(1-q)$	$(1-p)\pi(1-q)$
M2c	$(1-p)(1-\pi)q$	$p(1-\pi)q$	$p\pi q$	$(1-p)\pi q$
m2C	$(1-p)\pi q$	$p\pi q$	$p(1-\pi)q$	$(1-p)(1-\pi)q$
m2c	$(1-p)\pi(1-q)$	$p\pi(1-q)$	$p(1-\pi)(1-q)$	$(1-p)(1-\pi)(1-q)$

Assuming, further, random syngamy and absence of post-syngamic elimination, we calculate frequencies of all the F_2 genotypic classes as functions of three parameters (p, q and π), pool them over the phenotypes and, based on the observed frequencies of 16 phenotypic classes, obtain the ML estimates $\hat{p}$, $\hat{q}$ and $\hat{\pi}$. Provided segregation of markers is normal, we can test the hypothesis of chromosome affinity as the cause of quasi-linkage based on the residual value of χ^2 (d.f. = 12).

Consider a situation corresponding to case (b) (see Figure 2.2). Let p_1, p_2 and c_1 denote recombination frequencies $rf_{aw-(\alpha\beta)_1}$, $rf_{(\alpha\beta)_1-d}$ and coefficient of coincidence in the *aw–d* region, and q_1, q_2 and c_2 denote $rf_{m2-(\alpha\beta)_2}$, $rf_{(\alpha\beta)_2-c}$ and coefficient of coincidence in the *m2–c* region, with $rf_{(\alpha\beta)_1-(\alpha\beta)_2} = \pi$. Then, the expected frequencies of gametes and phenotypic classes depend on seven parameters $(\pi, p_1, p_2, q_1, q_2, c_1, c_2)$, but only five of these can be estimated. Coefficients of coincidence c_1 and c_2 cannot be calculated from data like these. In order to obtain the estimate $\hat{\theta} = \{\hat{\pi}, \hat{p}_1, \hat{p}_2, \hat{q}_1, \hat{q}_2\}$, it is therefore necessary to proceed from some a priori assumptions about interference within the segments in question, i.e. to set specific numerical values for c_1 and c_2 and then obtain an ML estimate for the vector θ. In particular, one may set c_1 and/or c_2 equal to 1, or set $c = c(p)$ according to a certain interference metric, for example that of Kosambi. In the latter case $c(p) = 2p$. In line with Kosambi's rule for combining recombination fractions in the presence of interference we obtain:

$$c_1 = c(p) = \frac{2(p_1 + p_2)}{(1 + 4p_1p_2)}, \quad c_2 = c(q) = \frac{2(q_1 + q_2)}{(1 + 4q_1q_2)}.$$

These formulae are substituted into expressions for phenotypic class frequencies. In a similar way, a situation corresponding to case (c) (see Figure 2.2) is considered.

Gamete selection

For simplicity, it is assumed that each of the two marker chromosomes carries genes or gene blocks, $(\alpha, \beta)_1$ and $(\alpha, \beta)_2$, such that viability of parental gametes ($\alpha_1{-}\alpha_2$ and $\beta_1{-}\beta_2$) is higher than that of recombinant gametes ($\alpha_1{-}\beta_2$ and

$\beta_1-\alpha_2$), i.e. blocks $\alpha_1-\beta_1$ and $\alpha_2-\beta_2$ are 'co-adapted'. The relative positions of the blocks in question with respect to marked segments are the same as in the previous model (see Figure 2.2). As an example, consider case (b) when loci $(\alpha\beta)_1$ and $(\alpha\beta)_2$ are located within segments *aw*–*d* and *m2*–*c* respectively. Using the previously employed symbols for recombination parameters $p_1, p_2, c_1, q_1, q_2, c_2$, we can calculate frequencies of various combinations of alleles α and β and of marker genes *aw*, *d* and *m2*, *c*:

	AwD	*Awd*	*awD*	*awd*
α_1	s_1	s_2	s_3	s_4
β_1	s_4	s_3	s_2	s_1

	M2C	*M2c*	*m2C*	*m2c*
α_2	t_1	t_2	t_3	t_4
β_2	t_4	t_3	t_2	t_1

where $s_1=\tau_1/2$, $s_2=(p_2-\tau_3)/2$, $s_3=(p_1-\tau_3)/2$, $s_4=\tau_3/2$, $t_1=\tau_2/2$, $t_2=(q_2-\tau_4)/2$, $t_3=(q_1-\tau_4)/2$, $t_4=\tau_4/2$, $\tau_1=1-p_1-p_2+c_1p_1p_2$, $\tau_2=1-q_1-q_2+c_2q_1q_2$, $\tau_3=c_1p_1p_2$ and $\tau_4=c_2q_1q_2$. Suppose that $\pi=0.5$, i.e. chromosomes segregate independently at meiosis. Then, the following procedure can be used to calculate a 16-dimensional set of frequencies of gametes at loci *aw*, *d*, *m2*, *c*, which carry alleles α_1 and α_2:

$$\Gamma\alpha_1\alpha_2=\begin{array}{ll|cccc|} & & AwD & Awd & awD & awd \\ & & s_1 & s_2 & s_3 & s_4 \\ M2C & t_1 & t_1s_1 & t_1s_2 & t_1s_3 & t_1s_4 \\ M2c & t_2 & t_2s_1 & t_2s_2 & t_2s_3 & t_2s_4 \\ m2C & t_3 & t_3s_1 & t_3s_2 & t_3s_3 & t_3s_4 \\ m2c & t_4 & t_4s_1 & t_4s_2 & t_4s_3 & t_4s_4 \end{array}=T_1\cdot S_1,$$

where $T_1=\{t_1, t_2, t_3, t_4\}$ and $S_1=\{s_1, s_2, s_3, s_4\}$. Similarly, $\Gamma\alpha_1\beta_2=T_2\cdot S_1$, $\Gamma\beta_1\alpha_2=T_1\cdot S_2$ and $\Gamma\beta_1\beta_2=T_2\cdot S_2$, where $T_2=\{t_4, t_3, t_2, t_1\}$ and $S_2=\{s_4, s_3, s_2, s_1\}$.

By the hypothesis under consideration, quasi-linkage between marker loci of two non-homologous chromosomes is due to differences in viability (competitive ability) of parental ($\alpha_1\alpha_2$ and $\beta_1\beta_2$) and recombinant ($\alpha_1\beta_2$ and $\beta_1\alpha_2$) gametes. Having the above gamete frequencies before selection and specifying the mode of selection we can calculate the expected phenotype proportions as functions of the unknown parameters. Then, based on the observed proportions the ML estimates of the parameters can be calculated.

Departure from random fertilization (assortative syngamy)

It is generally assumed that, in spite of differences in the rate of pollen tube growth, the actual union of male and female gametes (syngamy) is random, i.e. probabilities of zygotes *aa* and *Aa* being formed following a cross of *aa* × *Aa* type are proportional to frequencies of pollen tubes a and A that have 'won the competition' in the style. However, the upsetting of equiprobability is also possible at the stage of gamete union. This effect can, in principle, imitate quasi-linkage even under perfectly random assortment of chromosomes and in the absence of interaction of genes for viability (Korol, Preygel and Nyutin, 1989).

Consider a model for quasi-linkage resulting from non-random (assortative) syngamy. Let assortativeness be due to special loci, $(\alpha\beta)_1$ and $(\alpha\beta)_2$, whose possible positions in relation to the markers correspond to the cases discussed above. If the indicated loci act in a multiplicative way and their effects are x_1 and x_2 respectively, then the probabilities of formation of the F_2 zygotes carrying loci $(\alpha\beta)_1$ and $(\alpha\beta)_2$ can be written as a symmetrical matrix A:

$$A = \begin{array}{c|cccc|} & \alpha_1\alpha_2 & \alpha_1\beta_2 & \beta_1\alpha_2 & \beta_1\beta_2 \\ \alpha_1\alpha_2 & \delta_1 & \delta_2 & \delta_3 & \delta_4 \\ \alpha_1\beta_2 & \delta_2 & \delta_1 & \delta_4 & \delta_3 \\ \beta_1\alpha_2 & \delta_3 & \delta_4 & \delta_1 & \delta_2 \\ \beta_1\beta_2 & \delta_4 & \delta_3 & \delta_2 & \delta_1 \end{array}$$

where $\delta_1 = (1 + x_1)(1 + x_2)$, $\delta_2 = (1 + x_1)(1 - x_2)$, $\delta_3 = (1 - x_1)(1 + x_2)$ and $\delta_4 = (1 - x_1)(1 - x_2)$. What we need is the expressions for phenotypic class frequencies at marker loci. To derive these, it is necessary to calculate frequencies of combinations $\alpha_1\alpha_2$, $\alpha_1\beta_2$, $\beta_1\alpha_2$ and $\beta_1\beta_2$ for each of the 16 classes of gametes (with respect to loci *aw*, *d*, *m2* and *c*). This can be easily accomplished by using the expressions for probabilities of various types of associations of alleles at pairs of loci *aw–d* with $(\alpha\beta)_1$, and *m2–c* with $(\alpha\beta)_2$. Thus, a vector of frequencies corresponding to combinations $\alpha_1\alpha_2$, $\alpha_1\beta_2$, $\beta_1\alpha_2$ and $\beta_1\beta_2$ takes the form

$G_1 = (s_1t_1, s_1t_4, s_4t_1, s_4t_4)$ for the *AwDM2C* group
$G_2 = (s_2t_1, s_2t_4, s_3t_1, s_3t_4)$ for the *AwdM2C* group
$G_3 = (s_3t_1, s_3t_4, s_2t_1, s_2t_4)$ for the *awDM2C* group
....
$G_{16} = (s_4t_4, s_4t_1, s_1t_4, s_1t_1)$ for the *awdm2c* group,

where coefficients $s_1, \ldots, s_4$ and $t_1, \ldots, t_4$ have the same meaning as in the gametic selection model discussed earlier. The frequencies of zygotes z_{kl} $(k, 1 = \overline{1, 16})$ can now be represented in the matrix form: $z_{kl} = G_k'AG_1$. Coefficients z_{kl} depend on unknown parameters (e.g. $p_1, p_2, c_1, q_1, q_2, c_2, x_1, x_2$) some of which must be specified (c_1 and c_2) and the rest can be calculated from the observed frequencies of phenotypic classes by the ML method.

Differential viability of zygotes

The model allows for interaction effects both between marker loci themselves and between co-adapted blocks $(\alpha\beta)_1$ and $(\alpha\beta)_2$. Consider the latter case in more detail. Under the same conditions as in the model for gametic selection, we have four matrices $\Gamma\alpha_1\alpha_2$, $\Gamma\alpha_1\beta_2$, $\Gamma\beta_1\alpha_2$ and $\Gamma\beta_1\beta_2$ which determine frequencies of associations of various types of alleles at the observed (*aw*, *d*, *m2*, *c*) and selected $[(\alpha\beta)_1, (\alpha\beta)_2]$ loci. Based on these, we can calculate frequencies of various types of allelic associations at loci $(\alpha\beta)_1$ and $(\alpha\beta)_2$ for each zygotic class prior to selection. Then, by specifying some form of selection we obtain eventually the expected proportions of marker classes as functions of the model parameters which can be estimated from the observed frequencies.

Differential penetrance

It cannot be ruled out that the observed cases of quasi-linkage may sometimes be due to incomplete penetrance of a marker locus on one of the chromosomes (e.g. locus *m2* of chromosome 6) which depends on the allelic state of a locus on another chromosome (*d*). Indeed, in crowded stands, dwarf seedlings *dd* receive less light than do the normal plants *D*. Then, recombinants *m2d*/*m2d* are less likely than parental genotypes *m2D* to be classified as *M2* (because *m2* marker expression is dependent on illumination intensity). This will distort phenotypic ratios, resulting in increased numbers of parental phenotypes and, consequently, in quasi-linkage. It is easy to formulate a model allowing for differential penetrance based on classification errors (δ_i) depending on locus *d*:

	D	d
$P\{Aw \to aw\}$	δ_1	δ_2
$P\{m2 \to M2\}$	δ_3	δ_4

In our case, usually $\delta_2 > \delta_1$ and $\delta_4 > \delta_3$, since the degree to which anthocyanin coloration (controlled by the *Aw*/*aw* locus) and leaf variegation (controlled by the *M2*/*m2* locus) are expressed depends on illumination intensity, but in a segregating progeny seedlings *d* tend to be shaded.

2.5.3 Examples of application of quasi-linkage models

The data used were those from an experiment with the tomato hybrid *AwDm2c* × *awdM2C* (Korol, Preygel and Nyutin, 1989). Analysis was performed on the F_2 progeny of individual inflorescences of F_1 plants. Some results from this experiment are listed in Table 2.5. The largest deviations from the expected value of $rf = 50\%$ to be observed were those for $rf_{d\text{–}m2}$ and $rf_{aw\text{–}m2}$.

The presented evidence disagrees with the model for 'ideal' segregation of markers. Some of the data can readily be explained on the assumption that departures from the 'ideal' segregation are due to affinity or to gametic or zygotic selection, with the 'affinity' model being more preferable from a formal standpoint since it incorporates the least number of estimated parameters (and, consequently, a greater number of degrees of freedom are left to test the adequacy). However, it should be taken into account that the same evidence can be equally well explained by different mechanisms. In particular, the model for gametic selection under equal viabilities of reciprocal classes (*AB* and

Table 2.5 Recombination frequency *rf* (%) for linked and unlinked loci in F_1 of the tomato chromosomes 2 (*aw–d*) and 6 (*m2–c*) (as estimated from different F_2 families)

Family number	F_2 *size*	*Linked loci*		*Unlinked loci*			
		rf_{aw-d}	rf_{m2-c}	rf_{aw-m2}	rf_{aw-c}	rf_{d-m2}	rf_{d-c}
1	309	11.3 ± 3.6	34.2 ± 4.2	27.4 ± 5.3	22.3 ± 5.4	46.6 ± 4.8	54.7 ± 4.5
2	358	15.8 ± 3.4	33.7 ± 3.9	34.9 ± 4.7	26.1 ± 5.0	44.0 ± 3.7	43.5 ± 4.5
3	455	12.5 ± 2.9	32.9 ± 3.4	37.2 ± 4.2	34.3 ± 4.2	54.2 ± 3.7	48.0 ± 3.9
4	463	13.5 ± 2.9	27.6 ± 3.2	37.7 ± 4.1	37.7 ± 4.1	49.5 ± 3.8	47.7 ± 3.9
5	529	9.7 ± 2.7	31.4 ± 3.1	37.4 ± 3.8	30.0 ± 4.0	46.2 ± 3.6	50.1 ± 3.6
6	338	10.6 ± 3.4	28.5 ± 3.8	39.8 ± 4.7	37.4 ± 4.8	48.7 ± 4.4	45.7 ± 4.6
7	289	9.4 ± 3.6	31.2 ± 4.2	28.3 ± 5.5	33.0 ± 5.3	49.0 ± 4.8	56.7 ± 4.6
8	282	11.4 ± 3.7	25.9 ± 4.3	36.2 ± 5.3	38.4 ± 5.2	37.7 ± 5.3	33.7 ± 4.6
9	845	11.6 ± 2.2	25.4 ± 2.4	40.2 ± 3.0	35.3 ± 3.1	47.1 ± 2.9	45.2 ± 2.9
10	254	10.7 ± 3.9	34.4 ± 4.6	38.6 ± 5.5	36.2 ± 5.6	53.3 ± 5.0	55.5 ± 4.9
11	256	8.5 ± 3.8	28.7 ± 4.4	34.9 ± 5.6	32.3 ± 5.7	52.2 ± 5.0	57.4 ± 4.8

ab, *Ab* and *aB*) is equivalent to the affinity model. Models for gametic selection and zygotic selection yield closely spaced expected values of class numbers, with more precise 'adjustment' being possible for the zygotic selection model since it permits a greater number of parameters (3–8 versus 1–3) to be estimated. Clearly, formal criteria alone are insufficient for choosing the best model: the proposed hypotheses should be tested with additional experimentation.

The data on segregation of markers in F_2 (Table 2.5) are better explained by a model allowing simultaneously for chromosome affinity and gametic selection. In the majority of cases, the most probable position of affinity points is outside the marker sgements: distally (relative to *d*) on chromosome 2 and proximally (relative to *m2*) on chromosome 6.

The composite model 'affinity + gametic selection' is equivalent to a model in which either of the two mechanisms is realized based on the interaction of more than two points on both chromosomes. This is in agreement with the above-mentioned hypothesis of the possible role of affinity of telomeres (along with centromeres) in quasi-linkage (Sved, 1966). Blocks of intercalary heterochromatin can also play a role, in which the net effect of quasi-linkage may depend on the sum of individual effects and their interactions. It is also necessary to take account of the fact that quasi-linkage of a pair of non-homologous chromosomes in question may be affected by their associations with other chromosomes which have been ignored (not marked) in this particular experiment. Moreover, it is probably these associations that induce the observed departure from independence: it may be that each of the chromosomes in question (2 and 6) exhibits affinity to a third one, and in the absence of heterozygosity for the third chromosome's centrotype (when the centromere is the point of affinity) random segregation would be observed. In general, the behavior of any pair of marker loci of two non-homologous chromosomes seems to be the result of 'averaging' all the pairwise, tripartite, etc., interactions. Nevertheless, the expected dihybrid segregations resulting from model fitting agree very well with the observed ones, with the residual χ^2 values (for the trihybrid segregation) being largely due to discrepancies at the three-locus level.

In formulating the affinity hypothesis, only two points (one on each chromosome, $(\alpha\beta)_1$ and $(\alpha\beta)_2$) were assumed. Consequently, estimation of model parameters should yield somewhat averaged values of their positions with respect to marker loci. In view of this and the variability (biological and stochastic) of factors determining the pattern of polyhybrid segregation, model parameter values can be expected to vary between F_2 families. Indeed, our estimations yield different positions of affinity points (Korol, Preygel and Nyutin, 1989). This is reminiscent of a situation with the 'moving' affinity point established for a six-marker tomato hybrid (*c a j* × *br y wt*) (Korol, 1976). A similar reasoning holds when interactions between points $(\alpha\beta)_i$ and $(\alpha\beta)_k$ determine gametic or zygotic selection rather than chromosome affinity, with selective elimination of recombinant combinations being one of the manifestations of co-adaptation at the interchromosomal level (Dobzhansky, Spassky and Anderson, 1965).

On the whole, analyzing the F_2 data leads to the conclusion that the observed quasi-linkage between markers on chromosomes 2 and 6 is due to peculiarities of kinetic behavior of chromosomes during meiotic division I and/or to selective elimination of recombinants. In view of the estimates obtained, the involvement of such mechanisms as assortative syngamy and differential penetrance of marker genes has a very low probability. The number of points on each of the marked chromosomes whose interactions determine affinity or interchromosomal co-adaptation seems to be greater than one. This leads to some predictions about the behavior of markers *aw*, *d* and *m2*, *c* in later generations, for example in the F_3 progeny of various heterozygous F_2 genotypes:

1. Quasi-linkage should persist among individuals which are homozygous for only one of two marker loci per chromosome, for example *d* (since it is not the differential penetrance that is the cause of quasi-linkage).
2. Recombination between marker loci and $(\alpha\beta)_i$ points of affinity and/or differential elimination will produce the following effects:

   (a) extension of the range of quasi-linkage values obtained in the F_3 progenies of various F_2 genotypes (since in F_2 genotypes a broader range of states of interaction points is possible than in F_1);
   (b) a higher, than in F_2, incidence of families with an altered quasi-linkage phase ($rf > 50\%$, the F_2 analysis yielded predominantly values of $rf < 50\%$);
   (c) 'restoration' of random assortment in individual families (when at least one of the two marker chromosome pairs is homozygous for genetic content of affinity points) (see also Wallace, 1957);
   (d) increased diversity of estimates for parameters describing coordinates of affinity points.

These predictions were tested experimentally in collaboration with Dr. Yu. I. Nyutin using the F_3 progenies from selfing phenotypically dominant (with respect to marker loci *aw*, *d*, *m2* and *c*) F_2 genotypes (Table 2.6). In families 1–9, one or two markers do not segregate, nevertheless quasi-linkage persists, i.e. prediction 1 is confirmed. Predictions 2a and 2b also hold true: values were in the range of 17.6–62.2% for rf_{aw-m2}, from 28.0 to 51.5% for rf_{aw-c}, from 13.3 to 85.1% for rf_{d-m2} and from 22.8 to 70.3% for rf_{d-c}, i.e. in all the cases the range of values is wider than in F_2, with values of $rf < 50\%$ in families 2, 6 and 14, $rf > 50\%$ in families 3, 4 and 13 and $rf \approx 50\%$ in the other families.

Summarized in Table 2.7 are the hypotheses showing the best fit to the data on the behavior of marker loci in families 10–18 in which all four loci were segregating. As with F_2, the observed F_3 segregation ratios can be best

Table 2.6 Recombination frequency *rf* (%) for linked and unlinked loci in F_2 (as estimated from different F_3 families)

Family number	F_3 *size*	*Linked loci*		*Unlinked loci*			
		rf_{aw-d}	rf_{m2-c}	rf_{aw-m2}	rf_{aw-c}	rf_{d-m2}	rf_{d-c}
1	226	7.2 ± 4.0	—	—	28.0 ± 4.7	—	28.3 ± 4.7
2	416	—	30.1 ± 3.5	62.2 ± 4.6	51.5 ± 4.0	—	—
3	472	8.7 ± 2.8	—	46.4 ± 3.7	—	44.8 ± 3.6	—
4	404	13.6 ± 3.2	—	—	50.2 ± 4.3	—	55.3 ± 4.2
5	370	—	23.3 ± 3.5	—	—	34.8 ± 3.8	36.3 ± 3.9
6	396	7.1 ± 3.0	—	60.6 ± 4.4	—	60.6 ± 4.4	—
7	406	7.2 ± 3.0	—	17.6 ± 3.2	—	13.3 ± 3.1	—
8	490	—	—	—	32.3 ± 3.4	—	—
9	500	—	—	41.3 ± 3.4	—	—	—
10	402	11.7 ± 3.1	35.5 ± 3.7	43.1 ± 3.9	44.4 ± 3.9	40.9 ± 3.8	53.3 ± 4.2
11	391	7.5 ± 3.1	21.0 ± 3.2	28.1 ± 3.6	28.2 ± 7.0	20.0 ± 3.6	22.8 ± 3.4
12	409	4.9 ± 3.0	17.4 ± 3.2	28.7 ± 3.5	30.5 ± 3.5	25.3 ± 3.4	29.6 ± 3.5
13	412	11.9 ± 3.1	26.6 ± 3.4	45.6 ± 3.9	43.4 ± 3.8	45.4 ± 3.9	53.5 ± 4.1
14	202	18.8 ± 4.6	25.3 ± 4.8	55.8 ± 5.9	47.1 ± 5.6	85.1 ± 4.7	70.3 ± 6.5
15	379	9.7 ± 3.2	32.6 ± 3.7	39.9 ± 3.9	38.5 ± 3.9	43.5 ± 4.0	42.0 ± 4.0
16	481	15.8 ± 2.9	23.1 ± 3.1	40.0 ± 3.5	40.6 ± 3.5	42.0 ± 3.5	40.9 ± 3.5
17	449	17.1 ± 3.1	25.0 ± 3.0	58.7 ± 4.2	39.4 ± 3.6	34.2 ± 3.4	32.9 ± 3.4
18	353	15.2 ± 3.4	17.2 ± 3.5	38.7 ± 4.0	40.3 ± 4.1	36.9 ± 4.0	40.5 ± 4.1

Table 2.7 Quasi-linkage models showing the best fit to the data of Table 2.6

Family number	χ^2 *under standard assumptions*	*Formulation of the hypothesis*	*Residual* χ^2 *(d.f.)*
10	68.5***	Affinity of *D* and *M2* + selection of *AwC* gametes	21.4* (9)
11	140.3***	Selection of *DM2* gametes	14.7 (10)
12	124.0***	Affinity of *D* and *M2* + selection of *awc* gametes	14.8 (9)
13	66.9***	Affinity of *Aw* and *C* + selection of *DC* gametes + $m2 \xrightarrow{\delta} M2$	6.2 (8)
14	45.4***	Affinity of *Aw* and *C* + selection of *dM2* gametes	16.1 (8)
15	70.6***	Affinity of type (*c*) of *d* and *m2c* + selection of *d* gametes	11.7 (10)
16	28.7**	Affinity of type (*b*) of *awd* and *m2c* + $d \xrightarrow{\delta_1} D + m2 \xrightarrow{\delta_2} M2$	5.4 (8)
17	66.3***	Gamete selection of type (c)	8.5 (8)
18	67.0***	$d \leftrightarrow m2c + Aw \xrightarrow{\delta} aw$	10.4 (8)

Deviation of the model from the actual segregation is significant at $^*P < 0.05$, $^{**}P < 0.01$ and $^{***}P < 0.001$. (b) and (c) as in Figure 2.2.

explained based on the 'chromosome affinity + gametic selection' model. In line with prediction 2d, a considerably wider, than in F_1, range of values of parameters describing positions of chromosome affinity points or interchromosomal co-adaptation has been established among the F_2 genotypes studied. This supports the assumption of complex interactions of genes (gene blocks) responsible for the departure from independent segregation of loci of non-homologous chromosomes.

In conclusion, we would like to touch once more on the problem of quasi-linkage occurrence. If departures from randomness affect only some pairs of non-homologous chromosomes, then, in view of relatively small deviations from the 50% level (generally no more than 10–20%), one has to admit that the contribution of this factor to the restriction of recombinational variation is minor. To obtain an answer to this question, we have carried out an extensive retrospective analysis of evidence on recombination in the literature on tomato. A wealth of evidence on recombinational behavior of markers of various chromosomes has been accumulated in works concerned with the construction and refinement of genetic maps of tomato (primarily those appearing in annual issues of *Reports of the Tomato Genetics Cooperative*). Analysis of this material leads to the conclusion that quasi-linkage in tomato is the rule rather than the exception (Korol, 1987; Korol, Preygel and Nyutin, 1989). The summarized evidence presented in Figure 2.3 indicates that practically all chromosome pairs of the tomato genome show deviations from random assortment. The assumption that deviations of *rf* from the 50% level are only due to chance is disproved by numerous data: first, in a number of

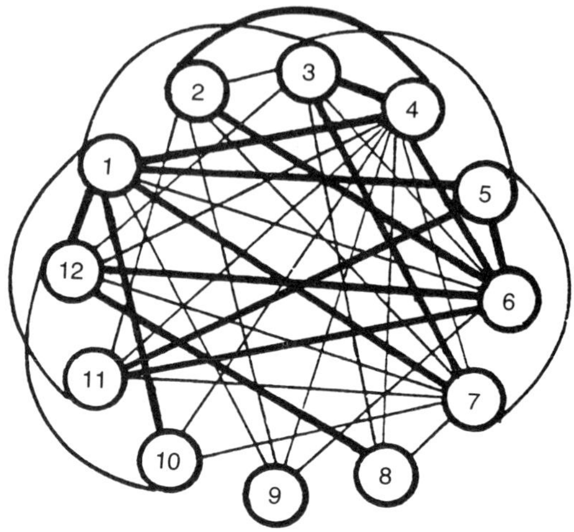

Figure 2.3 Pairs of nonhomologous chromosomes of tomato showing quasi-linkage (from Korol *et al.*, 1989). Deviations of *rf* from 50% are significant at $P < 0.05$ (—)and $P < 0.01–0.01$ (—).

samples the deviations are so high that they could appear by chance alone only in one case in a thousand or even ten thousand; secondly, test for independence of multiple pairs yielded no significant deviations in many experiments. Finally, it should be noted that under segregation some of the linked markers often exhibit concordance in recombinational behavior (quasi-linkage) with respect to markers on other chromosomes. This rules out incomplete penetrance as a possible trivial explanation for the observed deviations.

In our opinion, quasi-linkage has been one of the important reasons for numerous controversies in the extensive work on the construction of genetic maps in tomato and, probably, in other organisms. The dependence of quasi-linkage level on the genetic background might lead to significant deviations of *rf* from 50% in some crosses and to the absence of such deviations in others. On the whole, our results and data in the literature suggest that in some species (e.g. the mouse and tomato) quasi-linkage is an important factor restricting genetic variation.

The material presented here may be regarded as an illustration of how an approach to estimating simultaneously a large number of parameters can be applied to genetic analysis. We believe that, with a wider use of molecular markers (RFLP, VNTR, RAPD markers) belonging to different chromosome pairs, the problem of quasi-linkage and of its genetic mechanism can be studied more rigorously. Moreover, prerequisites have been formed by now for taking into account possible effects of quasi-linkage on the segregational behavior of both qualitative and quantitative trait loci, based on the use of advanced molecular genetic techniques.

Marker analysis of quantitative traits

3

3.1 THE MENDELIAN APPROACH TO QUANTITATIVE VARIATION

Until quite recently biometrical genetics (whose foundations were laid by the known works of Galton, Fisher, Wright and Mather) has been most commonly used to analyze quantitative variation. The polygenes controlling quantitative traits (QTs) were believed to be analogous to Mendelian genes determining alternative traits, but much smaller in their effects. Small effects of individual genes and high degree of QT dependence on environment require methods other than conventional techniques of Mendelian genetics (variance and covariance partitioning, determination of heritability, etc.).

The thesis of a small individual contribution of each of the genes controlling quantitative traits was partly based on experimental data and frequently used as a convenient simplifying assumption in a number of biometrical genetic models. Gradually, it came to be regarded as something self-evident (for criticisms see Thompson, 1975; Lewontin, 1977; Thompson and Thoday, 1979; Ginsburg, 1984). In the past decade, these notions were used in a great number of population genetic models of selection–mutation balance (for a review see Turelli, 1988; Barton, 1990). However, genetics has accumulated a wealth of evidence for the segregation of a small number of major genes, oligogenes or gene blocks being responsible for a considerable fraction, sometimes the bulk, of genotypic variation in QTs present in a population (Sax, 1923; Karp, 1940; Wehrhahn and Allard, 1965; Law, 1967; Zhuchenko *et al.*, 1979; Nevo, 1987; Paterson *et al.*, 1988; Doebley and Stec, 1991, 1993; DeVicente and Tanksley, 1993).

That such evidence has been obtained implies restrictiveness of one of the basic premises of the biometrical approach, i.e. the thesis of weak effects of loci controlling QTs. This opens up possibilities for analyzing quantitative variation by Mendelian methods including identification of individual genetic factors with megaphenic effect, determination of their positions and interactions with other factors.

The Mendelian approach to genetic analysis of QTs emerged simultaneously with the biometrical one (Fisher, 1918; Payne, 1918). Payne seems to have been the first to try, using marker genes, to understand the genetic nature of quantitative differences among *Drosophila* strains accumulated as the result of artificial selection. Serebrovsky (1936, 1970) was one of the first to begin applying mathematical methods in this area. The Mendelian analysis of QTs, which had gained a fairly wide currency in the 1920s to 1930s, to be subsequently superseded by the biometrical approach, began to be revived in the 1960s, thanks mainly to the works of Thoday and his school (Thoday, 1961, 1967; Spickett and Thoday, 1966). This trend in quantitative genetics continues to attract increasing attention because of the abundance of molecular markers (isozymes, RFLP, RAPD, mini- and microsatellites) (Tanksley and Rick, 1980; Beckmann and Soller, 1983, 1986; Stuber, Edwards and Wendell, 1987; Paterson *et al.*, 1988; Tanksley and Hewitt, 1988; Lander and Botstein, 1989; Martin *et al.*, 1989; Knapp, Bridges and Birkes, 1990; Reiter *et al.*, 1991). In order to obtain some idea of the range of problems to be addressed, a list of some possibilities follows.

- Key parameters of the genetic system of population adaptation, such as crossover frequency and genomic distribution, mutation rate, mating system type, migration rate, show quantitative variation (Zhuchenko and Korol, 1985; and Chapter 4 of this book). Mapping the modifiers and assessing their role in determining the **norm of reaction** of these components to environmental changes are important prerequisites for constructing realistic models of population dynamics, testing the predictions of population genetics and for our understanding of the patterns of long-term genetic variability transformation. The available evidence supports this approach (Gale and Rees, 1970; Chinnici, 1971; Cederberg, 1985; Charlesworth and Charlesworth, 1985a, b; Zhuchenko and Korol, 1985; Ebinuma, 1987; Ozias-Akins *et al.*, 1993).
- Assessing the effects of various genomic segments on QTs and on their norm of reaction to diverse ecological conditions may provide new insights into the nature of frequently discussed concepts of genetic balance in natural populations and its changes in the course of adaptation to new conditions. This approach is essential for achieving a more detailed understanding of such general notions as genome integrity and co-adaptation within a gene block and chromosome (Stephens, 1950; Grant, 1966, 1975; Dobzhansky, 1970; Mayr, 1970; Ford, 1971; Carson, 1975, 1987; Zhuchenko and Korol, 1985).
- Although it is the organism that is the target of selection, the entire genome is unlikely to be a suitable unit for both describing and predicting the genetic consequences of selection. Marker (segment) analysis seems to be just the

approach which will make it possible, at least partially, to overcome the difficulties in estimating individual effects of multiple loci on fitness and in analyzing the dynamics of multilocus systems (Lewontin, 1974; Jain, 1990). One implication of this interpretation of QT genetic determination is the appearance of new models for the maintenance of polymorphism in natural populations based on genes with essentially unequal effects (Barton, 1986; Nagylaki, 1989; Hastings and Hom, 1990; sections 7.2 and 7.3).

- The Mendelian analysis of the genetic architecture of QTs of wild species and primitive forms, used as sources of resistance to diseases, pests and adverse environmental factors, is an important prerequisite for improving the efficiency of introgressive selection. Selection of individuals carrying marker alleles of the donor strain permits a more efficient (more rapid) introgression of adaptively significant blocks of genes, linked to them, from wild species and, conversely, a more rapid elimination of donor segments with unfavorable genes or gene combinations (Soller and Plotkin-Hazan, 1977; Tanksley, Medina-Filho and Rick, 1981; Nevo, 1987, 1992; Hillel *et al.*, 1990; Sarfatti *et al.*, 1991; Eshed *et al.*, 1992; DeVicente and Tanksley, 1993). This is particularly important in breeding organisms with long generation times when economic benefits resulting from a lower number of individuals and generations analyzed may compensate even for the considerable costs of identifying expensive markers. Identification of chromosomes, chromosome segments and gene blocks, with megaphenic effects may also be important in intervarietal crosses (Law *et al.*, 1981; Law, Snape and Worland, 1983; Soller and Beckmann, 1983; Beckmann and Soller, 1986). Combined selection on markers and phenotypes, based on linkage disequilibria, could also increase the efficiency of breeding programs in outbreeding populations (Lande and Thompson, 1990; Zhang and Smith, 1992).
- Marker analysis of QTs may prove to be a useful tool for identifying strains and varieties and even for making predictions about heterosis (Dudley, Saghai Maroof and Rufener, 1992). This is also important in solving problems associated with the breeder copyright protection (Soller and Beckmann, 1987; Messmer *et al.*, 1991).
- Identification of a factor with a strong effect on a QT opens up possibilities for further physiological and biochemical analyses of its effects (Thoday, 1967) and for improving the reliability of genetic counseling (Lander and Botstein, 1986; Thomson, 1988; Gershon *et al.*, 1990). For genes with strong effects, precise marker-based genetic mapping offers possibilities for physical mapping and chromosome walking or jumping with a view to their cloning (positional cloning strategy) and transfer, via genetic transformation, to recipient genotypes (Young *et al.*, 1988; Sarfatti *et al.*, 1991).
- Marker analysis of QTs permits a study of the effect of crossing-over variation within particular chromosome segments on the behavior of

quantitative traits in marker groups. This raises the possibility of taking into account segment-specific effects of recombinogenic agents on variation in adaptively significant systems of traits, and of elucidating the causes and ways of rearranging genetic correlations between traits (Korol and Zhuchenko, 1980).

- Marker analysis of QTs may prove to be an effective tool for studying the genetic basis of interspecific differences (both morphological and reproductive) (Coyne and Charlesworth, 1986; Doebley *et al.*, 1990; Paterson *et al.*, 1990, 1991; Naveira, 1992). Of interest are the questions concerning the genetic architecture of interspecific differences in qualitative and quantitative traits, the nature of genetic factors determining reproductive isolation of species (poor crossability, low viability and fertility of hybrids) and the 'hybrid breakdown' effect (Mayr, 1970; Grant, 1981; Templeton, 1981; Singh, 1990).

One of the modifications of the Mendelian analysis of quantitative variation both under controlled crosses and in studying pedigrees of arbitrary structure is to construct models based on the simplest hypotheses of the genetic nature of differences between the parental forms, estimate model parameters and choose, based on statistical tests, the most adequate description. Various modes of monogenic inheritance, of a single major gene and a number of weak polygenes, two major genes with equal or unequal contributions, etc. can be considered as working hypotheses (Karlin, Williams and Carmelli, 1981; Lalouel *et al.*, 1983; Elston, 1984; Ginsburg, Fedotov and Chepkasov, 1986; Goradia *et al.*, 1992). In spite of such restrictive assumptions as normal distribution, additive effects and others, good agreement is, in many cases, achieved with simple one- and two-locus models (Elston and Stewart, 1973; Morton and MacLean, 1974; Ginsburg *et al.*, 1987).

It should be taken into account, however, that statistical fit of a model to actual data cannot be regarded as proof of its validity. Discussing the feasibility of a one-locus model, Elston (1966; p. 89) pointed out that:

> Compatibility of data of the type described and a one locus hypothesis is very weak evidence for such a simple hypothesis, as a great many of other genetic hypotheses will give rise to exactly the same results. Only further breeding tests can distinguish between the involvement of one and more than one locus.

Therefore, a purely statistical test of fit of a one- or two-locus model is primarily regarded as a means of *excluding* such a possibility for QTs (Elston and Stewart, 1973). Nevertheless, this approach has found practical application in animal breeding, for example, in typing of sires for major genes or QTLs (Knott, Haley and Thompson, 1992a, b).

With the Mendelian approach to quantitative variation analysis, the following interpretation can be used:

- A trait may be polygenic; moreover, it may even be assumed that each of the genes present in the genome affects the trait to a degree (general pleiotropy). But if segregation for a few of the genes accounts for most (say, 80%) of the heritable variation within the population in a particular environment, it may be assumed that the trait is controlled by a small number of genes (Thompson, 1975; Zhuchenko *et al.*, 1975; Korol, 1976; Maletsky, 1976; Zuchenko, Korol and Andryuschenko, 1978; Diers *et al.*, 1992). For most of the above problems, this simplification seems to be justified. Identifying each of these oligogenes appears to be quite a feasible task. The magnitude of the contribution to genotypic variation due to segregation of megaphene loci depends not only on their effects but also on the allele frequencies (Boerwinkle, Chakraborty and Sing, 1986; Dudley, 1992). This observation, usually irrelevant in controlled crosses, may be very important and should be taken into consideration when analyzing populations of arbitrary structure (e.g. in man).
- Many QT genes may be distributed throughout the genome not uniformly but as compact blocks with low probability of crossing over within the block (Hanson, 1959; Grant, 1966; Darlington, 1971; Ford, 1971; Lewontin, 1977; Zhuchenko and Korol, 1985; Ukai, 1987; Dentine and Cowan, 1990). Such a block can be viewed as an 'effective factor'. The ultimate objective of mapping quantitative trait loci (QTL) may be identifying loci with a strong individual effect or even cloning the corresponding genes (map-based or position cloning). In the latter case, resolving the 'gene block/major gene' alternative is a necessary condition for continuing the study.
- When crossing two strains which contrast with respect to the trait in question, the F_1 hybrid often appears to be heterozygous for a number of loci. However, among the progeny of a backcross between the F_1 and parental strains, some forms that are heterozygous for a smaller number of loci can be selected (Wehrhahn and Allard, 1965; Zhuchenko, Korol and Andryuschenko, 1978) and then examined with the aid of marker genes. Moreover, an approach like this allows the localized megaphene blocks to be subjected to further, fairly precise, recombination analysis (Shrimpton and Robertson, 1988a, b; Edwards *et al.*, 1992; Naveira, 1992).

Utilizing markers (especially molecular ones) considerably improves the efficiency of the Mendelian approach, allowing the effects of chromosome segments to be estimated without any assumptions of the total number of loci determining variation of the QT in question (Korol, 1973, 1976, 1987; Soller *et al.*, 1976; Korol, Zhuchenko and Preygel, 1983; Beckmann and Soller, 1986, 1990; Korol, Preygel and Bocharnikova, 1987; Paterson *et al.*, 1988, 1991; Doebley and Stec, 1991, 1993; Reiter *et al.*, 1991; DeVicente and Tanksley, 1993). Multimarker analysis provides a good basis for replacing the a priori approach to interpreting quantitative variation by a precise experimental analysis. Estimation of the effects of whole chromosomes and of their arms is also possible based on cytogenetic techniques, i.e. interstrain chromosome substitution (Law, 1966). Besides, cytogenetic analysis may enhance the power of the marker technique because of the use of cytological markers, e.g. chromosome interchanges (Linde-Laursen, 1982; Weber and Helentjaris, 1989).

3.2 ESTIMATING QTL MARKER LINKAGE IN CONTROLLED CROSSES

3.2.1 Genetic assumptions and the basic model

The efficiency of employing marker genes to identify megaphene loci (polygene blocks) with strong effects on QTs was most clearly demonstrated on *Drosophila* (Karp, 1940; Spickett and Thoday, 1966; Serebrovsky, 1970). Because of the known advantages of this model organism, the marker technique has allowed QTL mapping with a precision close to 1–2 cM (Spickett and Thoday, 1966), which is similar to the accuracy of mapping marker genes themselves.

Serebrovsky (1970) developed a very simple procedure of QTL mapping in *Drosophila*, taking advantage of the fact that male crossing over is suppressed in this organism. Let a megaphene gene or gene block (A/a) affecting the trait under study be located some distance, $rf = p$, from the marker locus (S/s), with parental genotypes being homozygous. Consider a case of coupling. Then the putative structures of parental forms and F_1 are as follows:

P_1	P_2	F_1
a –| |– a	A –| |– A	a –| |– A
s –| |– s	S –| |– S	s –| |– S

If the relationship between alleles A and a is that of complete dominance, further analysis based on backcross (BC) data is possible in the coupling phase when the recessive allele a is linked to the recessive s regardless of whether a is responsible for a lower value of the trait or vice versa. But if A/a shows an intermediate expression in the heterozygous condition, analysis is possible in both phases (as/AS and aS/As) with the terms 'coupling' or 'repulsion' merely implying that recessive allele s is linked to the suppressor allele a or to the enhancer allele A respectively.

The case of crossing over being absent in one of the sexes, considered by Serebrovsky, is convenient in that it enables marker groups to be compared in terms of the QT in question in the BC progeny consisting of non-crossover genotypes only (resulting from the cross $sa/sa(♀) \times SA/sa(♂)$) and in the progeny containing forms of both types – crossover and non-crossover (resulting from the $SA/sa(♀) \times sa/sa(♂)$ cross). These comparisons yield a simple system of equations:

$$Y = Y_S - Y_s = x_{Aa} - x_{aa} = d \quad \text{(in backcrossing } sa/sa(\female) \times SA/sa(\male)\text{)},$$

$$X = X_S - X_s = (1 - 2p)d \quad \text{(in backcrossing } SA/sa(\female) \times sa/sa(\male)\text{)}.$$

Solution of these equations permits the effect of substitution d at locus A/a and the crossover rate p between A/a and S/s to be readily estimated. An example of an economically important organism with crossing over suppressed in one sex is the silkworm ($rf_{\female} = 0$).

The assumption that only one genetic factor responsible for the differences between the parental types for the QT studied is present in the marker region does not imply that the QT is under monogenic control. A number of factors may be located on other chromosomes and contribute to genetic variation of the population without significantly affecting the resultant estimates (provided segregation of loci of non-homologous chromosomes is independent). Of course, in the presence of strong non-linear interactions between these and the factor located on the chromosome in question, estimate bias is possible and its reduction requires more sophisticated analytical procedures.

Consider a general case of multilocus QT determination while assuming:

(a) in a pair of marker chromosomes, parental forms differ in only one factor, A/a, affecting the trait under study (this may be a locus or a compact block of genes); for the rest of the genome, the F_1 hybrid may be a heterozygote of any level of complexity;
(b) the marker has no pleiotropic effect on the QT;
(c) there is no selection;
(d) marker penetrance is complete, i.e. genotypes *ss* and *Ss* always express as phenotypes *s* and *S* respectively;
(e) the meiotic behavior of the marker chromosome is independent of other chromosomes, i.e. quasi-linkage is absent;
(f) there are no interactions of A/a with loci on other chromosomes resulting in rank changes for genotypes *Aa* and *aa*.

The F_1 hybrid produces four types of gametes in the ratio $(1-p)$ *SA*: *pSa*: *psA*: $(1-p)$ *sa*, where p is the recombination rate between the megaphene factor A/a and marker S/s. In the progeny of the backcross to the recessive parent (P_1) the distribution of the trait within the group s is a mixture of two distributions: $f_{aa}(x)$ with the mean value x_{aa} and $f_{Aa}(x)$ with the mean x_{Aa}. The proportions of the first and second components of the mixture are $1-p$ and p respectively (Figure 3.1). For group S, the situation is the same, with the only difference that the proportions of the components are p and $1-p$. Thus, one-marker problem of parameter estimation in the backcross case can be reduced to the resolution of two mixtures into two components with complementary proportions:

$$g_s(x) = (1-p)f_{aa}(x) + pf_{Aa}(x); \; g_S(x) = pf_{aa}(x) + (1-p)f_{Aa}(x). \qquad (3.1)$$

With the above assumptions (a to f) and co-dominance, the distribution of the

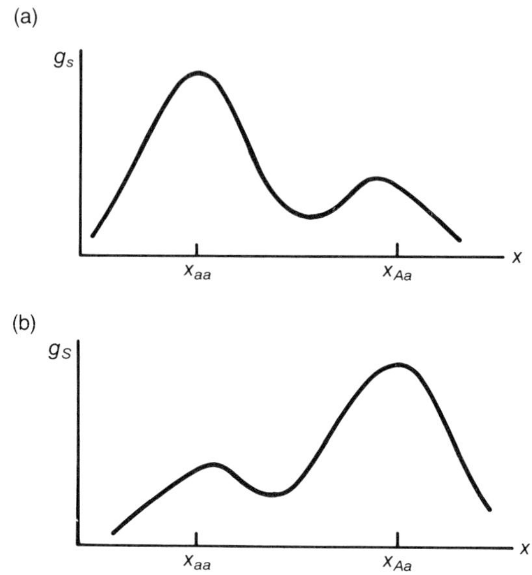

Figure 3.1 Distribution of trait x within marker groups of backcross progeny: (a) group s; (b) group S.

trait in marker groups of the F_2 progeny is a mixture of three components $[f_1(x) = f_{aa}, f_2(x) = f_{Aa}$ and $f_3(x) = f_{AA})$:

$$g_{ss}(x) = (1-p)^2 f_1(x) + 2p(1-p)f_2(x) + p^2 f_3(x),$$

$$g_{sS}(x) = p(1-p)f_1(x) + [1-2p(1-p)]f_2(x) + p(1-p)f_3(x),$$

$$g_{SS}(x) = p^2 f_1(x) + 2p(1-p)f_2(x) + (1-p)^2 f_3(x). \tag{3.1a}$$

A large statistical literature is devoted to resolution of one mixture into components (e.g. Titterington, Smith and Makov, 1985; McLachlan and Basford, 1988). The presence of several mixtures of the same set of components but with different proportions is a specific feature of the QTL mapping problem, which seems to give some advantage in the identification of the parameters, especially with small p. But estimation of p is a difficult problem in any case.

3.2.2 Statistical criteria used for parameter estimation

At different stages of a QTL mapping study one should make decisions about:

- the experimental design, i.e. the system of crosses, the way of accumulating data (parallel or sequential) providing the maximum power of statistical tests for the presence of a QT locus in the marked chromosome;

- the genetic model used to describe the situation (single or multimarker, one QT locus on the chromosome or several, etc.);
- how the data will be used in the analysis, i.e. separately for each of the QTs of interest or simultaneously, accounting for trait correlations;
- the statistical approach for estimation of genetic parameters and the calculation techniques;
- testing the results based on additional experiments.

The above questions are not independent and a choice at one stage could lead to some constraints at others. As mentioned above, QTL mapping can formally be considered as a statistical problem of mixture decomposition. It includes estimation of mixture proportions which depend on the unknown recombination rate(s) and parameters characterizing the postulated QT locus (loci) effects (mainly, on the trait mean and/or variance).

In most of the recent publications on marker analysis of QTs the method of maximum likelihood (ML) was used as a basic approach. This method has known advantages over other approaches, including generality, asymptotic efficiency, consistency and asymptotic normality of the estimates. Nevertheless, there could be numerous situations in which some of these advantages are not necessarily valid, e.g. with small sample size, violation of assumptions about component densities, non-additive gene action, etc. A proper choice of the basic distribution densities $f_i(x)$ in equation 3.1 or equation 3.1a in QTL mapping can be considered among the most difficult and yet unsolved problems. In most cases the normality is assumed, but this assumption contradicts the very essence of the marker approach, aimed at revealing QT loci with strong individual effects. Indeed, the presence of oligogenes in other chromosomes cannot be excluded. Because of this, and other factors as well, the components of the mixture distributions of the QT in the compared marker groups could be non-normal, leading to an increase in the biases of the estimates. In general, it is highly desirable to have linkage estimation procedures applicable to arbitrarily distributed components under moderate constraints. From the statistical viewpoint such a method can be referred to as distribution free (Kendall and Stuart, 1967). Three main approaches founded on rather different principles are used for parameter estimation from experimental data:

Methods based on probabilities (likelihoods) of the observed sample
These include the ML method and Bayes' approach. The component distributions of the QT in question should be specified from the beginning.

Methods based on minimization of the discrepancy between the observed and expected distributions
Mean square, mean absolute discrepancy or maximum absolute discrepancy between the observed and expected distributions of the studied QTs in marker

groups could be used as optimization criteria for identification of unknown parameters (reviewed by Titterington, Smith and Makov, 1985). Here, the component distributions of the trait should be specified too.

Distribution-free methods

These methods may also be based on minimization of discrepancies between observed and expected parameters, but the specification of the QT distribution is not necessary. The most important approach of this kind is the method of moments, which also demands information about trait distribution in marker groups. But in contrast to the above two groups of methods it is possible to utilize this information without specification of the densities $f_i(x)$.

One can expect that any method of estimation could have some advantages and drawbacks depending on the real genetic situation. We give here some more details about the method of moments because it is not described sufficiently in the literature. For a comparison of these approaches see Korol, Ronin and Kirzhner, 1993).

1. Maximum likelihood method

This approach is the same as the **lod score** method commonly used in estimating linkage between marker loci, particularly in human genetics. The lod score technique has also been applied in estimating QT marker linkage (Lange, Spence and Frank, 1976; Demenais, Lathrop and Lalouel, 1988; Lander and Botstein, 1989). For the backcross progeny ML estimation based on the model in equation 3.1 could be obtained by maximization of the log-likelihood:

$$\ln L(\theta) = \sum_{k=1}^{N_1} \ln g_1(x_k) + \sum_{j=1}^{N_2} \ln g_2(x_j) + \text{const.}, \tag{3.2}$$

where N_1 and N_2 are the *ss* and *Ss* group sizes and x_k is the observed value of the quantitative trait in question for the *k*th individual. Usually, the assumption of normality is used for the densities f_1 and f_2 from equation 3.1: $f_1 = f_{aa}(x) = N(x_{aa}, \sigma^2_{aa})$ and $f_{Aa}(x) = N(x_{aa} + d, \sigma^2_{Aa})$. Here, x_{aa} is the trait mean of the group *aa*, *d* is the effect of $aa \rightarrow Aa$ substitution and σ^2_{aa} and σ^2_{Aa} are the trait variance within the groups *aa* and *Aa* due to the environmental variation and segregation of loci on other chromosomes. Thus, in the case of normality, $\ln L(\theta)$ is a function of five-parameter vector $\theta = (p, x^{aa}, d, \sigma^2_{aa}, \sigma^2_{Aa})$. Given the forms of distribution of $f_{Aa}(x)$ and $f_{aa}(x)$, a set of parameters values, $\theta = \hat{\theta}$, for which the maximum of likelihood function (or, which is the same, of its logarithm) is attained is considered to be the solution (e.g. Weller, 1986; Jensen, 1989; Luo and Woolliams, 1993).

For many traits, the assumption of normality appears to be unacceptable (see Karlin, 1988 for a review). Despite some arguments to the contrary, the robustness of the QT genetic models (including those of QTL mapping) and their insensitivity to departures from normality may be insufficient for reasonable results to be obtained (Karlin, Williams and Carmelli, 1981; Korol, Zhuchenko and Samovol, 1981; Le Roy and Elsen, 1992; Korol, Ronin and Kirzhner, 1993). One possibility for extending the applicability of the ML method of QTL mapping problems is a more flexible specification of the component density distribution functions. For example, one can use an approximation (Trankle, 1975) that allows for deviations from normality both in asymmetry and curtosis:

$$f_{aa}(x) = N(x_{aa}, \sigma^2_{aa})\{1 + c_1 \tanh[(x - x_{aa})/(c_0 \sigma_{aa})] + c_2 \tanh^2[(x - x_{aa})/c_0 \sigma_{aa})]\},$$

$$f_{Aa}(x) = N(x_{Aa}, \sigma^2_{Aa})\{1 + d_1 \tanh[(x - x_{Aa})/(d_0 \sigma_{Aa})] + d_2 \tanh^2[(x - x_{Aa})/d_0 \sigma_{Aa})]\}.$$

In order to obtain a flexible description of a density $f(x)$ using its moments, a series approximation can be used:

$$f(x) = N(x) + N^{(3)}(x)c_3/3! + N^{(4)}(x)c_4/4! + \dots,$$

where $N(x)$ is the standard normal density, $N^{(i)}(x)$ is the ith derivative of $N(x)$ and c_i are the coefficients, which depend on the moments of the density $f(x)$. Two forms of the last approximation are mainly used, i.e. Gram–Charlier and Edgeworth series. Theoretically more preferable is Edgeworth's series, in which the members of the same order of n (n standing for the sample size) are grouped together (see Cramer, 1974). It has the following form:

$$f(x) = N(x) - [N^{(3)}(x)m_3/3!\sigma^3] + [N^{(4)}(x)(m_4/\sigma^4 - 3)/4! + N^{(6)}(x)(m_3/\sigma^3)^2 10/6!] + \cdots,$$

in which each of the large brackets represents a member of the same order of n, and m_i is the ith central moment of $f(x)$.

2. *Minimum discrepancy between the observed and expected distribution functions*

With this approach the following group of functionals can be used:

$$I_q = \sum_{k=1}^{N_1} \left| k/N_1 - \int_{-\infty}^{x_k} g_1(x)dx \right|^q + \sum_{j=1}^{N_2} \left| j/N_2 - \int_{-\infty}^{x_j} g_2(x)dx \right|^q. \quad (3.3)$$

With $q = 2$ we have the mean square (MS) method and $q = 1$ gives the mean absolute discrepancy method. Another version of this class is the minimax method, related to the Kolmogorov–Smirnov criterion. Clearly, the component distributions f_1 and f_2 of the expected mixtures g_1 and g_2 should be specified here. As a modification of these functionals one can use optimization criteria with different weights a_k and b_j for different parts of the observed distribution:

$$Iw_q = \sum_{k=1}^{N_1} a_k \left| k/N_1 - \int_{-\infty}^{x_k} g_1(x)dx \right|^q + \sum_{j=1}^{N_2} b_j \left| j/N_2 - \int_{-\infty}^{x_j} g_2(x)dx \right|^q, \quad (3.3a)$$

with $a_k > 0$ and $b_j > 0$. This allows one to check possible variation of different parts of observations with respect to the information content (see also Darvasi and Soller, 1992; Weller and Wyler, 1992).

3. Method of moments (MM)

According to this approach, introduced to statistics by K. Pearson (cited by Cramer, 1974), one should calculate expected moments of the trait in the groups *ss* and *Ss* as functions of parameters characterizing the postulated locus A/a. Then, some optimal estimates of these parameters can be found by minimizing the discrepancy between the observed and expected moments. In principle, no a priori information about the distribution densities f_1 and f_2 is needed, which makes the approach rather attractive when such information is not available.

It is worthwhile mentioning the following fact. A broad class of practically important distribution densities has been shown to obey the differential equation suggested by Pearson:

$$\frac{df(x)}{dx} = \frac{(x-a)f(x)}{b_o + b_1 x + b_2 x^2}.$$

The remarkable result is that Pearsonian densities are fully defined by the first four moments (for more details see Kendall and Stuart, 1963). Moreover, these densities can be used for approximation of other distributions if their moments are known.

The method of moments was employed in different situations of QTL mapping (Korol, 1973, 1976; Zhuchenko, Korol and Andryuschenko, 1978; Zhuchenko *et al.*, 1979; Korol, Zhuchenko and Samovol, 1981; Korol, Zhuchenko and Preygel, 1983; Korol, Preygel and Preygel, 1990; Weller, 1986; Preygel and Korol, 1989; Luo and Woolliams, 1993). Let us consider one of the versions based on the first four moments (Korol, Zhuchenko and Preygel, 1983).

Backcross data

The unobservable distributions $f_1(x)$ and $f_2(x)$ can be characterized by their means x_{aa} and $x_{Aa} = x_{aa} + d$, variances σ^2_{aa} and σ^2_{Aa}, third moments μ_{aa} and μ_{Aa}, and fourth moments λ_{aa} and λ_{Aa}. The expected first initial moment (the mean) and three central moments (from the second to the fourth) can easily be obtained for the two marker groups:

$$X_{ss} = (1-p)x_{aa} + px_{Aa} = x_{aa} + pd = E_1(\theta),$$

$$X_{Ss} = px_{aa} + (1-p)x_{Aa} = x_{aa} + (1-p)d = E_2(\theta),$$

$$\sigma^2_{ss} = (1-p)\sigma^2_{aa} + p\sigma^2_{Aa} + p(1-p)d^2 = E_3(\theta),$$

$$\sigma^2_{Ss} = p\sigma^2_{aa} + (1-p)\sigma^2_{Aa} + p(1-p)d^2 = E_4(\theta),$$

$$\mu_{ss} = (1-p)\mu_{aa} + p\mu_{Aa} - 3p(1-p)(\sigma^2_{aa} - \sigma^2_{Aa})d + p(1-p)(1-2p)d^3 = E_5(\theta),$$

$$\mu_{Ss} = p\mu_{aa} + (1-p)\mu_{Aa} - 3p(1-p)(\sigma^2_{aa} - \sigma^2_{Aa})d - p(1-p)(1-2p)d^3 = E_6(\theta),$$

$$\lambda_{ss} = (1-p)\lambda_{aa} + p\lambda_{Aa} - 4p(1-p)(\mu_{aa} - \mu_{Aa})d + 6p(1-p)[p\sigma^2_{aa} + (1-p)\sigma^2_{Aa}]d^2 + p(1-p)(1-p+p^2)d^4 = E_7(\theta),$$

$$\lambda_{Ss} = p\lambda_{aa} + (1-p)\lambda_{Aa} - 4p(1-p)(\mu_{aa} - \mu_{Aa})d + 6p(1-p)[(1-p)\sigma^2_{aa} + p\sigma^2_{Aa}]d^2 + p(1-p)(1-p+p^2)d^4 = E_8(\theta). \tag{3.4}$$

Here $\theta = (p, x_{aa}, d, \sigma^2_{aa}, \sigma^2_{Aa}, \mu_{aa}, \mu_{Aa}, \lambda_{aa}, \lambda_{Aa})$.

Some additional assumptions about the A/a mode of action are required to make further analysis possible. Consider genotypes of the groups *Aa* and *aa*. In addition to locus A/a linked to the marker, these may differ with respect to other chromosomes (see assumption (a) above). In view of assumption (e), each genotype in group *aa* has a counterpart in group *Aa* differing from the former only at locus A/a. When effects of the factor are purely additive (i.e. for any genotype k, $x^k_{Aa} = x^k_{aa} + \beta$), allele substitution $aa \to Aa$ results in a shift in the trait distribution, the shape of the density curve remaining unchanged. Most of the reports dealing with identification of major factors (oligogenes) controlling QTs are restricted to this case (Jayakar, 1970; Hill, 1975; Soller, Brody and Genizi, 1976; Weller, 1986; and others). However, in fact, the shape of the distribution curve may be altered when the substitution effect depends on the genetic background, as in the multiplicative case ($x^k_{Aa} = \alpha x^k_{aa}$). Now let us suppose that the relative change is the same for all genotypes, i.e. $x_{Aa}/x_{aa} = \alpha = \text{const}$. Combining these two cases, we obtain $x^k_{Aa} = \alpha x^k_{aa} + \beta$, i.e. locus A/a is regarded as a linear converter with x^k_{aa} as an input variable and x^k_{Aa} as an output one. Then, $\sigma^2_{Aa} = \alpha^2\sigma^2_{aa}$, $\mu_{Aa} = \alpha^3\mu_{aa}$ and $\lambda_{Aa} = \alpha^4\lambda_{aa}$. Therefore, the vector θ can be reduced to the set $\theta_\alpha = \{\hat{p}, \hat{x}_{aa}, d, \alpha, \sigma^2_{aa}, \mu_{aa}, \lambda_{aa}\}$.

In order to obtain the solution of the system 3.4 under linear assumption ($x^k_{Aa} = \alpha x^k_{aa} + \beta$) the least-squares method can be used. Let the seven-parameter set $\theta = \{\hat{p}, \hat{\alpha}, \hat{x}_{Aa}, \hat{d}, \hat{\sigma}^2, \mu, \hat{\lambda}\}$ minimizing the sum of squared deviations

$$F(\theta) = \sum_{i=1}^{8} [M_i - E_i(\theta)]^2 w_i = \sum_{i=1}^{8} [M_i - E_i(\theta)]^2/m^2_i,$$

where m^2_i is the variance of statistic M_i, be a solution of system 3.4. Thus, the task of identifying factor A/a is reduced to solving a standard optimization problem subject to certain, biologically justified, restrictions on parameter values ($0 \leqslant p < 0.5, \alpha > 0, \sigma^2 > 0$, etc.). While the search for the point $\hat{\theta}$ providing minimum $F(\theta)$ may present certain difficulties, the principal advantage of the approach described lies in its generality. Any QTL mapping model which is formulated based on a set of equations for moments may be reduced to system 3.4 with its vector of genetic parameters θ and a set of restrictions (Korol, Zhuchenko and Preygel, 1983). Comparative simulation analysis of the three methods (ML, MM and mean square discrepancy for the distribution functions) showed that it is much easier to obtain the global maximum for the MM functional than those for the other two methods (Korol, Ronin and Kirzhner, 1993).

The next step is to determine simultaneous confidence regions at least for the key parameters, for example p and d. It must be noted that determining simultaneous confidence regions in multivariate (even bivariate) problems of parameter estimation may be a far more complicated problem than estimation itself (Kendall and Stuart, 1967). In view of the asymptotic properties of sample moments (Cramer, 1974), it can be inferred that with increasing sample size each of the random variables $[M_i - E_i(\theta)]/m_i$ tends, when the model is adequate, to be normally distributed as $N(0, 1)$. Thus, the sum $F(\theta)$ is asymptotically distributed as χ^2 with the degrees of freedom being equal to the number of equations minus the number of parameters to be estimated, i.e. d.f. $= 8 - 7 = 1$. This provides a simple means for an approximate determination of simultaneous confidence domains for parameters specifying the identified factor A/a. The surface $F(\theta) = \chi_1^2(\gamma)$, produced by incrementing vector θ coordinates relative to point $\hat{\theta}$ until values are obtained ensuring the attainment of level $\chi_1^2(\gamma)$, serves as the domain boundary corresponding to confidence level $1 - \gamma$. Since assumptions underlying such a derivation are only satisfied under a large sample, it is also reasonable to have independent interval estimates based on a computer simulation technique.

In our opinion, the opportunity of working without the need to choose the proper form of distributions $f_{aa}(x)$ and $f_{Aa}(x)$ is a fairly sufficient compensation for a comparatively low statistical efficiency of the method of moments. Moreover, it should be stressed here, that this drawback, in comparison with more effective alternatives (such as ML), is valid only when the assumptions about the QT distribution, mode of gene action, etc. are not violated: otherwise MM can be superior (Korol, Ronin and Kirzhner, 1993). When necessary, the estimates derived by the method of moments may, because of their sufficiency (Kendall and Stuart, 1967), also be taken as a start point in the iterative procedure of the ML method (e.g. Luo and Woolliams, 1993). This can help in solving a rather difficult technical problem – ensuring the likelihood function convergence to the global maximum (Darvasi and Weller, 1992).

F_2 data

Model 3.4 has been constructed for analyzing backcross progeny data. Consider now an analog for F_2 (see Korol, Zhuchenko and Preygel, 1983). Assume that the genetic assumptions underlying equation 3.4 are satisfied and let also recombination frequency (p) between A/a and S/s be the same in male and female meiosis. We assume also complete dominance not only for the marker locus, but also for the identified factor, i.e. $x_{Aa} = x_{AA} = x_A$. By introducing parameter q, $q = (1 - p)^2$ in coupling phase and $q = p^2$ in repulsion phase, and setting $x_a = x_{aa}$, $d = x_A - x_a$ and $\tau_a^2 = \tau^2$, we obtain the equations for the trait moments:

$$M_1 = X_s = qx_a + (1 - q)x_A = E_1(\theta),$$

$$M_2 = X_S = (1-q)x_a/3 + (2+q)x_A/3 = E_2(\theta),$$
$$M_3 = \sigma^2{}_s = q\sigma^2 + (1-q)\alpha^2\sigma^2 + q(1-q)d^2 = E_3(\theta),$$
$$M_4 = \sigma^2{}_S = (1-q)\sigma^2/3 + (2+q)\alpha^2\sigma^2/3 + q(1-q)d^2 = E_4(\theta),$$
$$M_5 = \mu_s = E_5(\theta),\ M_6 = \mu_S = E_6(\theta),\ M_7 = \lambda_s = E_7(\theta),\ M_8 = \lambda_S = E_8(\theta). \qquad (3.4a)$$

Here $\theta = \{q, x_a, x_A, \sigma^2, \alpha, \mu, \lambda\}$, M_i and $E_i(\theta)$ $(i = \overline{1, 8})$ have the same meaning as in equation 3.4a. The explicit forms of $E_5(\theta), \ldots, E_8(\theta)$ are not provided because the resultant expressions are rather cumbersome. For calculating these it is enough to replace, in equation 3.4a, p by $1-q$ within the recessive group and by $(1-q)/3$ within the dominant one.

The above approach to the analysis of linkage between markers and quantitative traits have been used by Weller (1986) to solve a similar problem based on the F_2 data. The author has made an assumption of the genotypes *SS*, *Ss* and *ss* being distinguishable while supposing arbitrary allelic relations at locus *A/a* (i.e. assuming no complete dominance), and formulated his model in the form of a set of six equations for the first two moments with seven parameters. For parameter estimation, he has developed a special algorithm for approximate calculation combining the method of moments and ML method for the case of normal distribution of original components.

As an illustration, consider the application of equation 3.4a for identifying, on tomato chromosome 2 (marked by *aw*-lack of anthocyanin, and *d*-dwarfness), a factor affecting seedling weight (Korol, Zhuchenko and Preygel, 1983). The example chosen is a model one since it is known in advance that gene *d* affects a variety of quantitative traits including seedling size and weight. It is of interest to note that Weller (1986), by way of providing an illustration for his algorithm, also used an experiment estimating the effects of gene *d* in tomato. The expected distance of the factor *A/a* in question from the marker locus *Aw/aw* must coincide with the map distance between *d* and *aw*. The following values of the first four central moments within marker groups *Aw* and *aw* of the studied F_2 population have been obtained: $M = \{M_i, i = \overline{1, 8}\} = \{7.47, 5.51, 5.46, 3.91, 3.84, 5.28, 88.0, 49.6)$. Applying the mean square (MS) method, the following solution was obtained: $\hat{\theta} = (0.8359, 5.18, 7.60, 2.26, 1.67, 0.51, 10.5)$. The map distance between markers *aw* and *d* is $x = 11$ cM. Using Haldane's formula $p = 0.5(1 - e^{-2x})$, we find the expected rate of crossing over between the 'identified factor' and locus *aw* as $p = 9.87\%$. The value of $\hat{q} = 0.8359$ derived from the model corresponds to $\hat{p} = 8.57\%$ (coupling phase), which is close to the expected recombination frequency ($p = 9.87\%$). If locus *d* itself is used as a marker, then moments in marker groups *d* and *D* have the following values: $M = \{7.50, 5.34, 5.43, 3.45, 3.94, 2.84, 86.8, 28.1\}$. The LS method yields the solution $\hat{\theta} = (0.9916, 5.32, 7.50, 3.11, 1.33, 1.94, 26.9)$. The estimated value $q = 0.9916$ corresponds to $\hat{p} = 0.42\%$, while the expected value of p is, naturally, equal to 0. Thus, in both cases, we have determined rather accurately the position of locus D/d as the 'identified factor' A/a.

3.2.3 Resolution power of single-marker analysis

A detailed discussion of the issues concerning the power of tests for detecting linkage and designing experiments can be found in many publications (e.g. Soller and Genizi, 1978; Soller, Brody and Genizi, 1979; Demenais, Lathrop and Lalouel, 1988; Soller and Beckmann, 1988; Lander and Botstein, 1989; Soller, 1990; Weller *et al.*, 1990; Weller and Wyler, 1992). Based on the moment approach a simple test of linkage between the marker and a megaphene factor may be provided by the test of significance of difference between marker group means:

$$U = \frac{|x_S - x_s|}{\sqrt{(\sigma^2_S/N_S + \sigma^2_s/N_s)}}.$$

From this, the required sample size $N = N_S + N_s$ ensuring the desired power γ of the test at the given significance value δ may be found:

$$N > \frac{4(U_\gamma + U_{1-\delta})^2\alpha^2[1 + p(1-p)d/\sigma]^2}{(1-2p)^2(d/\sigma)^2},$$

where U_γ and $U_{1-\delta}$ are corresponding quantiles of the normal distribution. The calculated minimum sample size, N, is shown in Table 3.1 for some values of key parameters. These simple considerations show how the resolution power could depend on the effect of allele substitution measured in terms of mean differences d/σ. In order to improve the resolution of marker analysis by decreasing σ, each genotype can be subjected to progeny test or cloned (Thoday, 1967; McMillan and Robertson, 1974; Soller and Beckmann, 1990).

In deriving equation 3.4, we assumed groups *Aa* and *aa* to be different with respect to both the trait mean and two-, three- and higher order moments. With few exceptions (Korol, 1976; Zhuchenko *et al.*, 1979; Korol, Zhuchenko

Table 3.1 Dependence of the required sample size N on d/σ and on p at $\alpha = \sigma_2/\sigma_1 = 1$, to ensure $\delta = 0.05$ and $\gamma = 0.9$

	d/σ			
p	0.2	0.5	1.0	2.0
0.0	1050	168	42	11
0.1	1700	268	79	24
0.2	3106	544	157	53
0.3	7125	1282	384	139
0.4	2×10^4	5268	1615	602

and Samovol, 1981; Korol, Preygel and Preygel, 1990; Korol, Ronin and Kirzhner, 1993; Weller, 1986, 1987), such effects of QTLs have not been considered previously. However, as rightly pointed out by Weller and Wyler (1992), the effect of a QTL on the variance is at certain times likely to be economically more important than its effect on the mean (e.g. for flowering time, ripening time under machine harvesting, time to hatching in chicken). The same applies to QTLs related to fitness in natural populations. Equal variances among QTL groups is a usual simplifying assumption in marker analysis both in its single-marker and interval mapping (multimarker) versions.

In the case of equal variances in the trait groups *Aa* and *aa*, $\sigma^2_{Aa} = \sigma^2_{aa} = \sigma^2$, the ratio $d{:}\,\sigma$ is of primary interest when discussing the power of tests for revealing QT locus linked to the marker or the resolution capacity of algorithms for linkage estimation. Let, as before, the variance of group *aa* be $\sigma^2_{aa} = \sigma^2$. What can happen if the variability in the second group, *Aa*, grows? Intuitively, one would expect a decrease in the resolution capacity proportional to the increase in variance σ^2_{Aa}. Numerical simulations show that just the opposite can be true (Korol, Ronin and Kirzhner, 1993). Namely, with the same absolute distance *d* between the means x_{Aa} *and* x_{aa}, the mean deviation of the estimates of *p* from their 'real' (simulated) values may decrease with increased σ^2_{Aa} (Table 3.2). In other words, the precision in this case is negatively correlated with the relative distance between allelic groups. We can

Table 3.2 The effect of unequal variances in the QTL groups on the mean absolute errors of the estimates ($N = 1000$, $p = 0.1$, $\sigma_1 = 1$)

$d = x_{Aa} - x_{aa}$		0.0			1.0		
$\alpha = \sigma_2/\sigma_1$		1.5	2.0	2.5	1.5	2.0	2.5
ML	$\overline{\delta p}$	59 ± 2.6	29 ± 1.4	19 ± 1.0	31 ± 1.7	21 ± 1.1	16 ± 0.8
	$\overline{\delta d}$	40 ± 2.3	39 ± 2.3	37 ± 2.1	69 ± 3.5	48 ± 2.5	42 ± 2.3
	$\overline{\delta \sigma_1}$	34 ± 1.7	30 ± 1.6	25 ± 1.3	27 ± 1.4	25 ± 1.5	22 ± 1.4
	$\overline{\delta \sigma_2}$	23 ± 1.1	17 ± 0.9	16 ± 0.9	14 ± 0.8	14 ± 0.8	15 ± 0.8
MS	$\overline{\delta p}$	79 ± 2.9	48 ± 2.5	34 ± 1.9	48 ± 2.3	36 ± 2.0	29 ± 1.8
	$\overline{\delta d}$	47 ± 2.7	53 ± 2.9	62 ± 3.2	107 ± 5.3	78 ± 3.8	69 ± 3.4
	$\overline{\delta \sigma_1}$	48 ± 2.3	45 ± 2.2	42 ± 2.1	42 ± 2.1	41 ± 2.2	41 ± 2.2
	$\overline{\delta \sigma_2}$	49 ± 2.6	55 ± 2.9	61 ± 3.3	36 ± 1.9	54 ± 2.8	72 ± 3.8
MM	$\overline{\delta p}$	73 ± 3.6	47 ± 2.4	36 ± 1.9	47 ± 2.3	40 ± 2.0	36 ± 1.7
	$\overline{\delta d}$	51 ± 3.6	51 ± 2.9	54 ± 3.0	106 ± 5.3	86 ± 4.5	77 ± 4.0
	$\overline{\delta \sigma_1}$	60 ± 6.1	63 ± 4.4	68 ± 4.4	52 ± 3.4	72 ± 4.8	91 ± 5.4
	$\overline{\delta \sigma_2}$	37 ± 3.0	33 ± 2.0	32 ± 2.0	19 ± 1.1	27 ± 1.6	35 ± 1.8

Here and in Tables 3.4 and 3.5, the mean absolute errors of the estimates and their standard errors are multiplied by 1000. MS estimates are based on criterion (3.3) at $q = 2$.

see, that if $\sigma^2_{Aa} \neq \sigma^2_{aa}$, then a good solution to the problem is possible even with small effects of A/a on the trait mean value.

These results show that the normalized difference between trait means x_{Aa} and x_{aa} is not a good enough indicator to decide if the QTL mapping problem can be resolved. A more objective measure should take into account any differences between the distributions $f_{aa}(x)$ and $f_{Aa}(x)$ affecting the precision of identification of the postulated QT locus. For example, the maximum discrepancy between the distribution functions (like Kolmogorov–Smirnov statistics) seems to be a more sensible measure for this purpose. Let us consider the particular case of normal distributions:

$$f_{aa} \sim \sigma_{aa}^{-1}\exp[-(x-x_{aa})^2/2\sigma_{aa}^2];\ f_{Aa} \sim \sigma_{Aa}^{-1}\exp[-(x-x_{Aa})^2/2\sigma_{Aa}^2].$$

Then, to obtain the dependence of resolution power on the parameters specifying densities f_{Aa} and f_{aa} we should find

$$\max_x D(x) = \max_x |F_{aa}(x) - F_{Aa}(x)| = \max \left| \int_{-\infty}^{x} \{\sigma_{aa}^{-1}\exp[-(x-x_{aa})^2/2\sigma_{aa}^2] - \sigma_{Aa}^{-1}\exp[-(x-x_{Aa})^2/2\sigma_{Aa}^2]\}dx \right| = J(f_{aa}, f_{Aa}).$$

The solution of the equation $D'(x)=0$ will give the value of $x^* = x(x_{aa}, x_{Aa}, \sigma^2_{aa}, \sigma^2_{Aa})$, where D is maximal, and thereby $J(f_{aa}, f_{Aa}) = D(x_{aa}, x_{Aa}, \sigma^2_{aa}, \sigma^2_{Aa})$.

Having in mind different approaches to mapping one can ask for a specific measure of discrepancy between f_{aa} and f_{Aa} for each of the above three (or any other) methods. We assume that the best indicator for any employed functional is the corresponding measure of discrepancy between the component distributions. It is clear that an analog to the procedure just described for the Kolmogorov–Smirnov criterion can easily be constructed for the mean absolute or mean square discrepancy of the distribution functions $F_{aa}(x)$ and $F_{Aa}(x)$. For the maximum likelihood method the related discrepancy J between two densities $f_1(x)$ and $f_2(x)$ can be derived based on information approach (Kullback, 1959):

$$J(f_1, f_2) = \int [f_1(x) - f_2(x)] \log[f_1(x)/f_2(x)]dx. \tag{3.5}$$

When applied to the case of normal distributions $f_1(x) = f_{aa}$ and $f_2(x) = f_{Aa}$ this gives:

$$J(f_{aa}, f_{Aa}) = 0.5(\sigma^2_{Aa}/\sigma^2_{aa} + \sigma^2_{aa}/\sigma^2_{Aa} - 2) + 0.5(1/\sigma^2_{aa} + 1/\sigma^2_{Aa})(x_{Aa} - x_{aa})^2$$
$$= [(\alpha^2-1)^2 + (\alpha^2+1)(d/\sigma)^2]/2\alpha^2, \tag{3.5a}$$

where $\alpha = \sigma_{Aa}/\sigma_{aa}$ and $\sigma = \sigma_{aa}$. It is clear that J is monotonically increasing with d/σ, but the dependence on α is U-shaped with min (J) being achieved when

$$\alpha^2 = [1 + (d/\sigma)^2]^{1/2}, \quad \text{or} \quad \sigma^2_{Aa} = \alpha^2\sigma^2[1 + (d/\sigma)^2]^{1/2}.$$

Table 3.3 Dependence of the discrepancy measure $J(f_{aa}, f_{Aa})$ on the substitution effect of the QTL (d) and trait variability ($\alpha = \sigma_2/\sigma_1, \sigma_1 = 1$)

		Difference between the mean values (d)					
	α	0.0	0.5	1.0	1.5	2.0	2.5
MS	1.0	0.00	0.44	1.70	3.65	6.10	8.89
	1.5	0.18	0.52	1.53	3.13	5.23	7.70
	2.0	0.58	0.86	1.68	3.01	4.78	6.93
	2.5	1.12	1.38	2.07	3.18	4.69	6.54
	3.0	1.82	2.08	2.68	3.62	4.90	6.51
ML	1.0	0.00	0.63	2.51	5.64	10.03	15.67
	1.5	0.87	1.32	2.68	4.94	8.11	12.18
	2.0	2.82	3.21	4.39	6.34	9.09	12.61
	2.5	5.53	5.89	6.98	8.80	11.34	14.61
	3.0	8.91	9.26	10.30	12.04	14.48	17.61

Some numerical examples are given in Table 3.3. These results imply that the power for QTL marker linkage tests should be estimated allowing for simultaneous effects of the postulated QT locus on the trait mean, variance and other moments. To obtain such estimations the discrepancy measures could be applied to the expected distributions of the compared marker groups under given distributions in the trait groups (e.g. *aa* and *Aa*).

3.2.4 Robustness of the estimates of QTL marker linkage

Analyzing the resolution capacity of an algorithm, especially in the case of relatively close component densities, one can try to check how the deviations from the basic assumptions affect the results. We will consider here in some detail the case of the method of moments and then present shortly comparative results on the described three approaches (ML, MM and MS).

Departures from linearity

The linear model is based on the assumption that the QT value of any individual, say the kth, in group *Aa* may be obtained from that of its counterpart in group *aa* found in the same ecological conditions and only differing from the former with respect to locus *A*/*a*, by the following transformation:

$$x^{(k)}_{Aa} = \alpha x^{(k)}_{aa} + \beta. \tag{3.6}$$

Owing to linearity, the same relation holds for mean values

$$x_{Aa} = \alpha x_{aa} + \beta, \tag{3.7}$$

and we took advantage of this relationship for constructing the basic model of the method of moments. By equation 3.6, the effect of substitution ($a \to A$) is equal to $d^{(k)}_A = (\alpha - 1)x^{(k)}_{aa} + \beta$ and d_A increases with x_{aa} at $\alpha > 1$, i.e. the 'better' the genotype or environmental conditions, the stronger is the effect of a 'good' gene on the trait. However, various modifications of saturation effects of genes are possible, in which the effect of a 'good' gene declines with increasing trait value, i.e. by adding a 'good' gene it is easier to improve a 'bad' genotype than a 'good' one.

One way to describe the saturation effect of a gene is to use the relation (Korol, Zhuchenko and Samovol, 1981):

$$d_A^{(k)} = \alpha x^{(k)}_{aa} e^{-\beta x^{(k)}_{aa}}.$$

This may be viewed as a simple model of the effect of trait canalization. Instead of equation 3.6, we obtain in this case

$$x^{(k)}_{Aa} = x^{(k)}_{aa}[1 + e^{-\beta x^{(k)}_{aa}}].$$

However, a simple translation of mean values for groups *Aa* and *aa* similar to that of equations 3.6 and 3.7 is no longer possible owing to non-linearity. In the specific case of normal distribution $N(x_{aa}, \sigma^2)$, in group *aa*, the mean and variance in group *Aa* take the form:

$$x_{Aa} = x_{aa} + \alpha(x_{aa} - \beta\sigma^2)\, U,$$

$$\sigma^2_{Aa} = \sigma^2[1 + 2\,\alpha U(1 - \beta\, x_{aa} + \beta^2\sigma^2) + \alpha^2\, U^2\, e^{\beta^2\sigma^2}(1 - 2\,\beta\, x_{aa} + 3\,\beta^2\,\sigma^2)],$$

where $U = \exp(-\beta\, x_{aa} + \beta^2\,\sigma^2/2)$.

Clearly, estimation of parameters based on relations as complex as these (still more complex for moments of the third and fourth order) requires large bodies of data. Moreover, the requirement for normality in group *aa* may not be satisfied either. Therefore, some other approach is needed to assess the effects of departure from linearity. One of the linearization procedures commonly used in biometrical genetics is to alter the scale by transforming the original data. In cases like these, a logarithmic transformation (Wright, 1968; Zhuchenko *et al.*, 1979; Korol, Zhuchenko and Samovol, 1981; Weller, 1987) is frequently applied. In using the ML method, such a transformation serves the purpose of normalizing the distribution. Our analysis of actual data from tomato showed that a change in distribution pattern due to log transformation affects the estimates of key parameters relatively weakly, i.e. the proposed model based on the method of moments is rather robust and insensitive to violations of the linearity assumption (Korol, Zhuchenko and Samovol, 1981). This is a very important property since, with it, one need not be concerned about the correspondence of the model to the scale. At any rate, departure from linearity seems to be a less dramatic violation for the MM method than violation of the assumption of the distribution (e.g. normality) for the ML method.

Deviation from the assumed distribution

The dependence of the results of estimation on deviations in the form of the component distributions $f_{aa}(x)$ and $f_{Aa}(x)$ from the assumed density was studied in order to compare the described three methods (Korol, Ronin and Kirzhner, 1993). Consider some results of Monte Carlo simulations. In order to produce data with deviations from normality each of the densities $f_{aa}(x)$ and $f_{Aa}(x)$ was replaced by mixtures of two normal components:

$$f_{aa}(x) = (1-r)\,h_{11}(x) + rh_{12}(x) \quad \text{and} \quad f_{Aa}(x) = (1-r)\,h_{21}(x) + rh_{22}(x),$$

where the expected means of h_{11}, h_{12}, h_{21}, and h_{22} were 0.5, -1.0, 2.0 and 0.5 respectively. The variances of these components were equal, σ^2_h, but specific for each of the three chosen r: $\sigma^2_h = 0.7975$ for $r = 0.1$, 0.5275 for $r = 0.3$ and 0.4375 for $r = 0.5$. Parameters used in this presentation provide for any chosen r the same set of the basic parameters as in the initial, non-disturbed, situation with $r = 0$: $d = 1.5$, $\sigma^2_{aa} = \sigma^2_{Aa} = 1$, $p = 0.1$. Figure 3.2 represents the dependence of the mean absolute deviation of the estimates of main parameters from the simulated ones as a function of the 'noise' proportion r. In fact, the point $r = 0.5$ corresponds to the situation in which another oligogene affecting the trait of interest is segregating in the backcross population. It can easily be seen that with r up to 0.3 the ML method is rather robust to the disturbance of normality, while at $r = 0.5$ an appreciable reduction in precision occurs. The precision of the moment estimates in this case is better.

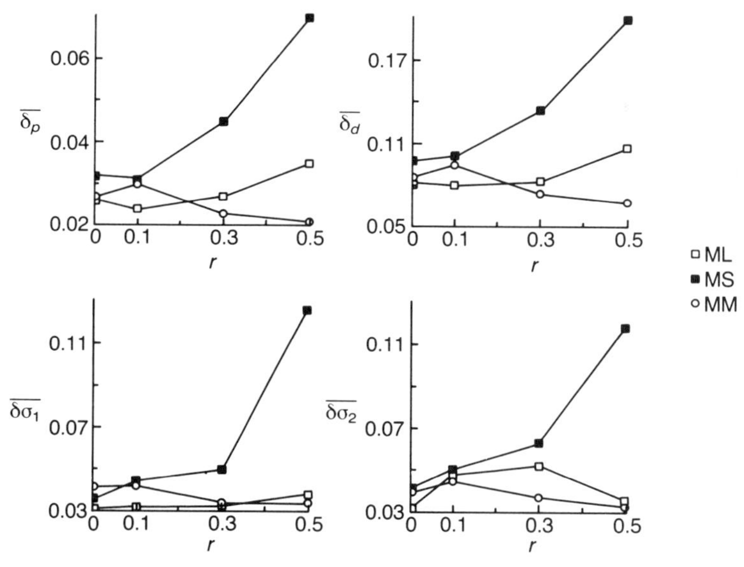

Figure 3.2 The effect of violation of the normality assumption on the precision of the parameter estimates. Instead of normal density each of the distributions f_{aa} and f_{Aa} was simulated as a mixture of two normals with the mixing proportion r ($d = 1.5$, $\sigma_1 = 1$, $\sigma_2 = 1$, $N = 1000$) (from Korol, Ronin and Kirzhner, 1993).

Table 3.4 The effect of violation of the 'equal variances' assumption on the precision of the estimates ($p = 0.1$, $d = 1$, $N = 1000$)

The situation		$\sigma_1 = 1$,	$\sigma_2 = 1$	$\sigma_1 = 1$,	$\sigma_2 = 1.5$
Assumption		$\sigma_1 = \sigma_2$	$\sigma_1 \neq \sigma_2$	$\sigma_1 = \sigma_2$	$\sigma_1 \neq \sigma_2$
ML	$\overline{\delta p}$	64 ± 2.5	65 ± 2.5	62 ± 2.6	31 ± 1.7
	$\overline{\delta d}$	149 ± 5.9	155 ± 6.1	165 ± 8.1	69 ± 3.5
	$\overline{\delta \sigma_1}$	40 ± 2.1	42 ± 2.1	261 ± 3.7	27 ± 1.4
	$\overline{\delta \sigma_2}$	41 ± 2.1	41 ± 2.1	239 ± 3.4	14 ± 0.8
MS	$\overline{\delta p}$	67 ± 2.7	65 ± 2.8	80 ± 3.0	48 ± 2.3
	$\overline{\delta d}$	167 ± 7.4	167 ± 7.8	227 ± 10.6	107 ± 5.2
	$\overline{\delta \sigma_1}$	48 ± 2.7	47 ± 2.9	194 ± 4.8	42 ± 2.1
	$\overline{\delta \sigma_2}$	49 ± 2.6	52 ± 3.0	306 ± 5.1	36 ± 1.9
MM	$\overline{\delta p}$	60 ± 2.6	58 ± 2.5	89 ± 3.2	47 ± 2.3
	$\overline{\delta d}$	144 ± 6.9	136 ± 6.3	245 ± 14.3	106 ± 5.3
	$\overline{\delta \sigma_1}$	41 ± 2.5	42 ± 2.5	167 ± 4.4	52 ± 3.4
	$\overline{\delta \sigma_2}$	41 ± 2.6	38 ± 1.9	346 ± 7.0	19 ± 1.1

The effect of non-equal variances in the QT locus groups on the obtained estimates
As mentioned above, the effect of the postulated factor (A/a) on trait variance may be no less important as a target for identification than the effect on the mean value. Moreover, we could see that non-equality of the variances (being taken into account) increases the resolution capacity of all the three methods, and especially that of ML (Table 3.2). One can also ask the opposite question: what will be lost if we ignore possible inequality $\sigma^2_{aa} \neq \sigma^2_{Aa}$? Table 3.4 shows that the assumption $\sigma^2_{aa} = \sigma^2_{Aa}$ when in fact the variances are different can dramatically affect the estimate precision of all parameters. This conclusion is independent of the estimation method. Especially sensitive to the above assumption are the estimates of the variances.

3.3 QUANTITATIVE TRAIT ANALYSIS WITH TWO AND MORE MARKERS

One could expect that the potential of the Mendelian approach will be sharply increased when at least two markers are used jointly. In particular, it is impossible to determine the orientation of the identified factor (to the right or to the left of the marker) when only one marker is present. Use of two or more markers allows not only this but also other, more interesting, problems to be solved: to assess the adequacy of one-locus QT description (with respect to the marker chromosome), to test various hypotheses of the number and distribution of megaphene loci and to estimate the recombination rates and interference. Nevertheless, if the available markers are rather closely spaced on the genetic map, single-marker analysis of QTs could provide nearly the same resolution power as interval procedures (Darvasi *et al.*, 1993).

Consider some examples of multimarker mapping of QTL. Using a complete genetic map covering all 12 tomato chromosomes and comprising 63 RFLP loci and five biochemical markers, Paterson *et al.* (1988) localized, based on the analysis of an interspecific backcross progeny, six factors affecting fruit weight, four affecting fruit solids content and five affecting fruit pH. Using RFLP markers in the F_2 progeny of a hybrid *Lycopersicon esculentum* × *L. pennellii*, DeVincente and Tanksley (1993) analyzed the genetic basis of transgressive segregation of several QTs. Some of the loci they mapped affected up to 10–30% of the phenotypic variance, with the joint effect of the loci constituting up to 88% of the genetic variance. Doebley and Stec (1991, 1993) studied the genetic basis of quantitative morphological differences between maize and teosinte. For some traits, they found that some of the individual QT loci could explain up to 20–50% of the phenotypic variance, while the combined action of the mapped QTLs accounts for up to 40–80% of the variation. The general conclusion is that the dramatic differences between the two species are largely controlled by five regions of the genome.

The above results were obtained employing the interval mapping approach (Lander and Botstein, 1989), one of the most promising advancements in QTL mapping in the last few years. However, the basic genetic idea of interval mapping can be found in much earlier publications (Thoday, 1961; Mather and Jinks, 1971; Korol and Zhuchenko, 1978). Most of the recent interval mapping algorithms are based on the ML method (e.g. in the form of lod score procedures). Nevertheless, for a broad spectrum of genetic situations in modern QTL mapping the available algorithms and software are still rather limited and not flexible enough. Serious difficulties remain unsolved. These include the sensitivity of the mapping results to restricting assumptions such as normality of the background densities, additivity of QTL action and equal variances, etc., though in multimarker analysis these problems are less troubling than in case of a single marker. As shown in section 3.2, there are situations in the single-marker case in which the method of moments has some advantages over the ML method, in spite of the higher statistical efficiency of ML. Therefore, it is worthwhile to have multimarker versions of the method of moments and some possibilities in this direction are presented below.

3.3.1 Method of moments

Let the marker chromosome of the F_1 hybrid have the following structure:

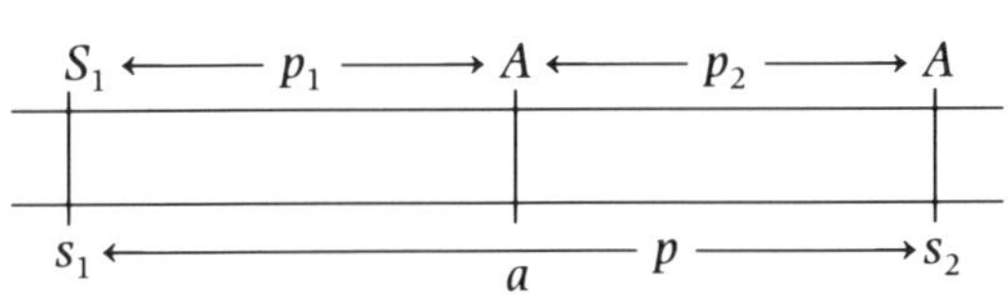

In a one-marker case, a significant difference between the means of S and s groups with respect to the QT in question is a usual test of linkage between A/a and S/s (but see section 3.2.3). However, it cannot be ruled out that a pleiotropic effect of the marker itself is responsible for the observed differences. Therefore, Thoday (1961) proposed the following variance test for situations with two markers: if the factor in question is located between the markers then, in the backcross progeny, the variance of each of the recombinant groups S_1s_2 and s_1S_2 will be greater than that of each of the parental groups S_1S_2 and s_1s_2. It turned out that the test provides a correct answer only for factors with additive effects in the absence of negative interference (Korol and Zhuchenko, 1978).

Assume that A/a acts as a linear transformer, i.e. $x_{Aa} = \alpha x_{aa} + \beta$. Let $\pi_1 = (1 - p_1 - p_2 + cp_1p_2)/(1 - p)$ and $\pi_2 = (p_2 - cp_1p_2)/p$, where c is the coefficient of coincidence in the region s_1–a–s_2. Then the composition of parental and recombinant (with respect to marker genes) groups can be represented as

$$S_1S_2 = \pi_1 Aa + (1 - \pi_1)aa, \quad s_1s_2 = (1 - \pi_1)Aa + \pi_1 aa,$$
$$S_1s_2 = \pi_2 Aa + (1 - \pi_2)aa, \quad s_1S_2 = (1 - \pi_2)Aa + \pi_2 aa.$$

The trait variance in each of the groups is

$$\begin{aligned}
\sigma^2_{S_1S_2} &= \sigma^2 + \pi_1(\alpha^2 - 1)\sigma^2 + 2\pi_1(1 - \pi_1)d^2, \\
\sigma^2_{S_1s_2} &= \sigma^2 + \pi_2(\alpha^2 - 1)\sigma^2 + 2\pi_2(1 - \pi_2)d^2, \\
\sigma^2_{s_1S_2} &= \sigma^2 + (1 - \pi_2)(\alpha^2 - 1)\sigma^2 + 2\pi_2(1 - \pi_2)d^2, \\
\sigma^2_{s_1s_2} &= \sigma^2 + (1 - \pi_1)(\alpha^2 - 1)\sigma^2 + 2\pi_1(1 - \pi_1)d^2, \qquad (3.8)
\end{aligned}$$

where $d = x_{Aa} - x_{aa}$, $\sigma^2 = \sigma^2_{aa}$.

Let $c = 1$. Then $|\pi_2 - 1/2| < |\pi_1 - 1/2|$ and $\pi_2(1 - \pi_2) > \pi_1(1 - \pi_1)$. Therefore, with the additive effect of A/a ($\alpha = 1$), values of $\sigma^2_{s_1S_2}$ and $\sigma^2_{S_1s_2}$ are greater than those of $\sigma^2_{S_1S_2}$ and $\sigma^2_{s_1s_2}$, which is the essence of Thoday's test. If $\alpha \neq 1$, then the variance in one of the parental groups may prove to be greater than that in any recombinant group. For example, when $\alpha > 1$ is large enough, the differences

$$\begin{aligned}
\sigma^2_{S_1S_2} - \sigma^2_{S_1s_2} &= (\pi_1 - \pi_2)(\alpha^2 - 1)\sigma^2 - 2(\pi_1 - \pi_2)(\pi_1 + \pi_2 - 1)d^2, \\
\sigma^2_{S_1S_2} - \sigma^2_{s_1S_2} &= (\pi_1 + \pi_2 - 1)(\alpha^2 - 1)\sigma^2 - 2(\pi_1 - \pi_2)(\pi_1 + \pi_2 - 1)d^2, \qquad (3.9)
\end{aligned}$$

may become positive. Moreover, Thoday's inference is then violated, but a weaker statement is satisfied: the sum of variances of recombinant classes is greater than that of parental classes. Indeed, the quantity

$$\Delta\sigma^2 = \sigma^2_{S_1s_2} + \sigma^2_{s_1S_2} - (\sigma^2_{S_1S_2} + \sigma^2_{s_1s_2}) = 4(\pi_1 - \pi_2)(\pi_1 + \pi_2 - 1)d^2 \qquad (3.10)$$

is positive since at $c = 1$ $\pi_1 > \pi_2$ and $\pi_1 + \pi_2 > 1$ for any p_1 and $p_2 \neq 0$.

Let $c \neq 1$. Equations 3.8 to 3.10 hold true, but with new values of π_1 and π_2; the formulated statements may, therefore, prove to be untrue. Indeed, it can be shown (Korol and Zhuchenko, 1978) that neither Thoday's test nor our generalization of it is satisfied under some level of negative interference.

In the two-marker case, use of the trait means is enough for assessing the effect of allelic substitution of the factor A/a and its chromosomal location. To this end, one may use the relations (Mather and Jinks, 1971):

$$D_0 = X_{S_1S_2} - X_{s_1s_2} = (1 - p_1 - p_2)d/(1 - p),$$
$$D_1 = [X_{S_1S_2}] + [X_{S_1s_2}] - [X_{s_1S_2}] - [X_{s_1s_2}] = (1 - 2p_1)d/2,$$
$$D_2 = [X_{S_1S_2}] - [X_{S_1s_2}] + [X_{s_1S_2}] - [X_{s_1s_2}] = (1 - 2p_2)d/2,$$

where X_k is the trait mean in a given class, and $[X_k]$ is the product of the trait mean in the class by its frequency. By setting $p \approx p_1 + p_2$, Mather and Jinks obtain

$$\hat{d} = D_0, \quad \hat{p}_1 = \frac{1}{2} - \frac{D_1}{D_0}, \quad \hat{p}_2 = \frac{1}{2} - \frac{D_2}{D_0}.$$

This solution is suitable for short segments, which is of interest when there are a large number of markers on the chromosome, e.g. PCR markers. Another application is in the analysis of near-centromeric regions with strong positive interference. In fact, the problem can readily be solved in the case of constant interference of any sign and level and the interference satisfying any metric, say, that of Kosambi.

The case of no interference

By Trow's formula for combining recombination fractions in adjacent intervals we obtain

$$p = p_1 + p_2 - 2p_1p_2, \quad \text{or} \quad 1 - 2p = (1 - 2p_1)(1 - 2p_2).$$

And since

$$(1 - 2p_1)/(1 - 2p_2) = D_1/D_2,$$

we can easily get

$$\hat{p}_1 = 0.5\{1 - \sqrt{[(1 - 2p)D_1/D_2]}\},$$
$$\hat{p}_2 = 0.5\{1 - \sqrt{[(1 - 2p)D_2/D_1]}\}, \quad \hat{d} = 2\sqrt{[D_1D_2/(1 - 2p)]}.$$

The case of arbitrary but constant interference

The formula for combining recombination rates takes the form $p = p_1 + p_2 - 2cp_1p_2$, where the coefficient of coincidence, c, can be any positive number. In the general case, c depends on the segment and its chromosomal location with respect to the centromere or telomeres, on genetic background and environment, and so on. Let c be constant. From the last equation, it follows that $1 - 2cp = (1 - 2cp_1)(1 - 2cp_2)$. Based on this, we derive an equation for finding p_1:

$$4cp^2{}_1 - 2[1 + c - (1 - c)D_1/D_2]p_1 + [1 - (1 - 2p)D_1/D_2] = 0.$$

Then, $\hat{p}_2$ and $\hat{d}$ are easily calculated. Kosambi (1944) used the relation $c(p) = 2p$. The formula for combining recombination fractions in this case is $p = (p_1 + p_2)/(1 + 4p_1p_2)$. From this we derive as before an equation for p_1:

$$8pp_1^2 - 2[1 + 2p - (1 - 2p)D_1/D_2]p_1 + 1 - (1 - 2p)D_1/D_2 = 0.$$

It would be interesting to identify the factor A/a when the degree of interference is not known. The corresponding estimates can be obtained by applying the mean square (MS) method to the equations (3.8) supplemented with the relation $p = p_1 + p_2 - 2cp_1p_2$ and with a set of equations for means:

$$X_{S_1S_2} = \pi_1 x_{Aa} + (1 - \pi_1)x_{aa}, \quad X_{s_1s_2} = (1 - \pi_1)x_{Aa} + \pi_1 x_{aa},$$
$$X_{S_1s_2} = \pi_2 x_{Aa} + (1 - \pi_2)x_{aa}, \quad X_{s_1S_2} = (1 - \pi_2)x_{Aa} + \pi_2 x_{aa}. \quad (3.12)$$

The estimate of p may be obtained by analyzing dihybrid segregations for markers s_1 and s_2. Thus we have nine equations for estimating seven parameters $\theta = \{p_1, p_2, c, x_{Aa}, x_{aa}, \alpha, \sigma^2\}$, eight equations being independent (since frequency-weighted means of marker groups should yield the overall mean of the population). As a modification of this method, p may be derived not from dihybrid segregations but from equations for moments in marker groups S_1, s_1 and/or S_2, s_2 with mixture parameters p_1 and p_2, respectively.

The problem is solved in a similar way for the F_2 population. Assuming complete dominance for the QT locus A/a, the expected frequencies of marker classes are determined based on the following table:

	S_1S_2	S_1s_2	s_1S_2	s_1s_2
A	$0.25 + \theta_1 + \theta_2 + \theta_3 - \theta_4$	$0.25 - \theta_2 - \theta_3 + \theta_4$	$0.25 - \theta_1 - \theta_3 + \theta_4$	$\theta_3 - \theta_4$
aa	$0.25 - \theta_1 - \theta_2 + \theta_4$	$\theta_2 - \theta_4$	$\theta_1 - \theta_4$	θ_4

where $\theta_1 = (1 - p_1)^2/4$, $\theta_2 = (1 - p_2)^2/4$, $\theta_3 = (1 - p)^2/4$, $\theta_4 = (1 - p_1 - p_2 + cp_1 p_2)^2/4$ (for the coupling phase).

As an example, consider the results of the character 'seedling height' in the F_2 population of tomato cross *m2c* × *L. pimpinellifolium* (markers *m2* and *c* are located distally on chromosome 6). The following values of moments were obtained for the marker groups:

Group	Numbers	Mean (cm)	Variance (cm^2)
M2C	736	11.30 ± 0.15	17.05 ± 3.53
M2c	104	9.10 ± 0.33	11.11 ± 3.94
m2C	135	10.83 ± 0.34	15.79 ± 7.15
m2c	118	9.11 ± 0.30	10.95 ± 5.26

Using the value of $\hat{p} = 29.85 \pm 1.71\%$ calculated from the marker segregation, we obtain the following estimates of parameters specifying the megaphene factor located within the *m2–c* segment: $\hat{p}_1 = 26.99\%$; $\hat{p}_2 = 7.54\%$;

$\hat{x}_{Aa} = 8.7$ cm; $\hat{d} = 2.6$ cm; $\hat{\sigma}^2_{aa} = 8.9$; $\hat{\alpha} = 1.43$; $\hat{c} = 1.1$. Similar results are obtained when p, like the rest of parameters, is estimated as the MS solution of a set of equations based on moments without invoking frequencies of marker classes: $\hat{p}_1 = 27.31\%$; $\hat{p}_2 = 8.19\%$; $\hat{x}_{Aa} = 8.7$; $\hat{d} = 2.7$; $\hat{\sigma}^2{}_{aa} = 7.6$; $\hat{\alpha} = 1.55$; $\hat{c} = 0.9$.

Consider now a situation in which the factor of interest is located outside the marker segment:

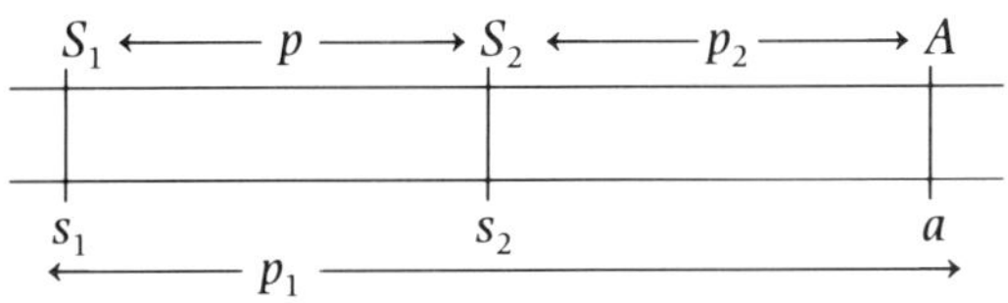

The means in marker groups are given by

$$X_{S_1S_2} = [(1 - p - p_2 + cpp_2)/(1 - p)]\, x_{Aa} + [(p_2 + cpp_2)/(1 - p)]\, x_{aa},$$
$$X_{S_1s_2} = cp_2 x_{Aa} + (1 - cp_2)\, x_{aa}, \quad X_{s_1S_2} = (1 - cp_2) x_{Aa} + cp_2\, x_{aa},$$
$$X_{s_1s_2} = [(p_2 - cpp_2)/(1 - p)]\, x_{Aa} + [(1 - p - p_2 + cpp_2)/(1 - p)]\, x_{aa}. \tag{3.13}$$

The QT distributions in recombinant groups S_1s_2 and s_1S_2 are distribution mixtures such that the contribution of each component depends on cp_2. Groups S_2 and s_2 are also mixtures, but with p_2 as the mixing parameter. Therefore, the equations for means and variances of groups S_1s_2, s_1S_2, S_2 and s_2 may be used to obtain MS estimates of parameters $\theta = \{p_2, c, x_{Aa}, x_{aa}, \alpha, \sigma^2\}$.

An important aspect in the analysis of situations like this is the assumption of interference. Indeed, with no interference, we obtain as few as two independent equations in (3.13); considering groups S_2 and s_2 adds no new information, and estimation of parameters using the indicated procedure will be impossible. Then, for estimating the location of A/a and its effect, it is necessary to apply the one-marker analysis described earlier in this chapter.

Consider a model example of estimating $\theta = \{p_2, c, x_{Aa}, x_{aa}, \alpha, \sigma^2\}$. An analysis of the character 'seedling height' in a backcross population, *Woawd* × (*Woawd* × *L. pimpinellifolium*), yielded the following values of moments for marker groups on chromosome 2:

Group	Numbers	Mean (cm)	Variance (cm^2)
Wo aw	20	8.55	13.6
wo Aw	20	12.15	9.6
Aw	85	12.56	16.5
aw	76	8.17	11.7

This example is a model one for the reason that involved in the cross, concurrently with *Wo* and *aw*, is mutation *d* (dwarf) which has a strong effect on plant growth. It is, therefore, possible to compare the estimates of recombination parameters obtained by the above procedure of the QT analysis with those from the standard marker analysis. The former approach yields $\hat{p}_2 = 5.04\%$; $\hat{c} = 2.10\%$; $\hat{d} = 4.72$ cm; $\hat{\alpha} = 1.06$; $\hat{x}_{aa} = 8.0$; $\hat{\sigma}^2 = 6.5$. With marker analysis the values are $\widehat{rf}_{aw-d} = 6.95 \pm 2.02\%$, $\hat{c}_{Wo-aw,aw-d} = 2.39 \pm 0.55$. Both approaches indicate the presence of negative interference in the *Wo–aw–d* region.

The above MM models can be extended to situations with two or more QT loci on a chromosome. A simplified case in which only mean values of the QT in the marker groups are included in the model is described briefly below in the section devoted to regression methods. One can consider MM versions when other moments will also be included in the analysis.

3.3.2 Method of maximum likelihood

In their multimarker interspecific QTL mapping in tomato Paterson *et al.* (1988) have employed a method similar to the lod score technique (which is usually used in linkage analysis of pedigrees and family data, in particular those in humans) and allowing the interval mapping (Lander and Botstein, 1989). In such cases, the lod score technique can also be used to test linkage between a marker and a QTL (Demenais, Lathrop and Lalouel, 1988).

The first step with this method is screening chromosomes for loci with strong effects on the traits in question. To this end, the lod score ratio is plotted against the hypothetical map position of the locus in question. If α is the effect of locus and σ^2 is the QT variance, then by the lod score one understands the log likelihood ratio

$$\text{lod} = \log_{10}[L(\alpha^*, \sigma^*)/L(0, \sigma^{**})]. \tag{3.14}$$

In order to calculate $L(\alpha, \sigma)$, it is necessary to make two essential assumptions: (a) the desired factor has an additive effect; (b) the trait distribution in the backcross is a mixture of two normal distributions (usually equal variances are also assumed). Then, for the kth genotype:

$$L_k(\alpha, \sigma) = (2\pi\sigma^2)^{-1/2}\left\{p_1 \exp\left(-\frac{x^2{}_k}{2\sigma^2}\right) + p_2 \exp\left[-\frac{(x_k - \alpha)^2}{2\sigma^2}\right]\right\},$$

where α is the effect of substitution at the QT locus, x_k is the value of QT in a given (kth) individual and p_1 and p_2 depend on the chromosomal location of the postulated locus and are calculated in specifying a corresponding mapping function (Lander and Botstein, 1989). For the entire population, the value of the likelihood function is computed as $L = \Pi L_k$. In calculating lod from equation 3.14, ML estimates of α^*, σ^* (the second estimate of the parameter σ, i.e. σ^{**}, is obtained assuming that $\alpha = 0$) are used for a given pair (p_1, p_2). By setting a critical value of lod exceeding which indicates the presence of a locus affecting the QT, it is easy to calculate the map position of the factor in question (i.e. the pair p_1, p_2 where the lod is maximum) and to determine the respective **supporting** interval (Lander and Botstein, 1989). The EM algorithm employing alternation of the **expectation** and **maximization** steps is considered a good calculation tool in optimization problems related to QTL mapping and mixture decomposition in general (e.g. Titterington, Smith and Makov, 1985; Lander and Botstein, 1989; Carbonell *et al.*, 1992).

The proposed approach is extremely attractive in that a complete genetic map of molecular markers (based, by the way, on the identification results from the same segregating population) is used in the analysis. However, the assumption of normality and of equal variances of components in the mixture of distributions forming the segregating population may not be satisfied. Nor is the situation necessarily improved by a scale transformation, for example by taking a logarithm (see, for example, Zhuchenko *et al.*, 1979). More promising in situations like these seems to be an analysis using more flexible specifications of component distributions, allowing for departures from normality (see p. 81). Nevertheless, some recent publications indicate that interval mapping could be more robust to deviations from normality than single-marker analysis (e.g. Knott and Haley, 1992).

A more serious problem arises when the multiple marked chromosome contains several QT loci. Let, for example, two QTLs be located in two non-adjacent intervals. Then, if one tries to fit a one-locus model, a QT locus can be found in an interval which does not contain QTs at all (Knott and Haley, 1992; Martinez and Curnow, 1992). Even more difficult is the problem of recognizing and treating situations with *trans*-effects of linked QT loci (Knott and Haley, 1992; Luo and Kearsey, 1992). That *trans*-association of QTLs could be a rather common phenomenon even in interspecific crosses has recently been demonstrated by DeVincente and Tanksley (1993) in the above-mentioned study with tomato: they found that up to 36% of the detected QTLs had alleles with effects opposite to the direction expected from the parental differences.

3.3.3 Multiple regression method

With a large number of markers, it is advantageous to use linear multivariate statistical techniques for screening the genomic regions involved in the control of the trait in question (Korol, 1973; Zhuchenko *et al.*, 1975; Stuber, Edwards and Wendell, 1987; Keim, Diers and Shoemaker, 1990; Keim *et al.*, 1990; Knapp, Bridges and Birkes, 1990; Diers *et al.*, 1992; Goddard, 1992; Haley and Knott, 1992; Martinez and Curnow, 1992). In particular, using stepwise regression, one can estimate not only the main effects of QT factors, but also interactions between them (Moreno-Gonzalez, 1992). The latter constitutes a serious and largely unsolved problem for some other models of QTL mapping.

Linear multivariate models have also been used for marker-based predictions of QT values in breeding programs (Cantet and Smith, 1991; Goddard, 1992). Regression estimations could also be considered as an initial approximation for the lod score interval mapping as well as an easy tool for evaluating the total effect of the identified QT loci (e.g. Doebley and Stec, 1993). A comparison made on simulated data shows the closeness of the results based on regression and the ML approach (Haley and Knott, 1992; Martinez and Curnow, 1992).

In the above-described interval versions of the method of moments one can confine the analysis to the mean values of the trait in the multiple marker groups, ignoring all other moments. Then, the obtained estimates can coincide with those of the multiple regression method. As an example consider a situation with two QT loci on a chromosome. Suppose that two factors have been detected in adjacent marked segments. Consider a simple algorithm for estimating their effects and positions. Let the genomic region in question be represented in the following form:

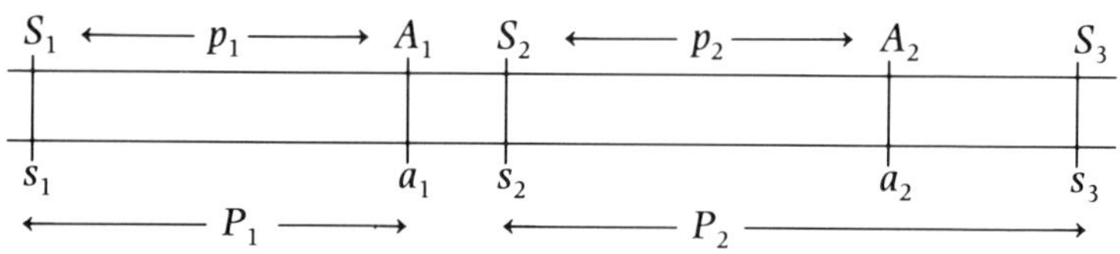

The number of marker classes for the progeny of a test cross here is 8. The distribution of a trait within each of these groups can be viewed as a mixture of four components (f_1, f_2, f_3 and f_4) with the means $x_{a_1a_1a_2a_2}$, $x_{a_1A_1a_2a_2}$, $x_{a_1a_1a_2A_2}$ and $x_{a_1A_1a_2A_2}$. With various types of interference within the region in question, a system of equations can be derived relating the observed (X_k) and the expected (E_k) mean values in marker groups:

$$E_k = \pi_{1k}[m + (d_1 + d_2)/2 + e_{12}] + \pi_{2k}[m + (d_1 + d_2)/2]$$
$$+ \pi_{3k}[m - (d_1 - d_2)/2] + \pi_{4k}[m - (d_1 + d_2)/2 - e_{12}].$$

Here, $\pi_{ik} = \pi_{ik}(p_1, P_1; p_2, P_2)$ is the proportion of the component f_i $(i = \overline{1, 4})$ in the kth $(k = \overline{1, 8})$ group, depending on known parameters P_1 and P_2 and unknown parameters p_1 and p_2; m, d_1 and d_2 are respectively the population mean and effects of heterozygous allele substitutions at loci A_1/a_1 and A_2/a_2; e_{12} is the effect of interaction between the identified loci. To estimate the vector of genetic parameters $\theta = \{m, d_1, d_2, e_{12}\}$, let us use the weighted mean squares (MS) method, i.e. the same approach that is usually employed in regression analysis. Consider the functional

$$S = \sum_{k=1}^{8} w_k(X_k - E_k)^2, \tag{3.15}$$

where weights, w_k, are the reciprocals of the squared standard errors of means X_k. For any point (p_1, p_2) from the domain $(0, P_1) \times (0, P_2)$, the set $\hat{\theta} = \{\hat{m}, \hat{d}_1, \hat{d}_2,$

$\hat{e}_{12}\} = \hat{\theta}(p_1, p_2)$ can be found ensuring a minimum of S. This will lead to a linear system of equations for the parameters m, d_1, d_2 and e_{12}:

$$m\Sigma\, w_k + d_1\Sigma\, w_k a_k/2 + d_2\Sigma\, w_k b_k/2 + e_{12}\Sigma\, w_k c_k/2 = \Sigma\, w_k X_k,$$

$$m\Sigma\, w_k a_k + d_1\Sigma\, w_k a_k^2/2 + d_2\Sigma\, w_k a_k b_k/2 + e_{12}\Sigma\, w_k a_k c_k/2 = \Sigma\, w_k a_k X_k,$$

$$m\Sigma\, w_k b_k + d_1\Sigma\, w_k a_k b_k/2 + d_2\Sigma\, w_k b^2{}_k/2 + e_{12}\Sigma\, w_k b_k c_k/2 = \Sigma w_k b_k X_k,$$

$$m\Sigma\, w_k c_k + d_1\Sigma\, w_k a_k c_k/2 + d_2\Sigma\, w_k b_k c_k/2 + e_{12}\Sigma\, w_k c^2{}_k/2 = \Sigma w_k c_k X_k,$$

where $a_k = \pi_{1k} + \pi_{2k} - \pi_{3k} - \pi_{4k}$, $b_k = \pi_{1k} - \pi_{2k} + \pi_{3k} - \pi_{4k}$, and $c_k = \pi_{1k} - \pi_{4k}$. Solutions of this system for various pairs (p_1, p_2) from the domain of permissible values $(0, P_1) \times (0, P_2)$ are compared based on Fisher's variance test, which characterizes the adequacy of the regression. It is reasonable to consider the set of genetic parameters $\hat{\theta}_p = \{\hat{p}_1, \hat{p}_2, \hat{m}, \hat{d}_1, \hat{d}_2, \hat{e}_{12}\}$ with which a minimum of S is achieved to be the general solution to the problem. One can also consider in a similar way the case when the QT loci in question are located in non-adjacent marker segments. Instead of the described scanning procedure the parameters p_1 and p_2 can be estimated by direct optimization of the functional equation 3.15, but a linear model is no longer possible in this case. Other moments could also be taken into account, but the resulting moment model can no longer be considered as a regression one.

The main advantage of the regression approach to QTL mapping is in its simplicity and the possibility of using standard statistical packages, and of including in the model any required number of QT loci (Haley and Knott, 1992; Moreno-Gonzalez, 1992). As mentioned above, no less important is the ease of estimating not only the main effects of QT factors but also the interactions between them.

3.4 MARKER ANALYSIS OF A SET OF QUANTITATIVE TRAITS

In section 3.2.3 we suggested a predictor for the expected precision of the parameter estimates for QTL mapping in the single-marker case. For this aim, some measures of the discrepancy $D(f_{aa}, f_{Aa})$ between the component distributions were considered. In particular, it was found that for the ML method the information measure predicts an increase in accuracy with increased deviation of $\sigma^2_{Aa}/\sigma^2_{aa}$ from unity when $x_{Aa} - x_{aa}$ is small. This expectation was confirmed numerically (see Table 3.2). An important question follows from these results: is it possible to find additional ways to increase $D(f_{aa}, f_{Aa})$, thereby providing more effective data utilization in the whole QTL mapping problem? A promising approach can be suggested employing simultaneous analysis of several QTs.

Earlier such a procedure was developed for the method of moments (Korol, Preygel and Bocharnikova, 1987; Korol, Preygel and Preygel, 1990; Ronin,

Kirzhner and Korol, 1993), using correlations between QTs in question. The idea of the multivariate approach is explained in Figure 3.3 for the two-dimensional case.

In certain conditions, one can expect to have a large discrepancy between the groups *Aa* and *aa* based on the joint distribution of two traits [say $f(x, y)$] even if the groups are strongly overlapping in their marginal distributions $f(x)$ and $f(y)$. The general multivariate case could be considered. Thus, the multivariate analog of the discrepancy measure J from equation 3.5 for the normal distribution has the following form (Kullback, 1959):

$$\begin{aligned} J(f_1, f_2) &= \int [f_1(x, y, z, \ldots) - f_2(x, y, z, \ldots)] \log [f_1(x, y, x, \ldots)/f_2(x, y, z, \ldots)]\, dxdydz \ldots \\ &= 0.5\mathrm{tr}[(\Sigma_1 - \Sigma_2)(\Sigma_2^{-1} - \Sigma_1^{-1})] + 0.5\mathrm{tr}[(\Sigma_1 + \Sigma_2)(M_1 - M_2)'(M_1 - M_2)]. \end{aligned} \tag{3.16}$$

Here, Σ_1 and Σ_2 are the covariance matrices of the trait set in groups *aa* and *Aa*, Σ_1^{-1} and Σ_2^{-1} are the corresponding inverse matrices and M_1 and M_2 are the vectors of the trait means in the same groups.

Consider the case of two dimensions. As with one trait, we can assume that allelic substitution $aa \rightarrow aA$ affects mean trait values and/or trait variances. The remaining variances were described above as a joint result of non-heritable variation and segregation of other (unlinked) loci. In addition, we

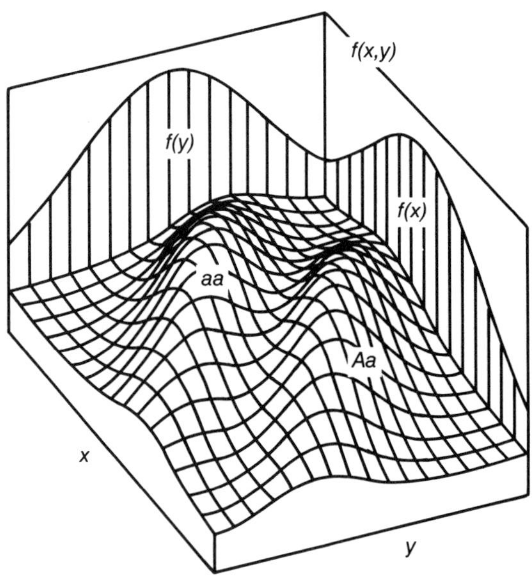

Figure 3.3 Joint distribution $f(x, y)$ of two correlated traits, x and y, in the backcross population. Even with a clear-cut bimodality of $f(x, y)$ [when the components $f_{aa}(x, y)$ and $f_{Aa}(x, y)$ are far enough and the correlation is high] the marginal distributions $f(x)$ and $f(y)$ are unimodal (from Ronin, Kirzhner and Korol, 1993).

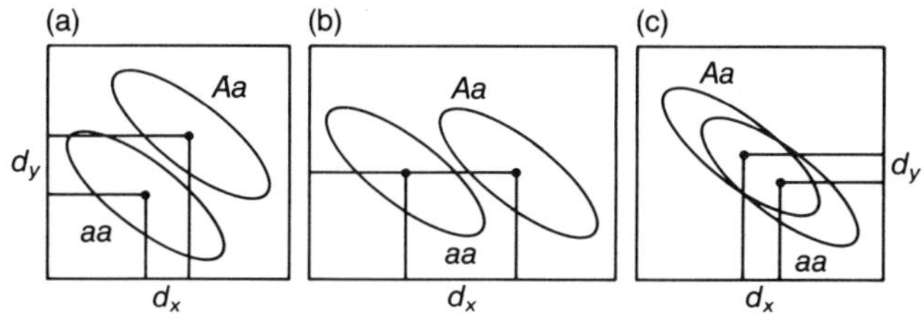

Figure 3.4 Some possible situations in two-trait mapping. An example is taken when the correlation between x and y within the groups is the same and negative. The effects of the locus A/a are: (a) an increase in x and in y; (b) increase in x and no effect on y; (c) decrease in x and increase in y.

assume here that the last two factors (other genes and non-heritable variation) can also affect correlation between the traits in the marker groups. Some possibilities are presented in Figure 3.4. It can easily be understood that with strong correlations between the traits in question this approach can give a serious increase in the resolution power even if the locus A/a influences only one trait. In a sense, the main idea is close to the classification methods based on the principal component methods.

To implement the multivariate mapping approach within the framework of the method of moments, system 3.4 should be complemented by the corresponding equations for the second trait and by new equations for the product moments, e.g. covariances (Korol, Preygel and Bocharnikova, 1987; Korol and Preygel, 1988a):

$$\mathrm{cov}_m = (N_m - 1)^{-1}\Sigma(x - X_m)(y - Y_m)(1 - r)\,\mathrm{cov}_{aa} + r\,\mathrm{cov}_{Aa} + r(1 - r)\,d_x d_y,$$
$$\mathrm{cov}_M = (N_M - 1)^{-1}\Sigma(x - X_M)(y - Y_M) = r\,\mathrm{cov}_{aa} + (1 - r)\,\mathrm{cov}_{Aa} + r(1 - r)\,d_x d_y,$$

where cov_{aa} and cov_{Aa} stand for the covariances in groups *aa* and *Aa* due to joint effect of non-heritable variation and segregation of other loci. If necessary, higher product moments, e.g. $(N_m - 1)^{-1}\Sigma(x - X_m)^2(y - Y_m)$, $(N_m - 1)^{-1}\Sigma(x - X_m)(y - Y_m)^2$, etc. could also be taken into consideration.

No special difficulties are expected in developing the multitrait analog of mapping based on the ML method. For example, in the case of two-dimensional normality the functional equation 3.2 can be written as before:

$$\ln L(\theta) = \sum_{k=1}^{N_1} \ln S_1(x_k, y_k) + \sum_{j=1}^{N_2} \ln S_2(x_j, y_j) + \text{const.}, \tag{3.2a}$$

where

$$S_1(x, y) = (1 - r)\,f_1(x, y) + rf_2(x, y),$$
$$S_2(x, y) = rf_1(x, y) + (1 - r)f_2(x, y),$$

$$f_1(x,y) = [2\pi\sigma_{1x}\sigma_{1y}(1-R^2{}_1)]^{-1/2} \times \exp\left\{-\frac{1}{2(1-R^2{}_1)}\left[\frac{(x-x_1)^2}{\sigma^2{}_{1x}} - 2R_1\frac{(x-x_1)(y-y_1)}{\sigma_{1x}\sigma_{1y}} + \frac{(y-y_1)^2}{\sigma^2{}_{1y}}\right]\right\},$$

$$f_2(x,y) = [2\pi\sigma_{2x}\sigma_{2y}(1-R^2{}_2)]^{-1/2} \times \exp\left\{-\frac{1}{2(1-R^2{}_2)}\left[\frac{(x-x_2)^2}{\sigma^2{}_{2x}} - 2R_2\frac{(x-x_2)(y-y_2)}{\sigma_{2x}\sigma_{2y}} + \frac{(y-y_2)^2}{\sigma^2{}_{2y}}\right]\right\},$$

x_i, y_i, σ_{ix} and σ_{iy} $(i=1,2)$ are the mean values and standard deviations of the traits x and y in groups *aa* $(i=1)$ and *Aa* $(i=2)$ and R_i are the correlations between x and y in the groups.

In this formulation we have now 11 unknown parameters: $\theta = \{r, x_1, y_1, x_2, y_2, \sigma_{1x}, \sigma_{1y}, \sigma_{2x}, \sigma_{2y}, R_1, R_2\}$. In spite of the relatively high dimensionality, the covergence to the solution is rather good, especially in cases with high correlations between x and y within the groups (Table 3.5). One can easily see how strong the effect of correlation between the traits is.

3.5 GENERAL DISCUSSION AND PROSPECTS OF THE PROBLEM

We considered above several methods of QTL marker linkage analysis. Until recently, applicability of such algorithms was limited by the number of available markers. With the discovery of new classes of molecular markers based on RFLP, mini- and macro-satellites, and particularly those based on polymerase chain reaction (PCR and RAPD markers), high-resolution analysis of QTs becomes applicable to almost any organism. It is obvious, however, that

Table 3.5 The precision of parameter estimates in the two-trait analysis as depending on correlation (R) between the traits ($p=0.1, \sigma_{11}=\sigma_{12}=\sigma_{21}=\sigma_{22}=1, N=1000$; in the ML method all σ_{ij} were assumed to be different)

$R_1 = R_2 = R$	0.6			0.9		
d_1	0.5	1.0	0.5	0.5	1.0	0.5
d_2	0.0	0.0	0.5	0.0	0.0	0.5
$\overline{\delta p}$	97 ± 3.5	45 ± 2.5	54 ± 2.5	53 ± 2.5	10 ± 0.6	10 ± 0.6
$\overline{\delta d_1}$	147 ± 8.5	94 ± 5.3	71 ± 3.4	68 ± 3.2	28 ± 1.6	27 ± 1.5
$\overline{\delta d_2}$	50 ± 3.3	38 ± 2.2	67 ± 3.4	38 ± 2.3	26 ± 1.5	27 ± 1.4
$\overline{\delta\sigma_{11}}$	29 ± 2.0	27 ± 1.5	15 ± 0.9	15 ± 0.9	11 ± 0.6	8 ± 0.5
$\overline{\delta\sigma_{12}}$	17 ± 1.2	12 ± 0.7	14 ± 0.9	11 ± 0.7	7 ± 0.4	7 ± 0.4
$\overline{\delta R_1}$	22 ± 1.3	19 ± 0.9	25 ± 1.2	7 ± 0.4	4 ± 0.3	4 ± 0.3
$\overline{\delta\sigma_{21}}$	25 ± 1.9	25 ± 1.5	15 ± 0.9	14 ± 0.8	10 ± 0.5	8 ± 0.4
$\overline{\delta\sigma_{22}}$	16 ± 1.1	12 ± 0.7	15 ± 0.8	12 ± 0.7	8 ± 0.4	9 ± 0.5
$\overline{\delta R_2}$	21 ± 1.6	19 ± 1.1	26 ± 1.4	7 ± 0.4	5 ± 0.2	5 ± 0.3

the need to develop new algorithms enabling one to analyze the obtained results as thoroughly as possible is also growing.

3.5.1 Opportunities for improving data analysis procedures

Among the recent promising achievements in this field one should mention the interval approach (Lander and Botstein, 1989; Carbonell *et al.*, 1992; Haley and Knott, 1992; Martinez and Curnow, 1992). Nevertheless, for the broad spectrum of genetic situations in modern QTL mapping, the available algorithms and software are still rather limited and not flexible enough. Serious difficulties remain unsolved. These include the restricting assumptions of normality of background distributions, additivity of QTL action, equal variances, relatively poor resolution capacity of the methods, calculation difficulties, etc. It is rather surprising how easily the assumption of normality is accepted by QTL mappers, in spite of its contradiction of the very idea of the Mendelian analysis of QTs.

The tendency to standardize the analytical techniques (e.g. based on the lod score approach) seems to be both attractive and progressive. But the extent of QTL mapping and high data cost make it worthwhile to search for new ways to extract information from experiments. We suggest, therefore, that it is essential to have for the QTL mapping problem a system of complementary (in terms of initial assumptions) methods and interactive algorithms, flexible at all the solution stages of the problem. The method of moments has, in our opinion, several advantages, if considered as a component of such a system. It could serve as a good adjunct to (and control of) the ML method, providing a first approximation for ML procedures and, thereby, partially overcoming the difficulties resulting from multiple extrema in the likelihood function (see also Darvasi and Weller, 1992).

As we could see, the assumption $\sigma^2_{aa} = \sigma^2_{Aa}$ when in fact $\sigma^2_{aa} \neq \sigma^2_{Aa}$ leads to serious bias of the estimates, especially of the variances. However, if $\sigma^2_{aa} \neq \sigma^2_{Aa}$, and this fact is taken into account, the resolution capacity increases. It is clear enough that the resolution capacity of any method should depend on the proximity of the distribution densities of QT locus groups. Usually, the proximity of the means is considered as the key factor in the power of QTL marker linkage tests and precision of the corresponding parameter estimates. In fact, any difference in the distributions could and should serve both for hypothesis testing and parameter estimation. For that, each test or estimation method should use an appropriate measure of discrepancy of the distributions (densities). Some examples, given in section 3.2.3, show why and how deviations from equality $\sigma^2_{aa} = \sigma^2_{Aa}$ affect the precision.

A strong increase in discrepancy between groups is expected when employing joint distribution of a set of correlated QTs. In principle, with high correlations between QTs, a good resolution is possible even if the QTL groups (say, *aa* and *Aa*) are strongly overlapping for their marginal distributions (see

Figures 3.3 and 3.4 and numerical examples in Table 3.5). The advantage of the multitrait analysis may be especially attractive when the QT factor (locus) in question influences several traits simultaneously. But this approach could be no less important in situations in which the trait of interest (say x) is dependent on the locus A/a and is strongly correlated with another trait (y), the latter being independent of A/a. Even in this case, the additional information provided by measurements of y can dramatically increase the mapping precision of the gene A/a. Such correlations may be caused by segregation of other genes, environmental effects and physiological limitations. As examples, trait pairs like 'grain weight–protein level' or 'milk production–fat content' could be mentioned. Because of the high cost of molecular marker typing, many QTs are usually measured in one experiment, so that the needed structure of data for such a mapping approach is not an exception.

3.5.2 Power of test in establishing linkage between a QT and a marker locus

Simultaneous consideration of segregation of many marker loci enhances the power of a test for detecting linkage between a marker and a QT locus. The sample size, N, required to reach the power of test γ at significance level δ depends on the value of $(d/\sigma)^2$ (where d is the effect of allele substitution at the QT locus and σ^2 is the QT variance due to environmental factors and to segregation at other loci affecting the QT). Since σ^2 includes the genotypic component, one may proceed, when applying multimarker analysis, as follows: first, identify genes with maximum effects, then consider the contributions of the genes identified at the first stage to σ^2 and repeat the procedure. The power of the test and accuracy of parameter estimation, incidentally, increase by virtue of the residual value of σ^2 being reduced. Such a procedure of identifying QT loci may, in principle, be an iterative one.

Reducing the number of individuals to be assessed with respect to marker loci so as to achieve the desired power of the test and accuracy of estimates is especially important in dealing with rather expensive molecular markers. Here, total expenses may be reduced by measuring a QT in a large population, with subsequent identification (with respect to markers) being confined to genotypic classes with the strongest and the weakest trait expression only. A procedure like this permits the desired power of the test and accuracy of estimates to be achieved with a significantly lower number of genotypes identified with respect to a set of markers (but, accordingly, with a higher total number of individuals assessed with respect to QT) (Lebowitz, Soller and Beckmann, 1987; Salmon *et al.*, 1988; Lander and Botstein, 1989). In some cases it is also advantageous to take into account a proportion of individuals from the modal class (Weller and Wyler, 1992). For further discussion of the issues concerning the power of tests for detecting linkage and designing experiments see Soller and Genizi (1978), Demenais, Lathrop and Lalouel

(1988), Soller and Beckmann (1988), Lander and Botstein (1989), Weller, Kashi and Soller (1990), Weller and Wyler (1992) and Darvasi *et al.* (1993).

A procedure based on selection of phenotypes with extreme QT values and subsequent marker-based identification of selected genotypes can also be used in the search for markers and marker combinations presumably linked to favorable QTLs. The detected marker combination can subsequently be used in artificial selection since the marker segment carrying a favorable gene block is readily traced among the progeny (Soller and Beckmann, 1983). A useful way of increasing the probability of detecting QTLs is to enlarge the number of parents involved in the analysis. This can be done, for example, by treatment of data from several segregating populations (Beavis *et al.*, 1991). Such an approach provides, in addition, information on multiple alleles controlling QTLs.

Efficient search for molecular markers closely linked to desirable genes can also be accomplished by using a procedure based on the production, via successive backcrosses and selection, of near-isogenic lines (NILs) differing only in the trait in question, and then looking for those markers with respect to which these lines differ (Tanksley and Hewitt, 1988; Martin, Williams and Tanksley, 1991). With this procedure, the differences between resulting lines in advanced generations are primarily localized in those genomic regions that control the differences in selected QTs. Thus, the markers bringing out these differences are located in these regions too.

When the analysis is based on segregation progeny of crosses between inbred lines, the number of individuals required for the detection of linkage between a QT and a marker locus is different for the F_2 and BC progeny; on average, the F_2 data are more informative (Soller, Brody and Genizi, 1976). An analysis based on recombinant inbred lines (RILs) requires larger samples than those used in F_2 or BC (Ellis, 1986), but the RILs can be repeatedly tested for their QTs (Soller and Beckmann, 1990). The detection and estimation of linkage between a marker and a QTL are also possible upon crossing two segregating populations (Beckmann and Soller, 1988), populations to inbred lines (Dudley, 1992) as well as based on pedigrees and family data (Jayakar, 1970; Smith, 1975; Cockerham and Weir, 1983; Geldermann, Pieper and Roth, 1985; Dentine and Cowan, 1990; Goddard, 1992).

3.5.3 Allowing for genotype–environment interactions: norm of reaction

Earlier we have seen that the efficiency of methods for detection and estimation of linkage between marker loci and a QT factor is strongly dependent on the magnitude of the QT effect relative to the overall phenotypic variance. Environmental variation and segregation for genetic factors on other chromosomes result in a reduced power of tests and in larger estimate variances for the parameters. Naturally, one tries various approaches in order to lower this variation (e.g. by vegetative propagation or by testing the progeny of each genotype). It is because of the well-known dependence of QT expression on the environment that the estimates obtained based on the identification of segregating progeny under some ecological conditions may be entirely different from those obtained under other conditions. Thus, on chromosome 3 of *Drosophila*, factors have been found which are responsible for interstrain differences in sternopleural bristle number in flies developing at normal temperatures, and on chromosomes 2 and 3 factors controlling the sensitivity of this trait to temperature changes (Caligari and Mather, 1975; Schnee and Thompson, 1984). In other words, the phenomenon of genotype–environment interaction can be analyzed not only based on conventional techniques for analysis of variance but also within the framework of the Mendelian marker approach. The dependence of Mendelian analysis of QTs on the conditions under which the traits are formed, on the sequence of action of limiting environmental factors in ontogenesis, implies, however, that the problem of QT mapping needs to be generalized. As is customary in genetics, instead of speaking of genetic factors controlling individual traits in a particular environment, one should rather speak of the genetic determinants of norms of reaction with respect to these traits in response to environmental changes and of the genetics of life histories in varying environments (Glotov and Tarakanov, 1985; Dingle, 1990).

It can be suggested that an appropriate analysis will be detecting factors (QTLs) affecting:

1. The 'developmental potential' of traits under minimum effects of limiting conditions.
2. The stability of traits in the face of limiting conditions (with little or no effect under optimum environmental conditons).

As applied to crop yield, these aspects of genetic determination of QTs are similar to the concepts of potential productivity and ecological stability (Zhuchenko, 1988).

3. Both the 'developmental potential' and stability simultaneously, the level of these effects depending on genetic background and environment.

If one can measure trait values for a set of genotypes of a mapping population in different environments, then the multitrait marker analysis could be used to reveal effects of individual QT loci on reaction norm, rather than on the trait value.

It follows, of course, that relative contributions of various genomic regions to within-population genetic variation should depend on particular ecological conditions. In general, such an approach is theoretically more justified and better reflects real genotype–environment interactions (see also Maletsky, 1976). It seems to be more capable of meeting the needs of applied breeding, on the one hand, and population genetics, ecology and evolutionary theory, on the other.

3.5.4 A possible experimental procedure for marker analysis of the QT norm of reaction

In view of the above, it seems worthwhile presenting here, by way of an example, a hypothetical experiment targeted to analyze the genetic nature of differences between two strains (species) with respect to a set of QTs allowing for the latter's dependence on the environment. Work of this kind may be carried out on various organisms. Let us take, for example, the tomato. This is a convenient organism to use, primarily for three reasons: availability of a detailed molecular linkage map including more than 1000 loci (Tanksley *et al.*, 1992); ease of vegetative reproduction; possibility of estimating a large number of genotypes under controlled conditions (in a climate chamber).

By virtue of being relatively complex, the experiment should be a multi-purpose one. The main objectives of such a study may be as follows:

- ascertaining the genetic nature of interspecific differences in morphological traits, biological and/or agronomical performance, resistance to stressful conditions, to diseases and pests, etc.;
- detection of genomic regions with specific and systemic influence on the traits under study, delimitation of segments responsible for only valuable (donor) or only unfavorable properties, and for correlations between desirable and undesirable traits;
- using the data obtained in 'calculating optimum genotypes', in predicting limits to selection, in providing a basis for map-based QTL cloning.

It is advantageous to perform the experiment on backcross progeny and/or F_2 of an interspecific cross. Although schemes based on the analysis of F_2 could be more informative, we prefer the backcross since in the latter the effects of developmental canalization appear to be less pronounced (Zhuchenko and Korol, 1985). In other words, what is evaluated here are the consequences of introgressing different genomic segments of one species into the genome of another. Therefore, it would be a good practice to make not one but two backcrosses. In view of the high cost of molecular marker analysis, it is desirable to apply sequential analysis of linkage as suggested by Morton (1955), Goradia and Lange (1990), Korol, Preygel and Preygel (1990), Motro and Soller (1993) and others.

With this approach, both the number of genotypes and the set of markers should not be fixed from the beginning but, on the contrary, change depending on the current results. For a start, a sample of 200–400 BC genotypes may be sufficient for estimations with respect to 30–50 molecular markers. With this number of markers, an initial resolution of no more than 20–40 cM can be assured.

It has already been noted earlier that under expensive marker-based identification of genotypes an improved efficiency of the method can be achieved by evaluating plants with extreme expression of the QTs in question (i.e. plus and minus groups) rather than a random sample (e.g. Darvasi and Soller, 1992). The same idea is applicable to the analysis of traits based on estimation of individual viabilities under stressful conditions (Soller and Beckmann, 1983) (here, instead of a minus group, a random sample from the initial population is used for comparisons). With this approach, it is necessary to establish an initial BC population of a significantly larger size (up to several thousand individuals) and to form, according to the traits in question, plus and minus groups, say a few tens of plants each. Thus, a random sample of about 200 BC plants and plants of extreme value groups are analyzed with respect to 30–50 markers and a set of QTs. Each of these genotypes reproduces vegetatively, so that repeat estimation with respect to QTs under different conditions will enable one to deal with the norm of reaction rather than a trait. Such an analysis of vegetative clones can be performed recurrently in a variety of environments, thereby providing new information on genetic topography of ecological stability features without additional expensive estimations with respect to molecular markers. For self-pollinating plants comparatively tolerant of inbreeding, a promising alternative to clonal propagation of genotypes from BC or F_2 may be the use of recombinant inbred lines (RILs) or dihaploids (Burr *et al.*, 1988; Jensen, 1989; Burr and Burr, 1991; Graner *et al.*, 1991; Zivy *et al.*, 1992). Such material can be subjected to comparative estimation under different external conditions.

Following the analysis of the data obtained, segments will be identified which are responsible for differences between parental strains in the traits concerned or in the norms of reaction. However, the resolution of 20–40 cM assured by utilization of 30–50 markers may prove to be insufficient and additional markers may be required, for example, in order to achieve a resolution of 2–5 to 10 cM. If such an accuracy is specified from the outset, then it will be necessary to consider 150–200 or even more than 300 markers, which is a rather laborious and inefficient procedure, since many markers will turn out to be 'inactive' (if the corresponding genomic regions lack the desired genes and gene blocks). To take an example, Paterson *et al.* (1988) have found, in tomato, as few as six genomic regions significantly affecting genetic variation in fruit weight in an *L. esculentum* × (*L. esculentum* × *L. chmielewskii*) backcross, and four regions affecting differences in fruit solids content. Similar results have been obtained in estimating the genetic basis of differences in QTs between maize and teosinte (Doebley and Stec, 1991, 1993). Therefore, as mentioned above, it is advantageous to apply sequential analysis.

With the information obtained at the first stage, it is possible, based on the markers used, to detect regions carrying the factors of interest, and then to repeat the analysis with additional markers, but now only from these segments. This approach could be applied iteratively, to obtain the desirable precision of mapping. Moreover, even the first stage can be conducted sequentially in order to minimize the total number of 'marker × genotype × environment' measurements. To achieve an acceptable reliability in such a complex flexible design, it is reasonable to accompany the real experiment with a massive parallel computer simulation. Then, each portion of the information obtained could be used to adjust the computer model, providing better possibilities to make decisions about the next step.

3.5.5 Prospects for cloning genes controlling quantitative traits

The detection of genomic regions carrying genetic factors controlling the traits in question is a most important, but not the final, step in a sophisticated genetic analysis. Basically, a study on the genetics of a QT, in case a major gene has been mapped, should involve identification of the corresponding physical DNA regions (map-based cloning), analysis of their regulation, use in genetic transformation, etc. In addition to cloning the QT loci mapped owing to marker analysis, another possibility also exists based on insertional mutagenesis.

Studies carried out over the past 10 years indicate that an important source of genetic variability in higher organisms is provided by mobile genetic elements or transposable elements (TEs). Transpositions of TEs may result in structural gene inactivation phenotypically manifested as a mutation process (insertional mutations). Based on this type of mutation, it is possible to clone the corresponding genes (Fedoroff, Furtek and Nelson, 1984). It is in this way that many genes controlling qualitative traits have been cloned in maize, *Drosophila* and other organisms. This approach is, in principle, also applicable to molecular genetic analysis of QTs, in particular for isolating genes with storng individual effects (Soller and Beckmann, 1987). Solving a problem like this is of great theoretical interest in terms of further insights into the nature of quantitative variation, gene expression control and developmental genetics. The practical value of such a problem is associated with the possibility of altering QTs of economic importance by controlling gene expression and/or gene copy number.

Consider briefly some aspects associated with solving the problem of cloning QT genes using insertional mutagenesis (see also Robertson, 1985; Soller and Beckmann, 1987; Feldman *et al.*, 1989):

- In obtaining and isolating insertional mutants, it is advantageous to use inbred material so as to reduce genetic variation due to segregation of other genetic factors.
- Insertional mutants may be produced through 'mobilization' of own TEs, or via genetic transformation employing own or foreign DNA. The transformation can be performed using either mobile genetic elements (Osborne *et al.*, 1991; Healy *et al.*, 1993) or clones carrying selectable markers. For plants, these may be the *Ac* or *Ds* elements of maize and genes for drug resistance (like *Bar*, Km^R, etc.). In order to increase the incidence of insertional mutants

among the selected transformants, the selectable marker should be used without the promoter. When this procedure is adopted, resistant cells are only formed provided the transforming gene has come under the control of the promoter of some gene within the recipient genome, i.e. the probability of the selected resistant genotypes being insertional mutants is high (Andre *et al.*, 1986);

- In using cell and tissue culture techniques for producing transformants, an additional source of noise (somaclonal variation) may arise, tending to reduce the probability of correct identification of insertional mutants with respect to QTs and resulting in the appearance of additional RFLPs (see Cullis, 1990, for a review). Of special importance in solving this problem are, therefore, alternative methods of transformation (involving no *in vitro* culturing), for example based on the pollination pathway of transformation (Chesnokov, *et al.*, 1990), bombardment of generative structures with micronuclei coated with plasmid DNA (Klein *et al.*, 1988) or absorption of DNA by germinating seeds (Ahokas, 1989).
- To enhance the probability of detecting insertional mutants, it could be a good practice to compare the selected transformants with the control plants based on multitrait analysis rather than with respect to individual QTs.
- An improved efficiency can also be achieved by retransformation as well as by using multicomponent systems with different selectable markers in the *trans* position.
- Cloning of QT genes is better started in organisms with small genome size and relatively low content of highly repetitive DNA. In plants, *Arabidopsis* is best suited to insertional mutagenesis. Many results obtained with this organism can then be extended to other species (Griffing and Scholl, 1991).
- For organisms with high levels of transpositional activity (*Drosophila*, maize), comparative analysis of TE distribution within strains that have undergone long-term directional selection seems to be a promising tool in solving the problem. It cannot be ruled out that the response to selection on a key trait as well as correlated responses to selection for other traits are due in part to TE insertions and/or excisions. In some instances, spontaneous and induced TE transpositions may result in drastic heritable changes in phenotypes and fitness in general (Mackay, 1985, 1987; Vasiljeva *et al.*, 1987; Kaidanov *et al.*, 1991; Torkamanzehi, Moran and Nicholas, 1992). A proportion of transgenic genotypes with reduced fertility may result from insertions at loci controlling meiosis, e.g. because of disturbances in normal segregation (Chapter 6). While usually considered as a kind of 'junk' material (in projects concerned with cloning economically important genes), some of these transformants are of primary interest in studies of genetic control of recombination (Zhuchenko and Korol, 1987).

In conclusion of this chapter, we would like to note that several stages of mass 'attack' on the quantitative variation problem can easily be distinguished in the history of genetics. In general, all these, despite some isolated advances, have provided neither a true understanding of the phenomenon nor a solid basis for medical applications and breeding efficiency improvement, nor, finally, analytical methodology for solving problems related to evolutionary biology. Nevertheless, it is to be hoped that the lingering crisis of quantitative genetics will now be eventually overcome through the synthesis of advances in molecular genetics, recombination analysis and computer modeling.

PART TWO

Population Genetic Models of Interaction Between Selection and Recombination

The genetic system and recombination

4

4.1 THE 'GENETIC SYSTEM' CONCEPT

A characteristic feature of most higher organisms is that of combining, within meiosis–fertilization processes, functions of propagation and genetic organization inheritance with those of generating novel variation. A set of structures and mechanisms ensuring a balance between genetic constancy and variation and determining the dynamics, qualitative and quantitative characteristics of this variation is referred to as the genetic system (Darlington, 1939, 1958; Mather, 1943; Stebbins, 1950; Grant, 1958; Carson, 1975, 1987). The concept covers control of the life cycle length and of relative duration of vegetative and reproductive phases, of mutation rate and spectrum, recombination rate and distribution, fertility level, pattern of gamete and seed dispersion, as well as chromosome number, ploidy level and features of the breeding system. The ideas of the genetic system (GS) and of its leading role in the formation of adaptations were developed by Darlington (1939) in his book *The Evolution of Genetic Systems*. This treatise is rightly called one of the most important biological books of the century (see Preface to Volume 7 of *Chromosomes Today*, 1981).

According to Darlington, it is the GS that provides a trade-off between the opposing requirements for maximum fitness of an organism to current environmental conditions and for genetic flexibility of a species with respect to future conditions. Such a trade-off is needed because maximum fitness to existing conditions does not exclude the high probability of a species becoming extinct as a result of changes in environment in the future (Darlington, 1981). The solution lies in generating trial (reserve) genetic variants adapted to a wide range of external conditions including other than the existing ones. Thus, the GS is a sort of predictor. Its changes are usually neutral for its direct carriers, but may confer long-term selective advantages (Darlington, 1939, 1958). This problem was further elaborated in the classical works of Mather, Huxley, Dobzhansky, Mayr, Stebbins and Grant.

However, the GS also performs another important function, that of controlling the development of organisms in a particular environment (Grant, 1958), determining their adaptive responses associated with survival. Therefore, one seems to be justified in recognizing as relatively autonomous systems two functional components of the organism's genetic constitution (Figure 4.1): the genetic system proper, determining the pattern of genetic variability conversion from the potential to the free state over the generations within a species (the system of population adaptation, or *R*-system) and the system controlling the development and physiological responses to environmental changes (the system of ontogenetic adaptation, or *F*-system) (Zhuchenko and Korol, 1982, 1983, 1985). A tendency to distinguish two components of the GS has already taken shape in Darlington's works. Thus, he speaks of interaction between the 'genetic system' and the 'vegetative system', and of the possibility of restructuring the former via natural selection on the latter. Later on, however, this distinction of functions is blurred and the GS comes to be identified with the entire genetic complex (e.g. Grant, 1975).

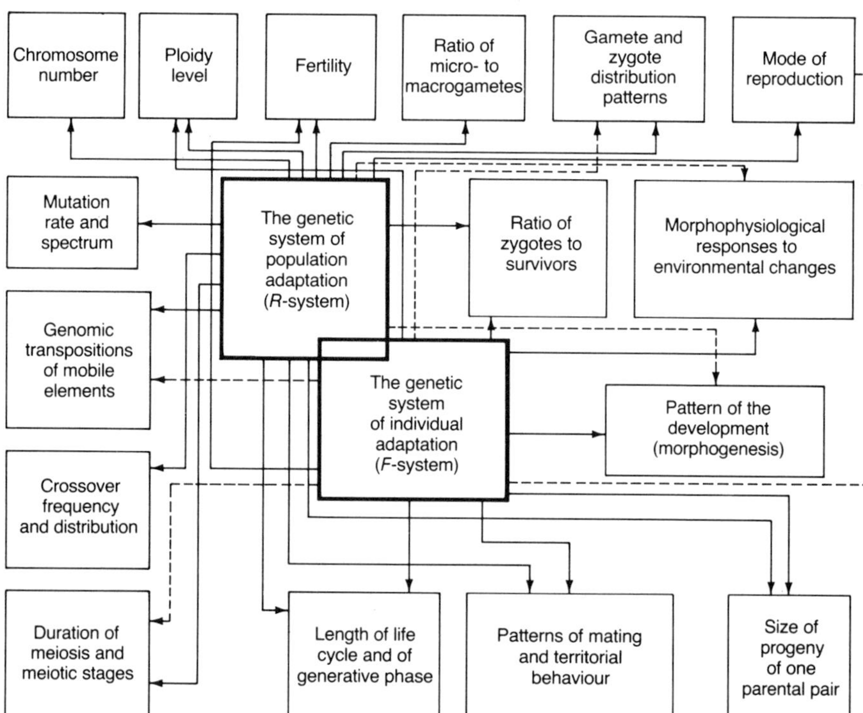

Figure 4.1 Functions of the two overlapping components of the genetic system. Solid arrows indicate primary functions, broken arrows point to secondary functions.

Of course, many components and features of the genetic complex may only rather arbitrarily be assigned to one group or another. Thus, in predominantly outcrossing plants a high rate of selfing occurring in some genotypes may confer an advantage at the boundary of the range where the population density is low. At the same time, the mean value and variation pattern of this trait are key parameters of the *R*-system. The multigene superfamily of immunoglobulin genes seems to be one of the most clear examples of overlapping functions of the *F*- and *R*-systems. An enormous diversity of immune responses to infection are ensured by highly efficient somatic cell recombination and hypermutation processes (Tonegawa, 1983; Schatz and Baltimore, 1988). Here, both functions – mutations and recombination – referred to as the *R*-system ensure high ontogenetic fitness (i.e. an *F*-property).

Recombination also plays an important role in such an essential (in terms of ontogenetic adaptation of eucaryotes) process as gene amplification (Passananti *et al.*, 1987). Another example of this type is the results from estimations of the effects of mobile elements on fitness in *Drosophila* (Kaidanov *et al.*, 1991). Upon termination of strong artificial selection for male low sexual activity, mass directional transpositions of mobile elements are observed together with a rapid increase in fitness. Note that the term 'controlling elements' introduced by McClintock (1984) is in itself suggestive of their influence on gene activity and, hence, on ontogenetic adaptation along with their role as sources of mutations (McClintock, 1984; Schwartz-Sommer and Saedler, 1987).

The expediency of the proposed division of the GS into components is obvious. Darlington's statements about the GS describe adequately the first component of the GS, but they would simply be incorrect when applied to the entire genetic complex. Distinguishing within the genetic system subsystems of individual and population adaptation corresponding to different levels of selection may help clear up the confusion which, as Maynard Smith (1982) puts it, has arisen in the minds of researchers with the appearance of Darlington's *The Evolution of Genetic Systems*. Similar to our dichotomy of the genetic system is also the notion (Altenberg and Feldman, 1987) of all the deterministic forces governing the population dynamics being explainable within the framework of two categories, **transmission** and **selection**, the first being connected with mutation, segregation, recombination, gene conversion and migration, and the second with differential viability, fertility and mating success.

There are numerous relationships between the components of the GS, which depend on environment and direction of selection. Thus, the nature of genetic changes in the *F*-system and particularly their combinations are determined by the *R*-system transferring the action of selection from the level of elementary chemical changes (mutations of individual genes) to the level of biological adaptations controlled by integrated gene complexes (Darlington, 1939, 1958; Mather, 1943; Mayr, 1963; Schmalhausen, 1968).

In an attempt to characterize the primary function of the GS – control of genetic variability within a population – Carson (1957) introduced the term 'recombination system'. The recombination system is, in his opinion, that fraction of the entire genetic system which regulates variability through mechanisms of meiosis and fertilization. The recombination system of each species provides a trade-off between complete genetic constancy (i.e. maximum immediate fitness) and the species potential for variation (i.e. genetic flexibility). Higher plants exhibit a variety of realized 'optima': from lack of recombinational variability in clonally propagated forms (see Antolin and Strobeck, 1985; Cullis, 1987; Hastings, 1991) to high recombination in obligate outcrossers.

Although variability of recombination systems is of quantitative nature, three basic types of these are distinguished: closed, restricted and open (Carson, 1957; Grant, 1958, 1975). Apomixis, maximum heterozygosity, but usually low intrapopulation polymorphism are characteristic features of closed recombination systems. An open recombination system exhibits low rates of selfing, higher levels of heterozygosity and a wide variety of polymorphisms. Restricted systems are distinguished by their predominant self-fertilization, high interpopulation differentiation and low mean heterozygosity of individuals. The difference between open and restricted systems lies not only in the number of possible recombination events, but also in the uniformity of 'variability flow' itself. While individuals in a restricted system generate novel variability sporadically in a few generations, an open system ensures a slow but steady 'flow of variability' over the course of generations (Grant, 1975).

Between the main types of recombination system all kinds of intermediate conditions are possible; various combinations of the three types are found in nature. According to Carson (1975, 1987), closed and open genetic subsystems may be components of one and the same GS: the genomic regions in which recombination is strongly reduced (because of a localized pattern of chiasma distribution, balanced lethals, etc.) constitute the first component and those with free recombination the second one. He believes that reorganization of components of a closed subsystem (blocks of co-adapted polymorphic genes) may trigger speciation.

Practical use of the term 'recombination system' has resulted in its broader meaning and, for example, as used by Grant (1975), it almost coincides with the term 'genetic system' in the narrow sense, i.e. is identical to the *R*-system concept (Zhuchenko and Korol, 1982). The recombination system, as defined by Carson, depends on the mode of reproduction and on factors controlling the rate and genomic distribution of recombination events. However, it is for the latter that the term 'recombination system' should probably be reserved since factors controling the organism's mode of reproduction are generally referred to as the 'breeding system'. Therefore, by the 'recombination system' (*rec* system) we will understand only that part of the *R*-system that determines the pattern of gene recombining as a result of the features of chromosomal machinery organization and of recombination mechanisms.

The relations between the components of the *R*-system, presumably, make it adaptive and internally adjusted (Grant, 1958, 1975). The total number of mutations and recombination, the spectrum of genotypes as well as the dynamics of accumulation and release of genetic variability within a population are the product of the combined actions of complex regulatory factors which seem to be dependent on ecological conditions and fitness (section 9.1). The net result of the combined actions of these factors, the 'point of balance', is the *R*-system of one type or another, providing a trade-off between ontogenetic adaptation and genetic flexibility (Mather, 1943; Grant, 1958).

The species' life conditions and its origin determine the strategy for adaptation and, hence, the optimum type of *R*-system (closed, restricted or open) (Grant, 1958). Thus, in long-continued studies on locusts, Vysotskaya and co-workers have demonstrated that species exhibiting larger amounts of variation in recombination (as measured by chiasma frequency) are ecologically flexible and less specialized. By contrast, species with stable chiasma frequencies occupy narrow ecological niches (Vysotskaya, Bugrov and Stebaev, 1983; Bugrov and Vysotskaya, 1988). These authors point out that low recombination capacity is characteristic of genera comprising one or few species, and high recombination capacity is characteristic of those with large numbers of species.

Clearly, the processes controlled by the *R*-system determine, in the first place, **the rate of evolution** (Endler and McLellan, 1988). Campbell (quoted by Endler and McLellan, 1988) has even termed the factors corresponding to the *R*-system 'the drivers of evolution'. By contrast, the *F*-system components act as factors determining **the direction of evolution**.

Comparision of *R*-systems of different types reveals that their adjustment results from the fact that various subsystems, acting in opposing directions, compensate the effects of one another but do not cancel them out (Stebbins, 1950, 1957; Grant, 1958, 1975). Thus, herbaceous plant species with short life cycles (a factor promoting variability) have, generally, low basic chromosome numbers (a factor restricting variability). The long-lived woody plants, by contrast, have high basic chromosome numbers. In mammalian species with low reproduction rates, longer life cycles and small progeny sizes, recombination (crossing over) seems to be higher than in species with short generation times and large progeny sizes (Burt and Bell, 1987; Sharp and Hayman, 1988). Koella (1993) found that in plant species with animal-dispersed seeds recombination is higher than in other plant species; perennials show a higher recombination than annuals when compared among genera, and an opposite tendency is displayed at the species level.

Of particular interest are possible relationships between the *rec* system (in the narrow sense of the term) and the other components of the *R*-system as well as those among the subcomponents of the *rec* system. Stebbins (1950) suggested that within herbaceous plant species with identical chromosome numbers positive correlation is possible between chiasma frequency and the rate of selfing. A consequence of this correlation is the differences between patterns of variability release in self-pollinating (especially pioneer) and cross-pollinating species: 'bursts' of variation following rare pollinations in the former and regular formation, albeit with lower frequencies, of new crossovers in each generation in the latter (see also Sun and Rees, 1964; Crowley, 1969; Arroyo, 1973; Brown, 1979). These results are probably due in part to ecological peculiarities. Spatial variation in environment provides an advantage for increased recombination under inbreeding occasionally interrupted by outcrossing. This promotes rapid formation and long persistence of combinations of co-adapted genes 'finely tuned' to local conditions. Temporal variation in environment produces an advantage for recombination under outcrossing, ensuring necessary genetic responses to continually arising environmental challenges. In the above-mentioned comparisons of Koella (1993), the variations in the mating system have not been taken into account. It is not only rapid release of variability that increased crossing over promotes. The results of computer simulations carried out by Ukai (1988) suggest that a higher rate of exchanges in selfers considerably increases the mean time to fixation.

The above subdivision of the genetic complex into two components is of major methodological importance for analyzing the evolution of factors and mechanisms controlling genetic variation, as well as for a better understanding of the role of environment in evolution and breeding. Evolutionary relationships between components and subcomponents of the genetic system considered by Darlington, Mather, Grant and Stebbins in group selection terms can and should be reformulated and studied at the individual selection level too (Charlesworth, 1993). However, many of the generalizations about variation of the *R*-system and its components are still largely unknown and untested in nature. Such tests, being conducted at local, regional and global ecological scales in different phylogenies, based on molecular markers and modern cytogenetic techniques, could make the theoretical models more realistic and provide new insights into the genetic basis and logic of the evolutionary processes.

4.2 RECOMBINATION NON-RANDOMNESS

4.2.1 Non-uniformity of genetic maps

The efficiency of crossing-over as a process ensuring recombination of genetic material strongly depends on between- and within-chromosome patterns of exchange distribution in the nucleus. Here, two different (albeit related) problems are generally distinguished: (1) crossover distribution in relation to major components of the chromosome–the centromere, telomeres, euchromatin and heterochromatin; and (2) relative positions of crossovers with respect to one another. Studies on exchange distribution within the genome have revealed its high non-uniformity. The range of spatial variation in exchange frequencies is extremely wide: long stretches, accounting for the major part of the cytological chromosome length, in which crossing over is almost blocked, and comparatively short segments containing recombination 'hotspots' have been observed. Functional significance of such non-uniformity is still unclear. Perhaps, the biological meaning of this non-randomness is in ensuring some expediency, a balance of processes of conversion of adaptively valuable variability from a potential to a free state and vice versa (Grant, 1975; Zhuchenko and Korol, 1985).

Recombinational non-uniformity of the chromosome was first established in *Drosophila* in studies of dependence of crossing over on age and temperature. Comparisons of genetic and cytological maps have shown that recombination in the near-centromeric regions of *Drosophila* chromosomes is strongly inhibited (reviewed by Smirnov, 1991). Similar results have been obtained with tomato (Rick, 1971), barley (Kunzel, 1982), wheat, rye and triticale (Devos *et al.*, 1992; Lukaszewski, 1992) and many other organisms. Thus, in the B-genome of wheat the number of DNA base pairs per chiasma varies more than 150-fold along the chromosome length (Lukaszewski and Curtis, 1993).

The recombination length of the near-centromeric euchromatic segment *ru–p* of *D. melanogaster* chromosome 3 constitutes 1% of the map length and 15% of the overall euchromatin length of this chromosome (Dennell and Keppy, 1979). Evidence for the inhibitory effect of the centromere is provided by the fact that relocation of a segment into the near-centromeric region or a deletion in the near-centromeric heterochromatin results in lowered *rf* (Beadle, 1932; Yamamoto and Miklos, 1978). The centromere also seems to inhibit gene conversion events (Lambie and Roeder, 1988).

The centric heterochromatin constitutes about 40% of the total length of the *D. melanogaster* X chromosome, while it is only 1 cM in terms of recombination length (Roberts, 1965; Hilliker, Appels and Schalet, 1980). In chromosome 2, the crossover and physical lengths of the heterochromatin differ by more than two orders of magnitude and constitute 0.1 and 25%, respectively, of the total chromosomal lengths (Smirnov and Smaragdov, 1979). It is worth asking whether these reductions in *rf* are not due to the same effects of the centromere. In order to ascertain that this really was the case, Roberts (1965) compared the responses of the heterochromatic and the adjacent euchromatic segments to an autosomal inversion. The value of *rf* in the euchromatin increased from 3.51 to 9.11%, whereas no differences were detected in the heterochromatin. Neither was *rf* increased in the heterochromatin of meiotic mutants exhibiting higher crossover frequencies in those euchromatic segments where crossing over is normally suppressed (Carpenter and Baker, 1982). That the properties of the heterochromatin itself are responsible for its low crossing-over frequencies may in general be considered to be a proven fact (John and Lewis, 1965; John and Miklos, 1979; John, 1990). However, the effect of heterochromatin on *rf* does not end there. In a variety of organisms, heterochromatic blocks have been shown to control the frequency and genomic distribution of crossovers (John and Miklos, 1979; John and King, 1985; Attia and Lelley, 1987; Loidl, 1987; John 1990).

As estimated by Rick (1971), the distribution of marker genes on tomato chromosomes significantly differs from the expected, e.g. calculated from the length at pachytene. Rick (1971) points out the tendency for genes to be clustered in blocks in chromosome maps. He also found that the majority of tomato, maize and barley chromosomes show a tendency for their marker genes to be clustered in the proximal region (nucleolus organizing chromosomes being an exception).

The distribution of markers within intervals on each chromosome and over the genome as a whole seems to be worth estimating as well. We have carried out such an analysis using the known data on the location of markers in tomato (Zhuchenko and Korol, 1985). The hypotheses of distribution of genes on chromosomes according to the total cytological length (at pachytene) and euchromatin length have been tested. The results obtained are consistent with Rick's conclusions. The main sources of heterogeneity are the excess of genes over the expected number on chromosome 11 and their deficit on chromosomes 5 and 12. It is of interest to note that the tomato genome only tolerates monosomy with respect to chromosomes 11 and 12 (reviewed by Zhuchenko, 1973), i.e. large differences between these chromosomes in the number of markers are difficult to relate to the similarity of their behavior with respect to aneuploidy. Significant differences between chromosomes with respect to the distribution of markers seem to be due in part to heterochromatin itself (e.g. chromosomes 5 and 12 with very low number of markers and a large proportion of heterochromatin), although this explanation does not provide a complete solution to the problem. Indeed, the chromosomal distribution of loci homologous with random cDNA clones differs significantly from the distribution pattern of the morphological markers. In particular, a large number of cDNA probes are assigned to chromosome 12 (Tanksley, 1987). In other words, the reliability of some estimates based on morphological mutations is an open question.

If the pattern of gene distribution in genetic maps reflects recombination properties of chromosomes, it is of interest to compare the latter with respect to these properties. Our estimations show a very high chromosome heterogeneity (Table 4.1): the chromosomes differ sharply in both mean distances between the adjacently mapped mutations (sixfold) and variance of these distances (45-fold). The recent high-density mapping studies in tomato based on molecular markers revealed the same general picture: very high non-uniformity of the maps (Tanksley *et al.*, 1992). This conclusion is true for both random DNA sequences and cDNA clones (i.e. transcribed genes). A natural explanation for

Table 4.1 Heterogeneity of tomato chromosomes with respect to morphological marker gene distances

Chromosome	Mean distance (cM) From the centromere	From the distal locus	Between adjacent markers	Variance of distances between adjacent markers
1	48.4	37.6	7.1	51.9
2	53.0	16.0	1.3	2.6
3	31.5	26.4	5.4	20.7
4	28.2	44.9	6.6	119.2
5	17.0	15.5	5.0	47.4
6	39.1	28.2	5.2	52.6
7	25.8	21.1	3.1	16.2
8	18.7	27.2	2.9	11.7
9	14.2	21.8	5.0	27.3
10	25.9	33.7	8.8	43.9
11	24.8	21.0	3.4	7.1
Averaged over the genome	32.6	27.7	4.5	37.4
Heterogeneity $\chi^2(10)$	106.7	54.9	140.7	113.0

Heterogeneity is significant at $P < 0.001$ for all of the characteristics.

the observed non-uniformity is the segment-specific crossover distribution, which is similar to the interpretation adopted for *Drosophila* and is confirmed by comparisons of genetic and cytological maps of tomato.

In animals, apart from *Drosophila*, similar analysis is still possible only for mouse (Lyon, 1976) and man (Hartley *et al.*, 1984). Lyon's data indicate a tendency for markers to form clusters; in this respect, different chromosomes of the mouse (like those of tomato) differ significantly from one another. The analysis of possible causes of such clustering leads to the conclusion that non-uniformity of gene distribution in mouse linkage groups is due to non-randomness of crossovers within the chromosome (also manifested in cytologically identifiable localized patterns of chiasma distribution (Lyon, 1976; Gorlov, Agulnik and Agulnik, 1987).

Hartley *et al.* (1984), using the method of direct *in situ* hybridization of specific clones with metaphase chromosomes, have compared physical and recombinational distances in the human X chromosome and demonstrated substantial non-uniformity of distribution. This is in agreement with the known data from the analysis of chiasmata at diakinesis. An extremely high degree of intragenomic non-uniformity in the extent of sex differences in *rf* has been demonstrated in human genetic studies carried out with the aid of molecular markers (White and Lalouel, 1988): generally $rf_{♀} > rf_{♂}$, although in some regions (more often in subtelomeric ones) $rf_{♀} < rf_{♂}$.

The structure of a chromosome segment as a **reacting system** may be responsible not only for suppression of recombination (as in heterochromatin) but also for a sharp increase in it, for the appearance of recombination **hotspots**. Recombination hotspots have been found in the yeast genome. One such segment is 14 cM long and spans the *SUP6* locus (DiCaprio and Hastings, 1976). Another hotspot is in yeast chromosome III, with a sixfold increase in *rf* compared to the genome mean density of the exchanges (Goldway *et al.*, 1993). Recombination non-uniformity (including hotspots) is also found in chloroplast genome (Newman *et al.*, 1992).

There is evidence for the recombinogenic effects of minisatellite sequences, e.g. *GT/CA* repeats from Z-DNA showing negative supercoiling, poly(dA:dT) sequences (Rich, Nordhein and Wand, 1984; Jeffreys, Wilson and Thein, 1985; Kmiec and Holloman, 1986; Treco and Arnheim, 1986; Chandley and Mitchell, 1988; White *et al.*, 1988; Schultes and Szostak, 1991; Pearlman, Tsao and Moens, 1992).

It is well-known that two marker sites located as closely as one nucleotide apart are separable by recombination. However, there is a minimum (specific for the organism and for the system under study) length of homology required for a recombination event to occur, e.g. 20–50 bp in *E. coli* and up to 300 in mammals (Nassif and Engels, 1993). In certain cases, highly specific interaction is possible between mutations in genes of the controlling system and the reacting system components, e.g. when mutations in *rec* genes result in a sharp increase in *rf* within strictly defined genomic regions (Rasooly and Robbins, 1991). These, in particular, may be Z-DNA regions interacting with mutations in topoisomerase genes (see Wallis *et al.*, 1989).

A class of genetic signals, *chi* elements (the sequence 5‘-GCTGGTGG-3’), have been discovered in *E. coli* and phage λ, which significantly increase recombination, their effects being polar and extending for up to 10 kb from their location site (Stahl *et al.*, 1980; Smith, 1987). Mobile elements may act as recombination inducers in eucaryotes too. Thus, with identical locations on homologous chromosomes, a pair of P elements of *D. melanogaster* can induce recombination at a frequency of 20% and more, including exchanges in males (Sved *et al.*, 1991).

Recombination hotspots have been found within the human genome. In the distal end (region q32) of chromosome 14, crossing-over frequency is five times as high as the expected rate corresponding to its physical length (Nakamura *et al.*, 1989). Interestingly, this segment exhibits increased frequencies of aberrations in somatic cells. In human genome, the recombination hotspots are the β-globin and insulin loci, and in mammals in general the *MHC* locus (Lebo *et al.*, 1983; Kobori *et al.*, 1986; Johnston and Rich, 1989). Thus, in laboratory mice, the DNA content per unit exchange within the *MHC* varies over the range of 300 kb to 30 Mb/cM, depending on the presence or absence of a recombination hotspot. The level of recombination in the region of the insulin locus (short arm of the human chromosome 11) is 30 times as high as the expected values. The GC-rich bands of human chromosomes, which represent only15% of all bands, contain 65% of the mapped genes and display six times higher chiasma frequency than G-bands (which are AT rich) (Holmquist,1992).

Finally, there is some evidence for non-uniform genomic distribution of gene conversion and intragenic recombination rates (Thuriaux, 1977). In yeast, it has been demonstrated that the features of nucleotide sequences may be responsible for differences in the rate of intragenic recombination, both meiotic and mitotic (Moore and Sherman, 1975). Recombination within the *Bz* gene of maize is 200 times higher than the average over the entire genome (quoted by Meagher, McLean and Arnold, 1988). In general, variation in intragenic recombination may be due to variations in physical length of genes, but it is also likely to result from variation in recombination frequencies in the genome, for example because of differential rates of hybrid DNA formation (Nicolas, 1979).

4.2.2 Recombination regulation and regional distribution of gene activity

The available evidence in general suggests high intragenomic variation in the frequency of recombination events per unit physical length of DNA (reviewed by Smirnov, 1991). This variation is under complex control in which several hierarchical levels can be distinguished:

1. organization of DNA sequences including the distribution of specific regulatory sequences of zygotene and pachytene DNA, minisatellites, etc. (Hotta, Chandley and Stern, 1977; Friedman, Bouchard and Stern, 1982; Stern, 1986; Pearlman, Tsao and Moens, 1992);
2. position of euchromatin (as the main target for recombination) relative to the centromere, telomeres and heterochromatic blocks (John and Miklos, 1979; Miklos, 1982; John, 1990);
3. segment-specific regulation of crossover frequency by genes of the 'fine control' system with relatively weak effects of individual components (Simchen and Stamberg, 1969);
4. combining the genomic and, simultaneously, chromosome and segment specificity of gene regulation within the system of 'coarse control' which includes regulation of the basic steps in the recombination (e.g. the pairing processes, DNA strand exchanges, recombinational repair) (Simchen and Stamberg, 1969; Carpenter and Sandler, 1974; Parker 1975; Tease and Jones, 1976; Carpenter, 1988; Zickler *et al.*, 1992);
5. control of the nucleus as a whole (Lukaszewski, 1992). It is manifested, for example, in the interchromosomal effect of rearrangements (Mather, 1936b; Lucchesi and Suzuki, 1968);
6. regulation based on cytonuclear interactions (Thoday and Boam, 1956; Moriwaki, Tobari and Matsuda, 1979; Lawrence, 1963; Kruleva *et al.*, 1992);

7. effects of mobile genetic elements (Slatko, 1978; Sved *et al.*, 1991; Vereschagina and Korol, 1991).

Previously, Thuriaux (1977) noted the discrepancy between macro-evolutionary variations in recombination frequency per unit physical length for the genome as a whole and for structural genes. In going from lower to higher organisms, the recombination density is reduced by 3–4 orders of magnitude, whereas for strucutral genes *rf* remains almost unchanged. He attributes this paradox to presumably preferential localization of recombination in the structural genes. As mention above, in human chromosomes, the concentration of mapped genes and meiotic chiasmata in GC-rich regions is several-fold higher the average for the genome (Holmquist, 1992).

Still earlier, Whitehouse (1966) suggested an operator model of recombination assuming some relationship between recombination and transcription initiation sites. Experimental evidence (both direct and indirect) has been obtained suggesting the existence of such a relationship (Ikeda and Kobayashi, 1977; Voelkel-Meiman *et al.*, 1987; Dul and Drexler, 1988a,b; Vizir and Korol, 1990; Voelkel-Meiman and Roeder, 1991; Yeom, Kuniya and Artzt, 1992).

A different interpretation of a possible relationship between the genomic distribution of structural genes and recombination initiation sites has been proposed by Holliday (1984). He believes it to be a manifestation of the meiotic recombination repair function ensuring elimination of epigenetic defects in DNA when passing to the next generation. However, if this meiotic function does play a critical role, it is unlikely to be associated with crossing over, since it is inconceivable how it can be realized by the heterogametic sex in species with achiasmate meiosis (e.g. *Drosophila* and silkworm). If one rejects this role of recombination, then the functional part of Holliday's explanation has no bearing on the problem of relationship between transcription and recombination, although the proposed regulatory mechanism seems to be rather plausible.

A remarkably high precision inherent in crossing over has remained largely unexplained for years. If in the course of evolution the distribution of recombination initiating points did really become associated with structural genes then a mechanism should have evolved protecting the coding sequences against possible recombination errors. The formation of non-coding regions (introns) can be regarded as constituting precisely this mechanism (Zhuchenko and Korol, 1985). In other words, we can speculate that the accuracy of the molecular mechanism of exchange may be only an apparent one. Actually, errors similar to intron sizes are permissible. However, these errors will not affect the protein structure if crossing-over events occur mainly within intron regions. In cases where the extent of the shift is greater than the intron size, gametes with duplications of one of the exons are formed. This formulation is close to the hypothesis of Gilbert (1978), who considered the exon–intron organization of structural genes in eucaryotes as a mechanism for a faster gene evolution through rearrangements of intragenic blocks (protein domains) (see also Darnell, 1978; Tonegawa *et al.*, 1978; Inge-Vechtomov and Soydla, 1978; Khesin, 1984; Kobori *et al.*, 1986; Bremermann, 1987; Doolittle and Stoltzfus, 1993).

Bremermann (1987), while in general sharing these views, believes that one of the main functions of the mosaic structure of eucaryotic genes is providing the possibility for rapid evolution of their proteins concerned with resistance to microparasites. He regards introns, also hypothetically, as possible 'trap' sites for the preferential insertion of viral sequences, unequal crossing over probably playing the role of a mechanism regulating the length of such inserts. Reanney (1984) considers the mosaic structure as a kind of anti-noise mechanism, which increases the stability of genetic information by screening the molecular deviates during RNA splicing (see also Altukhov, 1989). Our model coincides fairly well with the known data on very short periodicity in intron sequences of higher eukaryotes, i.e. purine–pyrimidine alternation, which is characteristic of some hotspots of recombination (Arques and Michel, 1987). Another interesting coincidence could be found if we relate the concept that transposable elements can behave as introns (see for review Purugganan, 1993) to the data that mobile elements can induce recombination (Sved *et al.*, 1991; section 6.2).

According to our hypothesis, comparisons of the intragenic recombination map and physical map of the gene can be expected to reveal alternation of regions of monotonic rise in *rf* with distance (within introns) with those of approximately constant *rf* (within exons). From the concept of introns being 'traps' for recombination errors, it follows that the divergence of intron regions should be more frequently (compared with intergenic non-coding regions) associated with deletions, duplications and insertions. As an example of such an evolution one can, possibly, consider the loss of the third exon of the *Adh* locus in *Drosophila willistoni* (Anderson, Carew and Powell, 1993).

All this suggests the possibility of a certain non-randomness in the formation of genetic variability within populations associated with genomic distribution of the template activity in a given environment. Of course, the type of recombination regulation considered here should be referred to as the system of 'fine control'. A simple model for the evolution of the dependence of recombination on environment will be discussed in section 9.1.

Comparison of recombinational behavior of chromosomes in meiotic mutants with that in the normal genotypes reveals that non-random genomic distribution of exchanges under regular meiosis is due, to a large extent, to evolutionarily established restrictions on recombination in certain regions which are under the control of meiotic (*mei*) loci. Meiotic mutants tend to lessen these restrictions. As a result, regions previously excluded from exchanges may become involved in crossing over, interference is reduced and, in spite of a general decrease in crossing-over frequency, the correspondence between physical and recombinational length of the chromosome segments becomes closer (Moens, 1969; Carpenter and Sandler, 1974; Lindsley and Sandler, 1977; Sandler and Szauter, 1978; Carpenter and Baker, 1982; Szauter, 1984). In a number of cases, *mei* mutations or *rec* mutations can significantly **enhance** crossing over (and not only in specific regions), or increase multiple exchange, intragenic recombination and/or mitotic crossing over frequencies many-fold (Moens, 1969; Rose and Baillie, 1979; Simchen *et al.*, 1981; Aguilera and Klein, 1988; Montelone *et al.*, 1988; Rattray and Rose, 1988; Wallis *et al.*, 1989; Palladino and Klein, 1992; Osman, Tomsett and Strike, 1993). Therefore, the high intragenomic variation in exchange frequency is due not only to features of structural organization of chromatin as a reacting system but also to the system of gene control. It seems justifiable to look for a functional (adaptationist) explanation for these and many other properties of the recombination system (Zhuchenko and Korol, 1985; Wohrmann and Jain, 1990).

4.3 LINKAGE OF CO-ADAPTED AND FUNCTIONALLY RELATED GENES

It is natural to assume that the biological function of recombination nonrandomness is to ensure a balance between the conversion of potential genetic variability into free, accessible selection, and the opposite process. It is therefore of interest to compare data on recombinational and adaptive organization of chromosomes. The problem has already been discussed in sufficient detail previously (Bodmer and Parsons, 1962; Ford, 1971; Grant, 1975; Zhuchenko and Korol, 1985). So, we will provide here only a short summary of data on linkage between co-adapted and functionally related genes.

There is abundant evidence (from both pro- and eucaryotes) suggesting that functionally related, or unrelated but co-adapted, genes are linked to one another, although, according to some authors, it is free recombination that is, rather, the general rule (Elston and Glassmann, 1967; Lande, 1980). In a number of cases, when adjacent genes control the successive steps in a biochemical chain, linkage is very close. Such groups of closely linked functionally related genes (clusters) have been studied in bacteriophages, bacteria and fungi (Perkins and Barry, 1977; Giles, 1978).

Examples of close linkage between functionally related genes are known in each genetically studied species of higher eucaryotes. In maize, the *Tu* (tunicate) locus is composed of two subunits separable by crossing over. The parental stocks have been found among maize ancestral forms. Upon crossing, these yielded the *Tu* form (Mangelsdorf and Galinat, 1964). In barley, the genetic loci *Est-1* and *Est-4*, standing far apart in the physical map of chromosome 3, are very tightly linked (Linde-Laursen, 1982). A group of closely linked loci (*T*, *F*, *Ki*) controlling tail length, vertebra size and other related characteristics are known in the mouse (Dunn and Caspari, 1945). In

Primula, heterostyly is controlled by a supergene composed of no fewer than seven functionally distinct, but co-adapted, loci (Ford, 1971). Other examples of co-adapted gene blocks are the Batesonian mimicry in butterflies, systems of recombinational lethals in *Drosophila*, a number of polymorphic complexes in man, loci controlling shell color and pattern in snails, etc. (Bodmer and Parsons, 1962; Ford, 1971; Grant, 1975; Hedrick, Jain and Holden, 1978; Shields, 1982; Zhuchenko and Korol, 1985; section 5.3).

A number of cases of linked functionally related genes have been found in *Drosophila*. These are both very tightly linked compound loci whose origin is associated with duplications and large groups of genes occupying comparatively long stretches of the chromosome (Golubovsky, 1977; McCartney, Renwick and Munday, 1977). One macrogroup is homeotic mutations localized in a segment 13 cM long on the third chromosome. For each of the three large chromosomes, examples are known of groups of closely linked functionally related mutations: genes for yellow body color at the map position 0.0 cM on the X chromosome; for a reduced number of bristles or extra bristles in the same position; for the orange, pink and scarlet eye color in the interval 54.3–56.0 on chromosome 2 (five genes) and 46.0–48.0 on chromosome 3 (four genes); for wing abnormalities in the interval 55.0–55.2 on chromosome 2 (six genes); and others. Several linked groups of related genes have been found in *D. virilis* (Korochkin, 1981). In particular, two groups of mutations affecting wing development are distinguished on the X chromosome: in the intervals 72.0–78.1 (four loci) and 94.5–104.5 (six loci).

Gene duplications via unequal crossing over seem to be a plausible explanation for a large portion of such data. New evidence that has been accumulated based on the use of molecular markers supports this view. Examples are provided by numerous duplications in *Brassica* (Slocum *et al.*, 1990; Song *et al.*, 1991), alfalfa (Brummer, Kochert and Bouton, 1991) and maize (Helentjaris, Weber and Wright, 1988).

As with other organisms, one can recognize groups of relatively closely linked genes with similar effects within the tomato genome. Thus, a cluster of esterase genes (*Est-1,- 5,- 6,- 7*) has been localized at map position 29 cM on chromosome 2, five mutations affecting plant growth (*Cu, ro, dil, prun, dry*) in the interval 8 cM long on the same chromosome, three such genes (*d-2, rv, sp*) in the vicinity of marker *c* on chromosome 6 and four (*re, cpt, bu, depa*) in the 12–18 cM region on chromosome 8. On the short arm of chromosome 4, six chlorophyll mutations (*alf, cm, syv, ver, ful, oc*) are located in the 6–24 cM region. With an increasing proportion of mapped mutations, the number of such groups will probably also rise.

Since the haploid chromosome number is rather high ($n = 12$) in tomato, it is reasonable to assume that some of the chromosomes can have certain functional 'specialization', at least with respect to the spectrum of arising mutations. To test this hypothesis, we characterized each mutation on a score of 1–3 or 1–4 according to the magnitude of its effect on plant morphology and reproductive system. Then, mean values of the indicated traits were calculated for each chromosome and for the genome as a whole, and the degree of homogeneity of these means and the extent of each chromosome contribution to the overall test of heterogeneity, χ^2, were determined (chromosome 12 was excluded because of the small number of mutations localized on it). The results obtained suggest high heterogeneity of tomato chromosomes with respect to the effects of their markers on some traits (Table 4.2).

Significant differences have been established for the characters 'morphological and cytological abnormalities of the reproductive system' and 'chlorophyll abnormalities'. At the same time, the chromosomes are homogeneous for the traits 'plant height' and 'leaf shape'. The presented data can be regarded as initial information in designing experiments on mapping of genes and gene blocks determining differences between genotypes in adaptive characters (Chapter 3). In any event, one can hardly agree with the assumptions that loose linkage of functionally related genes is a characteristic of higher organisms. Such an assumption leads to the conclusion of a small contribution of linkage disequilibrium to correlations between fitness-related traits. These conclusions are difficult to reconcile with the available evidence of the possible strong selection in natural populations.

A number of cases are known (including those in outcrossing organisms) of substantial linkage disequilibrium between markers, inversions and markers,

Table 4.2 Phenotypic effects of tomato morphological markers as measured for individual chromosomes

	Effects on				
	Generative system abnormalities		*Plant morphology*		
Chromosome	*Cytogenetic*	*Morphological*	*Height*	*Chlorophyll system*	*Leaf shape*
1	0.68 ± 0.24	0.52 ± 0.22	1.26 ± 0.19	0.36 ± 0.22	0.71 ± 0.18
2	0.61 ± 0.24	0.67 ± 0.24	0.76 ± 0.20	0.39 ± 0.18	0.85 ± 0.20
3	0.55 ± 0.18	0.30 ± 0.21	0.65 ± 0.21	0.60 ± 0.18	0.74 ± 0.25
4	0.56 ± 0.34	1.04 ± 0.33	0.90 ± 0.26	0.91 ± 0.23	0.74 ± 0.28
5	0.42 ± 0.23	0.59 ± 0.29	0.92 ± 0.38	0.01 ± 0.25	1.17 ± 0.42
6	0.68 ± 0.25	0.64 ± 0.22	0.72 ± 0.20	0.60 ± 0.22	1.00 ± 0.23
7	0.05 ± 0.07	0.05 ± 0.07	1.05 ± 0.25	0.85 ± 0.23	1.05 ± 0.28
8	0.39 ± 0.16	0.61 ± 0.23	1.04 ± 0.22	0.57 ± 0.18	0.79 ± 0.21
9	0.22 ± 0.13	0.06 ± 0.06	1.00 ± 0.27	0.72 ± 0.21	0.72 ± 0.28
10	0.12 ± 0.08	0.24 ± 0.18	0.71 ± 0.97	1.71 ± 0.27	0.47 ± 0.15
11	0.83 ± 0.24	1.00 ± 0.22	1.14 ± 1.61	0.53 ± 0.12	0.50 ± 0.16
As measured for the entire genome					
	0.51 ± 0.07	0.57 ± 0.07	0.95 ± 0.07	0.63 ± 0.06	0.76 ± 0.07
$\chi^2(10)$	79.73**	143.87**	9.24	29.34*	9.74

The overall heterogeneity is significant at $^*P < 0.01$ and $^{**}P < 0.001$.

and between inversions themselves (section 5.3). The important role of the relationship between epistatic interactions of loci and their linkage, as well as epistasis itself, is indicated by evidence on recombination (synthetic) lethals, numerous examples of supergenes and the phenomenon of hybrid breakdown (for review see Bodmer and Parsons, 1962; Ford, 1971; Grant, 1975, 1981; Hedrick, Jain and Holden, 1978; Zhuchenko and Korol, 1985). Many examples of linkage between functionally related genes are known in biochemical and molecular genetics. The results obtained in these areas not only substantiate the role of linkage and recombination in evolutionary reorganization of the genome but also reveal a diversity of mechanisms, sometimes unexpected ones, whereby a balance of constancy and flexibility is achieved in interactions between the genetic system and environment.

4.4 PARALLELISM OF SYNTENIC GROUPS

Of great interest in connection with the problem in question is the phenomenon of linkage parallelism. There is evidence for an extremely high stability of some linked gene complexes over long periods of time. The term 'synteny' was introduced to denote situations when two and more genes are located on the same chromosome but the degree of linkage between them is unknown (Renwick, 1971). The idea of possible chromosome homologies in gene locations in different species was put forward some 50 years ago (Sturtevant and Novitsky, 1941). However, as early as the 1920s, Morgan, Bridges and Sturtevant (1925) established the similarity of genetic maps of X chromosomes in some species of *Drosophila*. The chances of detecting parallelisms of groups of tightly linked genes have recently been increased as a result of the application of molecular genetic techniques, in particular physical mapping of genes based on pulsed-field electrophoresis of DNA (Brockdorff, Amar and Brown, 1989). In the last years, intensive studies of interspecific homologies in genetic material organization at the intralocus level have also been initiated (Hudson, Kreitman and Aguade, 1987; Aquadro, Lado and Noon, 1988; Weber *et al.*, 1990; Devos *et al.*, 1991; Green *et al.*, 1993; Hardison and Miller, 1993; Kliman and Hey, 1993; Ogawa *et al.*, 1993).

The largest number of cases of gene synteny parallelism concern the mammalian X chromosome which, in many respects, has remained unchanged for about 120 million years (Ohno, 1974). The dosage compensation effect, which is responsible for the functional haploidy of sex-linked genes in females and, probably, precludes the normal expression of these genes upon relocation into autosomes, has been assumed to be the cause of the freezing. But parallelism of syntenic groups is not confined to the X chromosome. The mechanisms for their conservation are not known. A large evidence of autosomal synteny parallelism have been accumulated in man, primates and other mammals including the mouse, rat, Chinese hamster, American mink, sheep, dog and pig (Stallings and Siciliano, 1983; Roderick *et al.*, 1984; Lalley and McKusick, 1985; Serov *et al.*, 1987; Zakharov and Valejev, 1988; Seldin and Kruh, 1989). It has been found that similar conservatism is also characteristic of mammalian oncogenes (Stallings *et al.*, 1985). These homologies in the genetic material topography occasionally prove so strong that they can be of practical value in mapping. Thus, in comparisons of the human and mouse genomes, the average length of conserved segments is 8 cM, reaching, in some cases, 20–25 cM (Nadeau and Taylor, 1984; Searle *et al.*, 1987; Nadeau, 1989; Scherer *et al.*, 1989).

Similarities in gene patterns have been found in insects: *D. melanogaster*, *Musca domestica* and sheep blowfly, *Lucilia cuprina* (Foster *et al.*, 1981). Attempts to analyze the causes of linkage group conservation in *Drosophila* have not yielded conclusive results (Hilliker and Trusis-Coulter, 1987).

The very fact of linkage parallelism suggests nothing about the significance of integrity of the gene groups. In principle, they may be a remnant of common ancestry which has not yet been destroyed by chromosome rearrangements. A rigorous analysis is required to make the argument for the adaptive role of syntenic groups sound. Early assessments of this kind show that the assumption of selective causes of synteny parallelism conservation in man and the mouse is not necessarily a correct one (Nadeau and Taylor, 1984; but see Searle *et al.*, 1987).

Particularly striking are the cases in which linkage is conserved during major evolutionary changes. Thus, structural genes of blood albumin (*Alb*) and vitamin D-binding protein (*Gc*) are closely linked in chicken; a similar situation has been found in man and horse (Juneja *et al.*, 1982). There are syntenic groups which are even older. In particular, comparative mapping revealed associations which are common among many vertebrates (e.g. fishes, frogs and mammals) and allow one to estimate the presumed genome organization of the ancient chordate ancestor (Morizot, 1990). Certain dehydrogenase loci are linked in such divergent groups as *Drosophila*, frog and mouse (see, for example, a review in Juneja *et al.*, 1982). Strong homology has been established in clustering of homeotic genes in insects and vertebrates (Akam, 1989). Finally, to complete the picture for animals, it should be mentioned here that there is evidence for parallelisms in the organization of mitochondrial genomes of various species, e.g. between man and mouse (Searle *et al.*, 1987).

Parallelisms of this type are also found in plants (Bonierbale, Plaisted and Tanksley, 1988; Smirnov, 1989; McGrath and Quiros, 1991; Binelli *et al.*, 1992; Devos *et al.*, 1992; Weeden and Muehlbauer, 1992; Bennetzen and Freeling, 1993). In particular, an extremely high similarity in positions of homologous genes has been shown in tomato and potatoes (Bonierbale, Plaisted and Tanksley, 1988). Using genomic copies and cDNA clones of tomato, the authors have performed recombination mapping of these loci in potatoes and compared the resulting map with the previously generated map of the same loci in the tomato. Nine chromosomes out of 12 exhibited identical gene orders in both genera, the differences in the remaining three being due to intrachromosomal rearrangements (paracentric inversions), i.e. syntenic groups are highly conserved. All these conclusions have subsequently been confirmed by Gebhardt *et al.* (1991).

On the other hand, within the same family of Solanaceae, the knowledge of gene linkage in tomato is a poor predictor of linkage in pepper, and vice versa (Tanksley, 1987; Tanksley *et al.*, 1988). It is of interest to note that, despite the four-fold higher genomic DNA content of pepper relative to that of tomato, the crossing-over level in comparable marker segments did not change significantly. Comparisons of wheat and rye homeologs have shown the relationship between the degree of syntenic group conservatism and crossing-over rate: regions with reduced exchange frequencies tend to exhibit higher parallelisms (Devos *et al.*, 1992).

In general, several types of genetic material integration can be recognized: co-adapted gene blocks (supergenes), gene clusters and multigene families, stable syntenic groups. The mechanisms for maintaining the integrity of these are currently unknown; it is more than likely that they are different. In the case of supergenes, it is co-adaptation (selection) and close linkage (crossing-over suppression) which play the key role. Gene linkage may be a critical factor determining the rate of evolution of resistance to toxic environmental agents, for example pesticides (Mani, 1985), as well as the pattern of antagonistic species co-evolution (Chapter 8). Another factor determining conservation of the chromosomal position of a locus can be its specific function. This is exemplified by the mating type locus which is linked to the centromere in many fungi, and can be viewed as an adaptation promoting heterozygosity (Zakharov, 1986). The fundamental cause of the evolutionary conservatism exhibited by groups of genes may be restricted randomness of chromosome rearrangements resulting from the spatial structure of the interphase nucleus and mitotic chromosome packing, as well as the pattern of attachment of germ cell chromosomes to the nuclear envelope (von Wettstein, Rasmussen and Holm, 1984; Stegniy, 1991). Like other karyotype characteristics, these restrictions can, in turn, be viewed as the product of evolution toward reduced aberrational losses, optimized functions of genes, of chromosome and cellular machinery (Lima-de-Faria, 1983; Imai *et al.*, 1986; Naora, 1987; Vogt, 1990; Trifonov, 1990; Foster and Whitten, 1991).

Another major force opposing the randomization of genetic material may be strong selection for a conserved order of loci successively activated during ontogenesis. Evidence for the 'temporal' principle of genome organization comes from microorganisms, maize, *Drosophila*, newt, mouse, and man (Grant, 1975; Korochkin, 1977, 1981; Krasilov, 1979; Li and Wu, 1984; Stubblefield, 1986; Hibner *et al.*, 1988). Transpositions of mobile elements may be a major factor behind evolutionary changes in the control of 'temporal' parameters of ontogenesis, the changes having the potential to be directional (provided insertion and excision sites are nonrandomly located) (Korochkin, 1983).

It is worthwhile to note that, unlike factors preventing co-adapted gene complexes from being broken up, the forces promoting the integrity of 'temporal' sequences of genes or multigene families can operate even in the absence of polymorphisms. Moreover, in view of the abundant evidence obtained in the past decade and suggesting a fairly high efficiency of the genome-destabilizing mechanisms (primarily transpositions of mobile elements), it can be assumed that the forces and processes involved in **conservation** of the genetic material organization operate even more effectively than previously suspected. Otherwise, the 'genome inconstancy' would have been converted from an important evolutionary potentiality to a source of high-frequency and all-embracing topographical transformations (chaotization) of genetic material. As we have seen earlier, this is not the case with gene locations, although much has yet to be done in terms of comparative analysis of structural and functional organization of the genome in various taxonomic groups for a more complete solution of the problem.

It is generally believed that complete suppression of recombination in a particular genomic region in all members of the population can result in this region being inactivated (Huxley, 1942; Darlington, 1958; Charlesworth, 1978). This may have been the cause of the Y chromosome heterochromatinization. Thus, inherent in the process of genome integration due to co-adaptation is an internal contradiction: integration of most important regions preconditions their future inactivation thereby weakening the repair functions (in a broad sense) of recombination. Similar arguments underline one of the most popular hypotheses (Muller, 1964; Felsenstein and Yokoyama, 1976; Kondrashov, 1984, 1988a; Charlesworth, 1990) accounting for the maintenance of sex and recombination in natural populations by selection against deleterious mutations (section 6.1). It is possible that the requirement for continuous correction of mutations in coding sequences of multiple copies is a key factor favoring clustered organization of multigene families. This effect may be due to exchanges during both meiotic and mitotic cell divisions (Robbins and Swanson, 1988; Petes and Hill, 1988).

High non-uniformity and non-randomness of crossing-over is characteristic of most eucaryotes studied. As suggested by genetic evidence, underlying this, along with specific nature of primary sequences (organization of the reacting system, in our terminology), are the functional features of genes controlling meiotic processes. In recent years, extensive studies have been initiated to identify these genes, clone them, analyze their primary structure, products and regulation pattern. Considerable structural and/or functional homologies have been demonstrated for some genes controlling meiosis–recombination–repair processes and cell cycle control genes between pro- and eucaryotes, and between lower and higher eucaryotes (Draetta, Beach and Moran, 1988; Spurr *et al.*, 1988; Lambert *et al.*, 1988; Bishop *et al.*, 1992; Shinohara, Ogawa and Ogawa, 1992; Hoeijmakers, 1993 a,b; Ogawa, Shinohara and Egelman, 1993; Scherly *et al.*, 1993). The possibility for complementation of repair gene mutations by transformation involving foreign genes and their products has been established, including cases of procaryote–eukaryote gene transfer (Arrand *et al.*, 1987; Lambert *et al.*, 1988; Yamaizumi *et al.*, 1988; Banga *et al.*, 1989). Hopefully, these findings will enable not only an understanding of the mechanisms for non-random recombination in higher eukaryotes but also the development of efficient methods for controlling this process (Zhuchenko and Korol, 1987).

Whatever the differences between the above systems with respect to their functions, mechanisms of origin and persistence, they have one feature in common – non-random distribution of genetic material – which is of fundamental importance in micro- and, probably, macroevolutionary processes. Rather little is currently known about the interrelations between functional and structural features of the genome organization. Yet, the generalization put forward more than 40 years ago that the genome is more than a mere 'bag of beans genes' can now be regarded as rather well substantiated experimentally. This is also true for intragenomic recombination variability considered above. Our main hypothesis is that the basic features of the recombination non-randomness have an adaptive value and result from adaptive evolution (Figure 4.2).

Recombination is non-random

- across the genome
- relative to genotype
- relative to environment

All these help to achieve a compromise between individual fitness and genetic flexibility

Hypothesis: The above features of recombination are a result of evolution for optimization of genetic variation (in quantity and spectrum)

Figure 4.2 Adaptationist view on the features of recombination in higher organisms.

5 Effect of recombination on population genetic structure

5.1 THE GENOME INTEGRITY AS A PROTECTION AGAINST EXCESS VARIABILITY

Increased fitness of organisms to existing conditions is one of the consequences of natural selection. In a constant environment (temporally and spatially) this could result in a certain maximum of fitness. Recombination, however, while ensuring the formation of new valuable gene combinations, at the same time continuously breaks these up. One of the possible strategies of a species is to achieve a maximum level of fitness and to switch to asexual reproduction or to another mode of variation restriction.

The majority of higher organisms pursued another evolutionary course – the development of systems ensuring a reconciliation of the opposing requirements for maximum fitness to current conditions and for flexibility with respect to future environmental changes (Darlington, 1958). A considerable decrease (by several orders of magnitude) in crossing-over rate per unit physical length of DNA is observed in going from lower to higher organisms (Catcheside, 1977; Thuriaux, 1977). The reasons for this seem to be, in general, clear. A sharp increase in the complexity and coordination of ontogenesis in higher organisms has offered them new opportunities for adaptation compared with lower forms (Schmalhausen, 1942). While in microorganisms the population adaptation is achieved under enormous overproduction of individuals by mutation (and partly recombination) of functionally relatively independent genes, in higher forms this overproduction is much lower, the value of each individual to the population is higher and adaptation to the environment depends more on ontogenetic (individual) plasticity (Schmalhausen, 1949, 1968; Bradshaw, 1965; Ratner *et al.*, 1985; Krasilov,

1986; Rodin and Rzhetsky, 1987). Nevertheless, the vital importance of recombination in RNA viruses is worth noting. The frequency of exchanges is very high in these organisms which, along with high mutability, plays a decisive role in their adaptive radiation (Steinhauer and Holland, 1987; Hahn, Lustig and Strauss, 1988; Strauss, and Strauss, 1988). Moreover, it is the high rate of exchanges (ensured, in particular, by their segmental replication) which probably serves to reduce the effects of deleterious mutations (Crow, 1988; Drake, 1993). There is evidence for adaptive significance of recombination in bacteria too (Redfield, 1988; Robertson and Meyer, 1992).

Higher complexity of multicellular organisms increases the danger of upsetting their internal balance and is, therefore, accompanied by a replacement of relatively free recombination by various adaptations serving to protect the most significant genomic region (zones of concentration of co-adapted gene systems) (Mayr, 1982; Grant, 1985; Carson, 1987).

A reduction in fitness associated with recombination may result from gene interactions within a single chromosome arm, between arms of the same chromosome or between different chromosomes, with the effect being manifested in both viability and fertility (Dobzhansky, 1946; Wallace *et al.*,1953; Gibson and Thoday, 1962; Dobzhansky, Spassky and Anderson, 1965; Templeton, 1979; Seager, Ayala and Marks, 1982; Shaw, Wilkinson and Coates, 1982; Thompson, 1986; Clark, 1987). Therefore, a general reduction in recombination rates has been accompanied, in the course of evolution, by an increase in non-uniformity of exchange distribution throughout the genome, which narrowed the range of genotypic variability down to 'reasonable' limits and, probably, resulted in smaller recombinational losses.

On the other hand, owing to higher 'regularity' of sexual processes (frequent outcrossing, presence of a synaptonemal complex at meiosis, the obligatory occurrence of at least one exchange per bivalent or per arm) and to increasing genome size, recombination in higher eucaryotes plays a greater role in short-term release of variability than mutation does (Zhuchenko, 1980, 1988; Severtsov, 1981; Zhuchenko and Korol, 1985). Although the rate of exchanges per unit physical length of DNA in lower organisms may be rather high (Catcheside, 1977), they generally do not seem to be of substantial evolutionary significance due to a small chance of sexual contacts between individuals (Maynard Smith, Dowson and Spratt, 1991). One reason for a lesser role of recombination than of mutations in generating variability in unicellular forms may be such an important component of the 'cost' of meiosis as a dramatic reduction in reproduction rate due to a longer duration of meiosis compared with two mitoses (Lewis, 1987).

As is known, the concepts of genetic material integrity, of the relationship between its structure and functions began to shape in the late 1930s to early 1940s. In his seminal book *Evolution. The Modern Synthesis*, Huxley (1942) wrote:

> If, as now seems established, it is the case with some or all genes that interaction with near neighbors in the same chromosome affects their expression in an important way, then it is clear that all the genes within a given chromosome must be delicately adjusted to each other so as to produce a harmoniously functioning whole. ... This involves a new conception of chromosomes. Up till quite recently, it was possible and usual to regard them as mere vehicles for the carriage and distribution of the hereditary constitution, without any functional organisation of the genes they contained. These were assumed to be arranged at random, like colored beads picked up haphazard by a blind man and threaded on a string; and their positions in the chromosomes were not supposed to have any relation with their effects on visible or other characters. ... But this randomness must then be given a functional polish: neighboring genes must be adjusted to each other by new mutation and by recombination. ... We must today consider chromosomes not as being purely mechanical gene-vehicles, but to a certain degree as organic gene-arrangements
>
> (cited from the 3rd edition, Huxley, 1974).

A similar meaning is implied in currently used concepts of 'gene ecology' or 'gene territories', within which optimization of functions of genes and gene clusters is ensured (Naora, 1987). The concept of 'chromosome fields' (Lima-de-Faria, 1983) includes the idea of the optimizing role of not only chromosomal locations of genes relative to the centromere and telomeres, but also of specific patterns of distribution of crossovers, aberrational chromosome breaks, chromatin changes induced by exposure to stress (low temperature), the formation of chromosome gradients at pachytene and of large heterochromatic knobs. Nonetheless, the problem of the proportions of the random and non-random in the genome organization of higher eucaryotes can by no means be regarded as solved. Apart from the question of the relationship between functions (effects) of genes and their positions, it includes mechanisms generating heritable variation (of mutational, recombinational or insertional nature) and a possible dependence of this on environmental conditions and on the degree of fitness attained by the genotype (section 9.1).

One point pertaining to the history of interaction of theoretical and experimental population genetics is worth noting. Mathematical population genetics (founded by Fisher, Wright and Haldane) is known to have played a critical role in the development of the synthetic theory of evolution. Mathematical geneticists were among the first to resolve the seeming contradictions between genetics and selection impeding the progress of the evolutionary theory. The result was the creation in the 1930s of a fairly well-balanced theoretical scheme of microevolution based on the notions of individual gene frequencies and individual gene effects. Underlying the scheme was the concept of the gene as a unit of selection. At about the same time, experimental genetics began to accumulate evidence for the genetic material integrity, which was summarized by evolutionists of the informal trend (Dobzhansky, 1937; Darlington, 1939; Huxley, 1942; Schmalhausen, 1949; Stebbins 1950; Mayr, 1963; Grant, 1964). Along with genes, entities of higher levels of organization – gene blocks (supergenes), chromosomes and even the entire genome – were regarded as units of selection. But these concepts hardly affected mathematical population genetics of the time. It was not until the 1960s that the problem of the evolution of genome organization became the subject of theoretical studies, which implied the need for developing a multilocus analog of the classical theory.

5.2 INTERACTION BETWEEN SELECTION AND RECOMBINATION

5.2.1 Survey of models

The very first studies along the new lines of enquiry demonstrated the problem to be an extremely difficult one even in the two-locus case. Over the past few decades, two-locus systems have been the subject of numerous studies aimed at elucidating the conditions of polymorphism maintenance under various types of selection (on viability, fertility, sex ratio, on the basis of sexual preferences), in constant and varying environments, with random mating and other breeding systems, under normal or distorted segregation of genes, allowing for mutations, migration and genetic drift (see, for example, Bodmer and Felsenstein, 1967; Ewens, 1968; Karlin, 1975; Karlin and Feldman, 1978; Akin, 1990; Gillespie, 1981a; Feldman and Eshel, 1982; Hastings, 1985; Nagylaki, 1989; Otto, 1991). The dynamics and fixed points of free (with no selection) multilocus infinite populations with arbitrary sexual structure, any number of alleles per locus and ploidy level, under arbitrary regimes of mutation, recombination and migration have been thoroughly studied by Geiringer (1944), Ellison (1966), Lyubich (1971, 1992), Kirzhner and Lyubich (1974). Starting in the 1970s, attempts have also been made to develop multilocus selection models, particularly in addressing one of the central problems of population genetics, viz. that of explaining the genetic variability observed in nature as well as inferring the patterns of phenotypic variation and natural selection from complex correlated systems of quantitative traits (Lande, 1976; Karlin and Avivi, 1981; Turelli, 1984; Barton, 1986; Gillespie and Turelli, 1989; Wagner, 1989; Hastings and Hom, 1990; Turelli and Barton, 1990; Nagylaki, 1992; Zhivotovsky and Feldman, 1992; and others).

In computer simulations of Lewontin and other authors, it has been established that the behavior of multilocus heterotic systems differs significantly from that of two-locus systems. In particular, interaction between linkage and selection can result in a greater store of potential variability and in a smaller one of free (available for selection to act upon) variability: random combination of alleles at various loci is severely restricted, the overwhelming majority of possible haplotypes are present at low frequencies, with 2–3 types prevailing (Lewontin, 1964, 1974; Sved, 1968; Franklin and Lewontin, 1970).

Consider a simple case of two diallelic loci (A/a and B/b). Let the population be a random-mating one. In the absence of correlation between the allelic states, the gamete frequencies $g(ab) = g_{00}$, $g(Ab) = g_{10}$, $g(aB) = g_{01}$ and $g(AB) = g_{11}$ satisfy the relation $D = g_{00}g_{11} - g_{01}g_{10} = 0$. Besides, $g_{ij} = P_i^A P_j^B$, $i, j = 0, 1$, where P_i^A and P_j^B are the frequencies of alleles i and j at loci A/a and B/b. Owing to selection, sampling effect, or other reasons, the coefficient of linkage disequilibrium D may differ from zero. Given $D = D_n \neq 0$ in a certain generation, n, then, in the absence of selection $D_{n+1} = D_n(1 - rf)$, or $D_n = D_0(1 - rf)^n$ (e.g. Lewontin, 1974). Clearly, the quantity D_n decreases fairly rapidly converging to the value $D_\infty = 0$.

Provided neither of the loci (neither A/a nor B/b) is neutral, then, under certain selection regimes, D_∞ can differ from zero, i.e. a population reaches a stationary state with a stable linkage disequilibrium. Such a state is characterized by a departure from random combining of alleles at various loci:

$$g_{00} = p^A{}_0 p^B{}_0 + D, \quad g_{01} = p^A{}_0 p^B{}_1 - D, \quad g_{10} = p^A{}_1 p^B{}_0 - D, \quad g_{11} = p^A{}_1 p^B{}_1 + D,$$

i.e. some gametes occur more frequently than they do under random combining, others less frequently. Of course, a stable disequilibrium may also arise in the absence of selection, due to quasilinkage resulting from genetic affinity (see Mike, 1977; Korol, Preygel and Nyutin, 1989; section 2.6), but this cause seems to be a fairly rare one. The two-locus case has most carefully been studied for three types of fitness matrix: additive (**A**), multiplicative (**M**) and symmetrical (**S**) (see Ewens, 1979):

$$\mathbf{A} = \begin{array}{c} \\ BB \\ Bb \\ bb \end{array} \begin{array}{c} \begin{array}{ccc} AA & Aa & aa \end{array} \\ \begin{bmatrix} \alpha_1 + \beta_1 & \alpha_1 + \beta_2 & \alpha_1 + \beta_3 \\ \alpha_2 + \beta_1 & \alpha_2 + \beta_2 & \alpha_2 + \beta_3 \\ \alpha_3 + \beta_1 & \alpha_3 + \beta_2 & \alpha_3 + \beta_3 \end{bmatrix} \end{array}; \quad \mathrm{M} = \begin{array}{c} \begin{array}{ccc} AA & Aa & aa \end{array} \\ \begin{bmatrix} \alpha_1\beta_1 & \alpha_1\beta_2 & \alpha_1\beta_3 \\ \alpha_2\beta_1 & \alpha_2\beta_2 & \alpha_2\beta_3 \\ \alpha_3\beta_1 & \alpha_3\beta_2 & \alpha_3\beta_3 \end{bmatrix} \end{array};$$

$$\mathbf{S} = \begin{array}{c} \begin{array}{ccc} AA & Aa & aa \end{array} \\ \begin{bmatrix} \alpha & \beta & \delta \\ \gamma & 1 & \gamma \\ \delta & \beta & \alpha \end{bmatrix} \end{array}.$$

The equilibrium value of D is known not to differ from zero under additive effects of loci A/a and B/b on fitness. For the multiplicative case, stable disequilibrium is attained under a certain relationship between selection intensity and recombination rate between the loci; the stronger the epistatic component (e) of selection and the tighter the linkage, the higher is the probability of disequilibrium and the more integrated the system. To characterize the degree of integration, Lewontin (1974) introduced the 'coefficient of integration', $c = e/rf$.

For some special types of fitness matrix specifying the multiplicative effects of loci, a maximum value of $rf = rf^*$ has been established at which selection still leads to permanent linkage disequilibrium in a two-locus system (Lewontin, 1964, 1974; Bodmer and Felsenstein, 1967). If $rf > rf^*$, then $D_\infty = 0$. In the general case, the situation is, however, an extremely complicated one. For example, the range of rf values for which $D \neq 0$ is not necessarily continuous, i.e. the range $0 \leqslant rf \leqslant 0.5$ can include more than one subinterval corresponding to $D_\infty \neq 0$ (Ewens, 1968). More than one stable stationary state differing in values of D_∞ can exist in one and the same population. Thus, even a two-locus population can, depending on the initial state, settle at values of $D = 0$, $D > 0$ and $D < 0$ (Franklin and Feldman, 1977; Karlin and Feldman, 1978; Lewontin and Feldman, 1988). Provided rf is sufficiently small, the surface $D = 0$ separates two zones of attraction with $D > 0$ and $D < 0$, and even starting at $D_0 = 0$ the system can move to one or the other set of $D_\infty \neq 0$ because of random fluctuations.

In a two-locus diploid model with constant symmetrical selection, up to seven polymorphic equilibria are possible, of which as many as four can be simultaneously stable (Karlin and Feldman, 1970; Hastings, 1985). If $\alpha = \delta$ then there exist two equilibrium points with $D_\infty \neq 0$ (see Lewontin, 1974):

$$D_{1,2} = \pm\sqrt{[1 - 4\,rf/(1 + \alpha - \beta - \gamma)]}/4,$$

provided $1 + \alpha - \beta - \gamma > 0$ and $rf < rf^* = (1 + \alpha - \beta - \gamma)/4$. In the case of a symmetrical matrix with $\delta = \alpha$ and $\beta = \gamma$, situations are possible with a simultaneous stability of the central polymorphic equilibrium and four monomorphic ones, provided that fitnesses of monoheterozygotes are lower than those of double homozygotes ($\alpha > \beta$) and recombination is sufficiently high ($1 - \alpha < r < 0.5$) (see Feldman and Liberman, 1979). The case of $\alpha \neq \delta$ has not been completely studied analytically.

Some two-locus models exhibit a still more complex dependence of dynamics on recombination. In particular, for certain values of rf, the possibility has been established of oscillatory regimes of the limiting cycle type under viability selection (Akin, 1979, 1990; Hastings, 1981). A similar pattern of behavior was found in the model of interaction between the determinant of sex in the mosquito *Aedes aegyptii* and of its segregation modifier (Lessard, 1987; Feldman and Otto, 1989). With changes in recombination, the system converges to a polymorphic equilibrium.

The principal distinctive feature of the results from studies of multilocus systems is a higher degree of integration compared with the two-locus cases. For five-locus models of multiplicative heterotic selection, Lewontin (1964) found an increase in the degree of integration of the system and a corresponding increase in the critical value of rf^* ensuring permanent linkage disequilibrium; the correlation between alleles at different loci is maximum in the center of the system (an insertion effect). It should be noted that the linkage disequilibrium between central genes disrupts random association of alleles at outside loci separated by a distance of $rf > rf^*$.

Further studies along these lines were carried out by Franklin and Lewontin (1970) with two-, five-, 18- and 360-locus models. The results are essentially as follows:

- A number of equilibria with different correlations between loci can occur within a system.
- Even when the two-locus theory points to the lack of linkage disequilibria effects (for example, under a weak selection), segregation **at many loci** results in a strong correlation between alleles (a cumulative effect).
- Allele frequencies at individual loci do not reflect their effects on fitness and are the product of **selection of the chromosome as a whole**. When recombination between the adjacent loci is lower than the critical value (which may be many times as high as the rf^* for a two-locus system), the entire chromosome behaves as if 'locked up', with only one pair of complementary gametes present at equilibrium.
- With increasing number of loci, the role of higher order interactions between these becomes greater.
- For a given degree of inbreeding depression per chromosome map unit, steady-state correlation between loci shows no dependence on their number but is determined by the map length. This implies that it is not the effects of individual genes (which are rather difficult to assess) that are actually important in population genetics, but the net result of interaction of many loci within whole chromosomes or their segments and recombination properties of chromosomes (Lewontin, 1974). This fact substantially increases the importance of techniques for assessing the effects of the whole chromosome and chromosome segment substitution including those based on marker analysis (Chapter 3).

Based on the linear programming technique, Hastings (1981, 1986) attempted to answer some questions associated with applying theory to experimental data interpretation:

- Provided rf is known and there is some disequilibrium, $D \neq 0$, then, what is the minimum intensity of selection, s, which results in the observed value of D?
- Given D and s, what is the maximum allowable value of rf?
- What is the maximum possible D for given s and rf values?

The authors have shown that the answers to these questions depend on population gene frequencies, but, in any case, the inequality $rf|D| < 0.1s$ holds. For example, when $p_a = 0.9$, $p_b = 0.5$, $s = 0.01$ and $D = D_{max}$, no fewer than 500 zygotes need to be tested so that the null hypothesis H_0:$\{D = 0\}$ can be rejected with a 90% confidence at $rf = 0.01$, and over 2300 zygotes at $rf = 0.025$. Thus, at permissible values of rf and s, one should, more than likely, expect to have small values of D in populations at equilibrium and, therefore, the data sets used should be large. However, in most publications on linkage disequilibrium, estimations are usually based on samples of 200 (rarely 300) individuals.

In general, multiplicative selection results in a comparatively low (of the order of s^2/rf) degree of linkage disequilibrium between the neighboring loci. A stronger correlation results from selection in hybrid zones, for example when two demes come in contact with each other, or in the case of continuous clinal variation (Barton, 1983). Here, as in the model of Franklin and Lewontin (1970), the selected loci exhibit a cooperative effect when the s/rf values are sufficiently large.

As follows from the model of Franklin and Lewontin, an important prerequisite for the formation of a small number of complementary gametes ensuring an increase in the equilibrium fitness (relative to the level expected under no correlation in the system) is generating an initial gametic disequilibrium in the population. This may, for example, arise by random drift. The initial disequilibrium is a sort of 'seed', from which a wave of increasingly stronger association spreads – the genome 'crystallization' occurs. Of course, this is not to say that the process involves changes in the chromatin properties. What is implied here is only a repatterning of genes in which the key role is played by the interaction of selection and linkage. The validity of these simulation results was verified analytically for a three-locus multiplicative model (Strobeck, 1976).

Thus, the analysis of multilocus models suggests that even not very strong interactions between the adjacent components can result in the integration of systems. To put it differently, co-adapted gene complexes within a population in a stationary state are expected to exhibit gametic disequilibrium. Therefore, one way of detecting co-adapted gene systems in natural or laboratory populations is to look for the cases with $D \neq 0$. It would seem that, apart from technical problems associated with the need to evaluate a large body of material, there should be no great difficulties in testing these predictions. Unfortunately, this is not really the case. As pointed out by Hedrick *et al.* (1978), ascertaining the cause of linkage disequilibrium is a difficult task even given the information on spatial and temporal variation of gamete frequencies. There are several reasons for the difficulties encountered in interpreting actual data. These are genetic drift, restricted panmixia, random mutations, population subdivision, etc. (e.g. Hedrick, 1982; Robinson, *et al.*, 1991; Takahata, 1993).

The problem of stationarity

Provided $D \neq 0$ in an experiment, can one interpret this as resulting from co-adaptation of the locus pair in question? Indeed, under sufficiently intense epistatic selection and tight linkage, the population can reach a steady-state value of $D = 0$. But is there always a means to ascertain whether a population is in equilibrium? On the other hand, genetic drift may result in a random deviation from $D = 0$ which will be maintained for a long period of time under close linkage even when both loci are completely neutral. Such a disequilibrium is referred to as transient. Nor can the problem be solved by comparing the states in which the populations find themselves for a few generations. The reason for this is that selection on a single locus can, for a number of generations, produce a disequilibrium between this locus and a neutral one linked to it, or between two neutral loci (Thomson, 1977; Kimura, 1983). Of course, a disequilibrium resulting from a hitch-hiking due to linkage is, sooner or later, reduced to zero by recombination. Given, however, strong selection and small values of recombination and of the selected allele frequency, D may even increase in value. At the same time, the value of D may, for a range of frequencies of the selected gene, decrease much more rapidly than it does under neutrality.

Under strong selection on a single locus, the results of selection on many weakly interacting loci are masked. In particular, the stationary state of a population with respect to some loci may be disturbed by its changes with respect to others. If the disturbance has occurred shortly before the data are obtained, this can result in a misinterpretation of the observed gametic correlations.

In general, larger values of D_∞ are only possible under strong epistasis and close linkage between loci (Hastings, 1986). One way to enhance the power of analysis is to consider a higher number (than 2) of alleles per locus. By lowering the sampling error of D, this enables selective effects to be distinguished from random mutations and drift (Thomson and Klitz, 1987). Useful information for elucidating the nature of two-locus disequilibria can be obtained by studying the constraints imposed on them by three-locus disequilibria (Robinson, Asmussen and Thomson, 1991). Incidentally, four-locus interactions already place no further constraints on two-locus values. Based on these approaches, gametic correlations within the *HLA* system in humans have been shown to result from selection (Thomson and Klitz, 1987; Klitz and Thomson, 1987; Robinson *et al.*, 1991). A detailed analysis of the distribution of different classes of *HLA* alleles in human populations may help in a better understanding of the pattern of interaction of natural selection and recombination during evolution (Hughes, Hughes and Watkins, 1993).

The effect of hitch-hiking can influence the rate of evolution by retarding the spread of beneficial mutations and speeding up the accumulation of deleterious ones. Consequently, by reducing linkage disequilibrium, recombination can accelerate adaptive evolution and slow down the rate of molecular evolution, since deleterious mutations occur at higher frequencies than beneficial ones (Birky and Walsh, 1988). Begun and Aquardo (1992) have compared the density of recombination events in 20 regions of *D. melanogaster* genome with the polymorphism at DNA level and found these variables to be positively correlated. Their assumption is that the effect is due to hitch-hiking caused by fixation of advantageous mutants.

Hitch-hiking due to disturbed panmixia

The analysis is further complicated by restrictions on random mating, e.g. in the form of selfing, a system characteristic of many higher plants and some animal species. In the absence of selection, the relationship between values of D in two successive generations is given by the relation (Karlin, 1969; Weir and Cockerham, 1973) in which rf and S appear symmetrically:

$$D_{n+1} = 0.5((1 + L + S)/2 + \{[(1 + L + S)/2]^2 + 2\,LS\}^{1/2})D_n,$$

where $L = 1 - 2\,rf$ and S is the rate of selfing. Consequently, in the system of neutral loci, hitch-hiking due to non-random mating is as effective as that due to linkage. Partial selfing reduces the effective population size, thereby increasing the contribution of genetic drift to the observed departures of D from zero. In particular, attributable to drift are gametic correlations, provided that $(1 - S)\,rf < 10^{-4}$ (Golding and Strobeck, 1980). Under selection for linkage disequilibrium, the effects of the breeding system (rate of selfing) and recombination are rather complex even in the two-locus case (Holden, 1979). In particular, a polymorphism with $D \neq 0$ may be observed either for small S or, conversely, for large or intermediate ones, depending on the values of rf and selection intensity. With partial selfing, the maintenance of a polymorphism for two multiplicative loci is possible under conditions which are insufficient for maintaining a polymorphism in each of them taken separately (Eshel, 1978).

Gametic disequilibrium between loci resulting from hitch-hiking due to partial selfing may also be observed at $rf = 0.5$. In this case, correlation is enhanced provided the effect of selection is partitioned among several selected loci (Hedrick and Holden, 1979; Hedrick, 1980). Restricted panmixia, as a cause of hitch-hiking, exerts stronger influence on gametic disequilibrium than linkage does, since it affects all the genomic regions simultaneously. Finally, it should be noted that disequilibrium may arise as a result of the mixing of the gene pools of two previously isolated populations differing in allele frequencies (Feldman and Christiansen, 1975; Christiansen, 1990). In such a case, the analysis of temporal variation in D may be very helpful in gaining insights into linkage between the genes under study (Chakraborty and Weiss, 1988; Lande and Thompson, 1990).

In conclusion of this section, it seems appropriate to mention some additional points related to the interaction between recombination and selection on multilocus systems. This issue has a direct bearing on our central theme – the *rec* system evolution – since without taking into account the details of the effect of recombination on the selected variability it is impossible to understand the reciprocal process induced by this selection, namely, the evolutionary dynamics of the *rec* system itself. It is clear that the effect of recombination depends on the sign of linkage disequilibrium between the selected loci (Felsenstein, 1988). But the situation is complicated by the fact that both selection and recombination cause a change in D. Furthermore, the selected loci in different genomic regions may generate linkage disequilibria which are opposite in sign.

If selection favors allele combinations with $D < 0$, then an increase in rf accelerates evolution until alternative combinations come to constitute a majority in the population (i.e. as long as selection is a directional one). High frequency of exchanges is also of great importance for rapid restoration of variability after the population has experienced a 'bottle-neck' effect (Dobzhansky, 1959; Ayala, 1981; Bryant, McCommas and Combs, 1986, Bryant and Meffert, 1988; Carson, 1990). However, an excessively high level of rf can lead to losses in variation, to superfine 'adjustment' of a population to a particular environment (G.C. Williams, 1975; Rees and Jones, 1977). The same will probably be the effect of too tight linkage. One would expect, therefore, long-term advantages to be generated at certain intermediate values of recombination and various patterns of exchange localization. Therefore, corresponding to different phases of population adaptation are different optima of the R-system. A trade-off can be achieved based on a more flexible response of systems controlling variability to current conditions (genetic and environmental).

5.2.2 Survey of experimental evidence

Protein electrophoresis studies of natural populations have revealed (and recent estimations at the DNA level have confirmed) the existence of high genetic polymorphism (Lewontin, 1974; Nevo, 1978, 1988a, b; Brown, 1979; Ayala, 1983; Nevo, Beiles and Ben-Shlomo, 1984; Nevo *et al.*, 1990, 1993; Altukhov, 1989; Gilbert *et al.*, 1990; Aquadro, 1992; Erlich and Arnheim, 1992; Jiang and Gibson, 1992). Two classes of hypotheses have been proposed for explaining the reasons for the observed variation: those based on selection (the balance concept) and the neutralist ones. Naturally, an adequate explanation should be based on models whose parameters are estimated from experimental data and can serve as criteria for testing the hypotheses (Lewontin, 1974). It turned out, however, that attempts to resolve the neturalist–selectionist controversy encountered considerable difficulties. In particular, extensive data accumulated by experimental population geneticists can be interpreted from both balance and neutralist perspectives (Lewontin, 1974; Kimura, 1983, 1991; Gillespie, 1986; Nevo, 1988a; Altukhov, 1989).

Strictly speaking, the problem of whether the genome is integrated and what the characteristic features of this integration are is both narrower and broader than the balance–neutralist alternative. Its solution requires techniques for elucidating the patterns of distribution of selective effects along the chromosome length and throughout the genome (Clegg *et al.*, 1976), which are just beginning to be developed. It is important to take account of interactions between different genomic components, especially linked ones. This is an even more difficult task. Estimating correlations between allelic states at linked loci (gametic disequilibrium) is but one of the indirect approaches to the problem.

Outbreeding species

An important consequence of balance models of genome integration of the kind proposed by Lewontin (Lewontin, 1964; Sved, 1968; Franklin and Lewontin, 1970; Slatkin, 1972) is the prediction of a high gametic disequilibrium in polymorphic systems subjected to epistatic selection. In general, judging by the evidence in the literature, significant disequilibrium is uncommon in random-mating populations and is fairly common in predominantly selfing species (Hedrick, Jain and Holden, 1978; Brown, 1979; Brown, Feldman and Nevo, 1980; Allard, 1988). The bulk of the evidence comes from *Drosophila* studies involving estimation of linkage disequilibrium between enzyme markers as well as between enzyme loci and inversions and between inversions themselves. An early work on the subject (Charlesworth and Charlesworth, 1973) reports significant correlations among five linked loci on chromosome 3 of *D. melanogaster*. Comparisons were made on 10 possible pairs of loci in samples from a natural population taken in two consecutive years as well as in a sample from a large laboratory population. In four cases out of 30, a significant deviation of *D* from zero has been established, the results showing correlation between years. Further estimations have, in general, shown that the proportion of cases of significant gametic disequilibrium between enzyme loci is small in natural populations of various *Drosophila* species, the value of *D* generally being also small (reviewed in Zhuchenko and Korol, 1985).

In an extensive experiment with samples from two *D. melanogaster* populations (Japan and the USA), only three cases of a small but significant *D* have been found among 36 pairs of linked markers (loci on chromosomes 2 and 3 have been examined, the number of gametes analyzed in each population being over 200) (Langley, Tobari and Kojima, 1974). In view of the large number of comparisons made (36), only one case out of the three should be considered as significant. Barker, East and Weir (1986) have found significant disequilibria between markers *Est-2*, *Aldox* and *Hex* on chromosome 2 of *D. buzzatii*.

Estimations by other authors based on enzyme systems show that linkage disequilibria within natural populations of *D. melanogaster* are weak or absent (Mukai, Mettler and Chigusa, 1971; Mukai, Watanabe and Yamaguchi, 1974; Langley, Ito and Voelker, 1977). Actually, the proportion of cases in which *D* is significantly different from zero is approximately the same for linked and unlinked loci (Langley, Smith and Johnson, 1978). An additional example of a possible association between alleles at unlinked loci is the data on gametic disequilibrium between the genes *Hk* (chromosome 2) and *Pgm* (chromosome 3) (Mont, Moya and Marinkovic, 1988).

A different picture is observed in laboratory populations of *D. melanogaster*. In their experiments with cage populations, Langley, Ito and Voelker (1977) observed a significant deviation of D from zero in pairs of linked loci nearly four times as frequently as in unlinked ones (37.5% and 10.3% respectively). A similar result (34.5% and 8.9%) has also been obtained in a work in which several laboratory populations of *D. melanogaster* were studied with respect to 17 enzyme loci (Laurie-Ahlberg and Weir, 1979). In cage populations of *D. melanogaster* with a significant $D_0 = 0$, Clegg, Kidwell and Horch (1980) observed return to the equilibrium $D = 0$ was twice as rapid as would be expected assuming no interaction of genes. Another study, with closely linked enzyme loci on chromosome 3, has revealed significant associations for some pairs of loci in samples from natural populations, a decrease in D upon switching to laboratory cultivation of these samples as well as the formation, under laboratory conditions, of new linkage disequilibria for some pairs of loci which were independent in initial populations (Smit-McBride, Moya and Ayala, 1988). The authors also observed non-random three-locus associations of alleles.

Thus, from the data on the behavior of protein markers in natural populations of *Drosophila*, it follows that non-random gene associations, while possible, are rather uncommon even under tight linkage; the amount of recombination and the degree of disequilibrium are not always correlated. The main reason for this may be low incidence of strong epistasis between closely linked loci in *D. melanogaster*. But there are at least two other factors which may be responsible for the correlation between linked loci becoming weaker and which have not been discussed previously in connection with the problem. These are the dependence of crossing-over rate between marker loci on ecological conditions and on the genetic background.

Indeed, the laboratory estimates of *rf* often differ significantly from the values characteristic of varying environments. Particularly sharp differences are observed for closely linked loci. As a result, the effective frequency of crossing-over in the natural population under study may prove to be substantially higher than the estimates obtained under laboratory conditions. Moreover, the laboratory estimates are usually based on analysis of the F_1 progeny from a cross between particular strains and, by virtue of *rf* being dependent on genotype, characterize the hybrids of precisely these strains. Since genes enhancing recombination in many cases are likely to be recessive, we shall obtain, in a laboratory experiment, a biased *rf* estimate compared with that in a natural population. All this is likely to be responsible for a serious underestimation of the amount of recombination and, hence, the overestimation of the expected D value. It also follows that errors are possible in estimating the correlation between *rf* and D. The above findings about the existence of critical values of *rf* do indicate that even a small increase in crossing-over rate may be sufficient for the population to move to a trajectory with an equilibrium value of $D = 0$.

According to the neutralist concept, the occasionally observed significant correlations between markers are due to random drift and should disappear with time (transient disequilibria). A convincing argument against the neutralist hypothesis would be the correspondence between *D* values in different populations of the same species, and such evidence has actually been obtained. Thus, nearly identical linkage disequilibria between enzyme markers have been found in two *D. montana* populations located 500 km apart (Baker and Kaeding, 1981). In a study of 15 electrophoretic loci of a mole cricket *Gryllotalpa gryllotalpa* in 11 localities in Israel, E. Nevo (unpublished) has revealed a high polymorphism accompanied by a marked deficit of heterozygotes and significant linkage disequilibria. For many pairs of loci, the *D* values were of the same sign across the localities.

Systems with stable disequilibria have also been found in other random mating organisms: four α-esterase loci in *D. virilis* and some isozyme loci in *D. obscura*; the *MN–Ss* pair and loci of the ABO system in humans; genes for shape and coloration of the shell in the snail *Cepaea nemoralis*; enzyme loci in a mussel *Mytilus edulis*, sea fish *Fundulus heteroslitus*, salamander *Plethodon cinereus* and ryegrass *Lolium multiflorum*; esterase loci in pine (*Pinus sylvestris*); genes for morphological traits in cats, etc. (reviewed by Hedrick, Jain and Holden, 1978; Laurie-Ahlberg and Weir, 1979; Zhivotovsky, 1984; Hedrick, 1985; Epperson and Allard, 1987).

The components of the mammalian major histocompatibility complex (MHC), including human *HLA*, are characterized by a large number of alleles, high polymorphism and significant disequilibrium with each of the other components and with autoimmune diseases and are likely to play a key role in defenses against some infections (Bodmer, 1972; Robertson, 1982; Tiwari and Terasaki, 1985; Hedrick, Thomson and Klitz, 1986; Klitz and Thomson, 1987; Hedrick and Thomson, 1988; Thomson, 1988; Hill *et al.*, 1991; Robinson *et al.*, 1991; Nevo and Beiles, 1992). As mentioned above, this genomic region is, at the same time, a recombination and mutation hotspot in mammals (see Kobori *et al.*, 1986; Meagher, McLean and Arnold, 1988; Johnston and Rich, 1989; Yeom, Kuniya and Artzt, 1992). According to Bremermann (1987), high heterozygosity at loci of this region may result from fertilization nonrandomness caused by sperm selection.

A number of cases of significant linkage disequilibrium between loci of hereditary diseases and RFLPs have been reported in human populations, which holds great promise for genetic counseling (Chakravarty, 1983; Estivill *et al.*, 1988; Strain *et al.*, 1988; Limborskaya, 1989). This approach could also be important in marker-assisted selection in animal breeding (Lande and Thompson, 1990; Zhang and Smith, 1992). Finally, with the use of RFLP markers, data on linkage disequilibrium between various sites **within** structural genes (e.g. *Adh*, *Est-6*, *Amy* and other *Drosophila* loci) began to be accumulated (see Laurie, Bridgham and Choudhary, 1991).

The above-mentioned example of an esterase system of pine is of particular interest: in spite of high recombination among the four markers studied, two complementary haplotypes, $A_1B_2C_1D_1$ and $A_2B_1C_2D_2$, account for nearly half of the number of gametes in the population, thus ensuring high heterozygosity (Zhivotovsky, 1984). Here, linkage disequilibrium tends to maximize fitness at later stages of ontogenesis owing to strong selective elimination at early stages.

It is likely that epistatic interactions (co-adaptation) between linked loci in other species are not as strong as that and generate weaker, if actually existing, gametic correlations. *Drosophila* seems to be just such a type of organism, and, in this respect, may be an inconvenient subject for this kind of population genetic work. This assumption is probably supported by the fact that in laboratory experiments with *Drosophila* the proportion of cases with significant D is fairly large.

If that is the case, the detection of disequilibria will require, as already noted, large volumes of material (according to Hastings' condition $|D| < 0.1s/rf$) or measuring the cumulative action of many weak disequilibria (Barton, 1983). This may call for other techniques of studying gametic integration, e.g. those based on QTs, including multivariate analysis of complex systems of traits (Bulmer, 1985; Zhivotovsky, 1984; Zhivotovsky and Gavrilets, 1990).

An important argument in favor of the balance hypothesis is the abundance of strong gametic correlations between isozyme loci and inversions. The establishment of these correlations was considered to be one of the greatest contributions of protein electrophoretic studies to population genetics (Hedrick, Jain and Holden, 1978). The cases of high degree of linkage disequilibrium between protein markers and inversions found in various species of *Drosophila* and other organisms are usually interpreted as realizations of the balance school predictions (Lewontin, 1974). Indeed, even if the condition for a stable linkage disequilibrium between two markers is not satisfied, the probability is high that it will be satisfied for a marker and at least one of the genes within the inversion, or for the inverted block as a whole. The probability is enhanced when the marker is located inside the inversion; if that is the case, the effective recombination frequency, rf_e, is close to zero. However, a neutralist explanation has been proposed for data like these as well, e.g. by an assumption that the gametic correlation is not due to selective effects (co-adaptation) but to a disequilibrium between neutral elements (a protein marker and a system of genes involved in the rearrangement) which arises at the moment of the inversion formation (Nei, 1975).

It is true that such an explanation is, in principle, rather difficult to disprove for a particular population at a given moment. However, a number of observations are at variance with the neutralist hypothesis. Langley, Tobari and Kojima (1974) established a significant correlation between markers and inversions in 11 cases out of 36 in *Drosophila*. What is most important in these findings is the geographic and temporal stability of gametic disequilibrium and the correlation with ecological conditions which are inexplicable in terms of the neutralist theory but are in good agreement with the co-adaptation hypothesis. Evidence for stable associations between enzyme markers and inversions and between inversions themselves has been obtained in other works as well (Prakash and Lewontin, 1968; Knibb and Baker, 1988; Kumar and Gupta, 1988). In view of the abundant evidence for the selective nature of inversion polymorphisms, the assumption of co-adaptation being the cause of disequilibrium in such systems seems most plausible (see also Day *et al.*, 1982; Fontdevila *et al.*, 1983; Peixoto and Klaczko, 1991; Levitan and Scheffer, 1993).

Studying the role of linkage disequilibrium as a manifestation of the effects of selective forces maintaining the integrity of co-adapted gene complexes is of particular interest in connection with the problem of 'hybrid zones' (Barton, 1983; Barton and Hewitt, 1985, 1989; Barton and Clark, 1990). The formation of a hybrid zone proceeds over hundreds and thousands of generations due to interactions among selection, recombination and migration. Here, gametic correlations act as barriers to gene introgression, and it is logical to assume that certain gradients of *rf* can be observed within a hybrid zone. Unfortunately, studies of this type have not yet been carried out with real populations, although mathematical models for the evolution of the *rec* system in subdivided populations have been proposed.

Indirect experimental evidence for the importance of recombination processes going on within hybrid zones is provided by the data on variation of ribosomal DNA segments in the grasshopper *Caledia captiva* (Arnold, Contreras and Shaw, 1988), which the authors attribute to gene conversion events. Attempts have also been made to analyze non-random cytonuclear associations of genes within hybrid zones and conditions for their stability (Arnold, Asmussen and Avise, 1988; Asmussen, Arnold and Avise, 1989). Scrinber and Avise (1993) found within a suspected hybrid zone of mosquitofish *Gambuzia* significant nuclear and cytonuclear disequilibria which can be interpreted as an indication of introgressive hybridization between *G. affinis* and *G. holbrooki*. Like nuclear–nuclear disequilibria, nuclear–cytoplasmic disequilibria may be due to crossing, random drift and the founder effect, gene flow and selection. Their analysis may provide valuable additional information about the population structure and evolution (Asmussen and Arnold, 1991; Scrinber and Avise, 1993).

Inbreeding species

The highest degree of association between the allelic states of linked loci has been established, as already noted, in species with high rates of selfing. As a result of studies of composite synthetic populations of barley (composite cross *CCII* and *CCV*) as well as of natural populations of wild oat and barley, Allard and collaborators have accumulated some evidence for selective value of the indicated associations. As pointed out by Allard (1975, 1988), predominant selfing promotes the maintenance not only of a high between-locus correlation but also of a considerable microgeographic variation at the within- and between-population levels. The former occurs due to a decrease in the proportion of heterozygotes and, hence, randomizing effect of recombination. An increased intragenomic and spatial genetic heterogeneity results in a finer 'adjustment' of the population to its environment, i.e. in a higher frequency of genotypes best adapted to a given environment. Environmental effects producing changes in the population genetic structure (population subdivision, fluctuation in numbers and the associated drift) have been termed 'indirect' effects as opposed to direct effects resulting from differential fitness (Pamilo, 1988). It has been assumed that indirect effects of environmental heterogeneity may also be responsible for the maintenance of variation at the molecular level.

According to the estimates of Allard *et al.* (1972), the distribution of natural populations of the wild oat *Avena barbata* with respect to five isozyme loci is largely non-random and closely associated with ecological conditions. Populations in xeric and mesic habitats are monomorphic with respect to opposite allelic combinations *12221* and *21112* (*1* and *2* stand for the alternative alleles), being polymorphic for all loci under intermediate conditions. Interestingly, under intermediate conditions, out of 32 possible gametic types as few as two haplotypes (*12221* and *21112*) account for the overwhelming majority of genotypes (57% and 11%) and, accordingly, most individuals in populations are either homozygous or heterozygous for all five loci. Similar data have also been obtained by Hamrick and Holden (1979). The authors view this as resulting from the formation of co-adapted gene systems due to natural selection.

Enzyme loci can either be 'components' of the selected system itself or act as neutral markers closely linked to the selected loci. Homozygotes for the above-mentioned combinations *12221* and *21112* corresponding to xeric and mesic ecotypes have been shown to differ with respect to a set of quantitative traits (morphological and phenological) as well as to the rate of outcrossing, the traits probably reflecting an adaptation to the environment. Moreover, the environmental gradient is correlated with the gradient of gamete frequencies and quantitative trait combinations (Allard and Kahler, 1974). This led the authors to conclude that quantitative differences are due to genes interacting with the other components of a multilocus system. It is significant that, in addition to the two alternative types *21112* and *12221*, the population also produces recombinant types (limited as they are in number) capable of colonizing new ecological niches.

In a series of studies with various organisms, both random mating and self-fertilizing, Nevo and co-workers have established highly significant correlations between polymorphisms for multilocus systems of isozyme or DNA markers, on the one hand, and ecological factors, on the other, which are also inexplicable in terms of the neutralist theory (Nevo, 1978, 1988a, 1991; Lavie and Nevo, 1988; Nevo and Beiles, 1988; Nevo *et al.*, 1990, 1993). In particular, population geographic microdifferentiation with respect to a multilocus system indicative of adaptive significance of the polymorphism has been demonstrated in self-pollinating wild barley, *Hordeum spontaneum* (Brown, Feldman and Nevo 1980; Nevo, 1992), and wild emmer wheat, *Triticum dicoccoides* (Golenberg and Nevo, 1987). For example, water, temperature and soil factors were very informative in classifying the distribution of *T. dicoccoides* multilocus patterns of amylase components across Israel (Nevo *et al.*, 1993). Based on multilocus marker systems (both enzyme and DNA) and ecological factors, an effective sampling strategy has been proposed for identifying rich genetic sources of plant adaptivity to stressful factors and of disease resistance, as well as for quality improvement (Nevo and Beiles, 1989; Nevo, 1987, 1992).

An interesting situation has been found in an autogamous annual legume, *Amphicarpaea bracteata*: an extremely high linkage disequilibrium for 9 out of the 11 enzyme loci studied (Parker, 1988a). More than 97% of the sample belong to two complementary classes differing also in resistance to pathogens. Non-random associations between genes for resistance to diseases and other ecologically important traits seem to be fairly common among selfers (Parker, 1991).

An experimental verification of the hypothesis of gene co-adaptation as a cause of gametic correlations has been done based on the data on the marker genotype frequency changes for three pairs of successive generations (8–9, 19–20, 28–29) in a population of the *CCV* barley (Clegg, Kahler and Allard, 1978). To determine various selection components, the authors assessed the genetic constitution of populations at two stages of life cycle (seedlings and adult plants). From one-locus comparisons, it follows that selection can produce drastic changes in allelic frequencies even in the course of a single generation, with the changes being generally opposite in sign in passing from zygotes to adult plants and then to zygotes of the next generation. The experimental design and analytical approach used in this study enabled estimates of fertility components to be obtained separately for female and male gametes. For some loci, the effects of a particular allele on fertility are nearly identical in both types of gamete; however, opposing effects of allele substitution with interactions between loci involved have also been observed. The authors performed the above estimations on the basis of three-locus models using data on eight homozygote classes for the three pairs of generations. Out of 24 viability estimates for the *EA–EB–EC* combination of linked esterase loci, 10 differed significantly from unity; viability ratios frequently reached sixfold and more, with as large as 24-fold differences having been registered for genotypic classes *221/221* and *122/122* in generation 19 (!). As with one-locus estimates, selective effects related to viability and fertility are opposite in sign. At least three- to fivefold differences in fertility, viability and total fitness have also been established for the three-locus combinations *ABD*, *ACD* and *BCD*, in which only two of the three genes are linked. That viability selection is apt to reverse its direction at different life cycle stages was also shown on *D. buzzatii* for genes of the inversion polymorphic system on chromosome 2 (Ruiz *et al.*, 1986).

An important result of co-adaptation studies is the conclusion that multi-locus selective effects cannot be predicted on the basis of single-locus effects alone. Interaction (co-adaptation) between system components is one of the critical factors determining the population behavior, with the interaction pattern (e.g. at the three-locus level) varying from generation to generation (see experiments of Clegg, Kahler and Allard, 1978). This seems to be because a three-locus system is but a component of the entire selected mutilocus complex, its characteristics serving as marginal effects for higher level systems. Changes in the latter lead, as it were, to changes in the 'context' (Lewontin, 1974) within which the effects of the lower dimensionality system under study are interpreted. In an evolving population, the 'context' is constantly changing, and changing with it is also the 'meaning' of each gene and gene combination.

When speaking of co-adaptation of genes and gene complexes, it is usually implied that certain combinations ensure a higher degree of adaptation to varying environmental conditions. But it is not only environmental conditions that vary. Mutations, gene flow from other populations, recombination, all these factors produce internal changes. Therefore, co-adaptation should also ensure homeostasis with respect to internal changes, a kind of stability in the face of continuous repatterning of the *F*-system. In this connection, the possibility cannot be ruled out that internal changes could be ranked according to the degree of their predictability. It is to frequent changes that co-adapted systems should exhibit the highest insensitivity, and including them, as components, in an appropriate subsystem of a higher hierarchical level seems to be the best way of solving the problem. It is probably in this that the explanation for the 'neutrality' of many mutations lies: they are neutral not *per se* but only because of the effects of certain compensatory complexes. Thus, the concepts of 'harmful', 'neutral' and 'beneficial' mutations appear to be very relative; their correct interpretation is only possible within the concept of co-adaptation, structural and functional integrity of the genetic material of higher organisms. In our view, this remark should be taken into consideration for proper classification and evaluation of various models for the *rec* system evolution (sections 6.1 and 6.4).

Both theoretical results and experimental evidence discussed earlier indicate that genomes of higher organisms exhibit a high degree of non-randomness that assures a considerable correlation in the combining of alleles within populations, thereby promoting the attainment of an appropriate trade-off between ontogenetic adaptation and genetic flexibility. Therefore, recombinational organization of the genome is one of the important factors determining fitness. But, can selection on adaptive traits be responsible for changes in the recombination system? If yes, how effective? Moreover, is recording these changes in the *rec* system experimentally a realistic task? In other words, it is necessary to ascertain whether selection on the *F*-system components can fairly soon (in a few dozens of generations) result in an appreciable repatterning of the *R*-system, including increased frequency of crossover exchanges, their altered interference and changes in allelic relations at the *rec* loci. We will try to provide answers to these questions in the following sections.

6 Evolution of the system controlling recombination

6.1 A REVIEW OF THEORETICAL EXPLANATIONS

6.1.1 The puzzle of recombination

The adaptively valuable gene complexes accumulated in a population are subject to constant erosion by mutation and recombination. Stabilizing selection is considered to be a factor maintaining fitness (Schmalhausen, 1949). Recombination can have a particularly destructive effect on fitness under strong epistatic interactions between polymorphic genes. Therefore, Fisher (1930) suggested that when a population has reached genetic equilibrium in a uniform environment, selection will favor closer linkage between loci since maximum population fitness is assured at the minimum possible recombination level. As pointed out by Darlington (1971), the necessity of protection against crossing-over is one of the two or three greatest paradoxes of genetics underlying the evolution of discontinuity both within and between species.

It is worthwhile adding, in connection with the last remark, that the problem of recombination evolution is closely associated with the more general, and also as yet unresolved, problem of accounting for the reasons of maintaining sex in general. Although we shall be touching upon specific aspects of this issue, its discussion is, in general, beyond the scope of this book. We refer the interested reader to the works of Williams (G. C., 1975), Maynard Smith (1978), Moore and Hines (1981), Bell (1982), Hoekstra (1982), Lewis (1983), Cohen and Zohary (1986), Manning and Dickson (1986), Stearns (1987a), Pamilo, Nei and Li (1987), Kelley, Antonovics and Schmitt (1988), Michod and Levin (1988), Nunney (1989), Uyenoyama and Bengtsson (1989), Vrijenhoek (1990), Melzer and Koeslag (1991), Szathmary and Kover (1991) and Maguire (1992).

In a series of theoretical works, Feldman and co-workers have advanced and discussed a general 'reduction principle' describing the evolution of selectively neutral modifiers of recombination, mutation and migration in a constant environment under panmixia (Liberman and Feldman, 1986; Altenberg and Feldman, 1987). The introduction into a population of a new modifier allele affecting one of the indicated processes is accompanied by an increase in its frequency if it reduces the parameter (recombination frequency, mutation or migration rate). The minimum of recombination ($rf = 0$) in a constant environment with panmixia corresponds to a local maximum of mean fitness (Lewontin, 1971). And if the evolution of a population (species) does proceed toward higher fitnesses, then why does the process not result in the attainment of a minimum (zero) recombination, i.e. 'Why doesn't the genome congeal' (Turner, 1967) and turn into a single, huge supergene? As we have already seen, chromosomes of higher organisms are highly heterogeneous with respect to recombination, and for certain regions this 'congealing' is an all but accomplished fact. In some species (e.g. in translocation heterozygotes of *Oenothera*) the genome can actually be regarded as being a single supergene (see Ford, 1971; Grant, 1975, 1981; Zhuchenko and Korol, 1985; Vřijenhoek, 1990). There is more than one answer to the question of why the process has never come to an end for many or most organisms. But before we proceed to analyse these, a remark must be made about a reduction in recombination being in principle possible in natural populations.

The results of experiments on artificial selection for altered *rf* suggest that almost every population has enough genetic variability stored to assure response to selection for changed *rf*, in at least one direction (section 6.3). Two conclusions emerge from this. Provided there do exist potential reasons for the 'genome congealing', then their realization should result in a detectable reduction in *rf* and in exhaustion of corresponding variability. Therefore, observed polymorphism at loci affecting recombination is either balanced (selected) or transient. There is experimental evidence for a possible negative correlation between individual fitness and recombination fraction (see, for example, Tucic, Ayala and Marinkovic, 1981; Zhuchenko and Korol, 1985; Zhuchenko *et al.*, 1986; Cvetkovic and Tucic, 1986). The existence of such a relationship can be viewed as a realization of negative feedback in regulation of genetic variability (Zhuchenko and Korol, 1982). But it implies also that the problem of identifying factors ensuring the maintenance of non-zero recombination within natural populations proves to be even more complicated than previously believed.

Let us for a moment accept the view that *rec* modifier polymorphism has no effect on population adaptive processes. Then, we shall have to refute the argumentation that led to the very question of 'Why doesn't the genome congeal?', account for extensive evidence on local 'genome congealing', sharp sex differences in *rf*, interrelationship between evolutionary changes in breeding system and in crossing-over rate and the known general tendency toward a reduction in recombination frequency per nucleotide by several orders of magnitude in going from lower to higher life forms (Catcheside, 1977). Even among eucaryotes, the density of recombination events within the genome varies widely: from 10kb/cM in yeast, 230, 280 and 570 kb/cM in *Arabidopsis, Brassica rapa* and *Drosophila*, to 1000, 4800 and 20,000 kb/cM in man, barley and Liliaceae (see for review: Meagher, McLean and Arnold, 1988; Nodari *et al.*, 1993).

Facts like these are sometimes attributed to the limited synthetic potential of cells: with increasing genome size, the amount of recombination enzymes per unit physical length of DNA must perforce decrease. If that is the case, then the recombinational length of the entire genome should be relatively constant over the evolutionary time scale (Thuriaux, 1977). However, there is increasing evidence that it is rather the average number of exchanges per chromosome that is relatively constant (Bell, 1982; Kondrashov, 1984; Cavalier-Smith, 1985; Rees and Durrant, 1986; Koella, 1993). Higher eukaryotes, with their chromosomal DNA content varying over three orders of magnitude, show, generally, no more than a two- to fourfold variation in per bivalent chiasma frequency, this index being in a fairly narrow range of 1–4 (one chiasma per bivalent is obligatory in the majority of species). This value may, however, be large in some cases see, for example, Ukai, 1988; Burt, Bell and Harvey, 1991). The total recombination length of the longest human chromosome is 600 and 300 cM for oogenesis and spermatogenesis respectively (O'Connell *et al.*, 1989).

All this suggests high 'regularity' of relatively rare crossing-over events, which is particularly striking considering the extremely low efficiency with which primary single-strand breaks (nicks) are transformed into crossovers (for example, 1 out of 10^4 in lily) (Hotta, Chandley and Stern, 1977; Stern, 1986). How well stochasticity and regularity like these agree with one another is an unsolved and probably as yet poorly recognized problem. In any case, it can be concluded that the per bivalent number and distribution of exchanges in higher eucaryotes are under the control of highly efficient and finely adjusted mechanisms, probably very ancient ones (Friedman, Bouchard and Stern, 1982; Stern, 1986; Zhuchenko and Korol, 1985; John, 1990; Pearlman, Tsao and Moens, 1992). Notice that the mean rate of mutations per genome per sexual generation also exhibits rather narrow variation across metazoa, as does the mutation rate per genome per DNA replication – across living organisms (Drake, 1991).

As pointed out by Miklos (1982), in analyzing recombination of eukaryotes, the discussion is sure to be more fruitful if one looks at the problem from the evolutionary rather than cellular (we would say, 'and not only cellular') standpoint. We have mentioned above that in a constant environment a reduction in recombination rate between tightly linked genes is accompanied, in the absence of mutation, by an increase in the mean fitness (W) of a population (Lewontin, 1971). Thus, it can easily be concluded that with genetic variation in *rf* selection will tend to suppress crossing-over. Such reasoning assumes, explicitly or implicitly, group selection (Maynard Smith, 1978, 1982; Lewis,

1987). Indeed, genetic differences in recombination ability among individuals may cause no differences in fitness. Nevertheless, by virtue of *W* being dependent on *rf*, populations with different amounts of crossing-over will have different chances of long-term survival, and groups (species) with an optimum *R*-system will eventually replace their less successfully organized competitors (Stanley, 1978; Stearns, 1987b; Nunney, 1989).

Note that the choice of mean fitness as a criterion for evaluating the evolutionary role of recombination is somewhat arbitrary since mean fitness is not the only and even probably not the best characteristic to assess the evolutionary advantage of one group of individuals or another (Cooper, 1984; Damuth and Lorraine, 1988; Barton and Clark, 1990). According to Cooper, for instance, the best formalization of what a biologist considers to be 'fitness' is not the mean fitness of a population but rather its expected longevity, i.e. the average number of generations to its extinction.

Group selection as a factor in the evolution of genetic systems can only act over a longer time scale: the rate of changes in gene complexes is inversely proportional to the mean group size. Therefore many evolutionists have been opposed to the idea of group selection (but see a discussion in Barton and Clark, 1990). Fisher believed sexual reproduction and recombination associated with it to be the only system to have arisen and evolved by group selection. The early models intended to account for the factors maintaining sex and recombination were analyzed in terms of 'advantage to the population'. Some of these were proposed as far back as the 1930s by Fisher (1930), Muller (1932) and Serebrovsky (1973) (it should be noted here that Serebrovsky completed the manuscript of this book in 1939). It was assumed that the advantage of sex and recombination results from the possibility of combining in a single organism several favorable mutations that otherwise might have been lost due to chance elimination. As Muller (1932) pointed out, 'Sexuality, through recombination, is a means for making the fullest use of the possibilities of gene mutations; thus it is itself an accessory process, accessory to the primary process of gene mutation'. Further analysis has shown that the issue is an extremely complicated one and no definitive answer can be provided to it, the result depending on population size, mutation rate, breeding system and mode and intensity of selection (Crow and Kimura, 1965; Eshel and Feldman, 1970; Felsenstein, 1974, 1988; G.C. Williams, 1975; Maynard Smith, 1978; Haigh, 1978; Kondrashov, 1984, 1988a; Bell, 1989; Charlesworth, 1990).

6.1.2 Recombination as a protective mechanism

Selection against harmful mutations

According to Muller, harmful genes tend to be fixed in a population without recombination (Muller's ratchet). To put it more exactly, the genomic regions in which recombination is, for some reasons, blocked **in every member** of the population may become genetically inactive owing to accumulation of harmful mutations (Darlington, 1939; Huxley, 1942; Muller, 1964; Charlesworth, 1978).

If u is the mutation rate per genome per generation and s the intensity of selection against mutants, then, in the absence of recombination, the proportion of individuals with the lowest number of mutations can be calculated as $n_0/N = \exp(-u/s)$, where N is population size, which is assumed to be constant (Haigh, 1978). Numerical experiments (Bell, 1988) show that a ratchet can complete a single turn in $10n_0$ generations, from which Bell calculates the level of recombination assuring an arrest of the ratchet: $r = -2\ln(1 - k/n_0)$ (here $k = 0.1$ is the per generation number of mutation-free genomes produced by recombination). Unfavorable mutant alleles arising independently at various loci of two homologous chromosomes at different times are placed by crossing-over on one of the recombinant chromosomes, leaving the reciprocal product free of mutations. Therefore, it is this effect, associated with the arrest of Muller's ratchet, that is considered to be one of the primary functions of sex and recombination and a factor in their evolution. Discussing two combinative functions of meiosis (associated with adaptation to varying environments and selection against harmful mutations), Crow (1988) assumes that independent chromosome segregation is a more important form of recombination than crossing-over. This assumption is not readily acceptable, particularly when the point at issue is selection against deleterious mutations. Indeed, even with a large haploid number, arguments about Muller's ratchet operating in the absence of crossing-over are simply transferred from the level of the genome to that of the chromosome (see also Bernstein, Hopf and Michod, 1988).

Some experimental evidence has recently been obtained demonstrating the possible role of recombination in systems with increased mutation rates. Experiments with the nematode *Caenorhabditis elegans* have shown that with increasing recombination rate due to crossing-over enhancer *Rec-1*, fitnesses of genotypes carrying the mutator gene *mut-6* increase (Zetka, Rose and Baillie, 1987). Consequently, the presence of a gene for increased recombination, which in itself has no effect on mutation, reduces the genetic load due to mutator system *mut-6*. Here, recombination performs repair functions through combinative ones. Bell (1988) has called this effect of sex and recombination 'exogenous repair' as opposed to common ('endogenous') systems of DNA repair, which are independent of selection. Complementation of deleterious mutations under outcrossing is also regarded as a repair function of sex (Bernstein *et al.*, 1985; Bernstein, Hopf and Michod, 1988) and a factor promoting an evolutionary increase in the duration of diplophase (Kondrashov and Crow, 1991; Perrot, Richerd and Valero, 1991; Otto and Goldstein, 1992).

Using an *E. coli*–phage T4 system, Bernstein (1987) has shown that genetic damage induced by UV irradiation of the phage reduces the resistance of host cells to a reinfection by phage particles, thereby providing an opportunity for damage repair by recombination. But in experiments with *Bacillus subtilis* and *Haemophilus influenzae*, Redfield (1993) found no increase in competence to transformation after treatment with DNA-damaging agents (UV light or mitomycin C). The author concludes that explanations other than DNA repair should be considered for the evolutionary functions of bacterial transformation.

It has been suggested that the high level of recombination exhibited by RNA viruses, which results, among other things, from their segmental replication, is an evolutionary adaptation improving the efficiency of selection against harmful mutations (Crow, 1988). These viruses show very high mutability, some of them (e.g. the flu virus) being, in addition, subject to bottle-neck effects due to the bulk of the host population developing viral resistance (Donis, 1991).

We think it is appropriate to make just here some critical remarks concerning the role of recombination in the efficiency of purifying selection:

- In asexual organisms, Muller's ratchet can, in principle, only operate in small populations with slightly deleterious mutations ($s \approx 10^{-3}$ to 10^{-2}), assuming that the rate of back-mutations is lower by an order of magnitude and more. But the last assumption does not necessarily hold for small mutations, which tends to reduce the difference between sexual and asexual reproduction (Pamilo, Nei and Li, 1987). Moreover, accounting for possible effects of compensatory mutations for fitness-related quantitative traits led Wagner and Gabriel (1990) to the conclusion that Muller's ratchet does not necessarily act in asexual forms, at least in short periods.
- The models showing the very possibility of harmful mutations accumulating in asexual populations consider constant population sizes regardless of the mean number of fixed harmful mutations, i.e. 'soft' selection is assumed. Melzer and Koeslag (1991) have studied the effects of this artificial assumption. They have shown, based on computer simulations, that with extinction of asexual clones caused by size fluctuation differences (resulting from their differential mutation loads) Muller's ratchet does not operate.
- The advantage of recombination should not be taken automatically as the advantage of sex. Indeed, let us assume that selection against deleterious mutations definitely favors a high level of recombination. But why then could it not lead to a sharp increase in the rate of mitotic exchanges (reciprocal ones and conversions) within asexual systems, of course, following the establishment of diploidy? As is known, these events may be increased in frequency by orders of magnitude in some *hyper-rec* mutants (Aguilera and Klein, 1988; Montelone, Hoekstra and Malone, 1988; Palladino and Klein, 1992). A clear example of a possible sharp increase in the rate of recombination in mitotic divisions is provided by the system generating variability of the mammalian immune cells. The question of a possible evolutionary role of genetic variability within germ-line cells is taken up by

Hastings (1991). Mutations, conversion and mitotic crossing-over of genes and chromosome segments involved in 'housekeeping' metabolism will all result in differential reproduction (selection) at the cellular level. Previously, this question was also discussed by Zhuchenko and Korol (1985). It has been argued that selective relations should arise, in principle, whenever there is self-reproduction, be it of molecules, cells, organisms or groups of organisms. Thus, purifying selection against deleterious mutations could, hypothetically, favor an increase in the rate of mitotic exchanges up to the level of meiotic ones. Of course, harmful effect of homozygotization of segments distal to the crossing-over point should be taken into account. But this problem seems to be not too serious to selfers. Moreover, it would be avoided by evolving non-reciprocal recombination.

All this leads one to wonder whether sex and meiotic recombination are the response of nature to other challenges, e.g. adaptation to environmental changes (both biotic and abiotic, in time and space).

Recombination and repair processes

In principle, it is not unlikely that recombination and sex involve something more than combinative functions alone (Bernstein *et al.*, 1985; Grafen, 1988) (Figure 6.1). Maynard Smith (1978) has called explanations of the role of recombination not associated with its combinative functions 'physiological'. One can probably class under this category the hypothesis of recombination as primarily a repair mechanism. It has been suggested that meiosis ensures a sort of 'rejuvenation' of genetic material in passing to the next generation by

Combinatoric effects and R-functions

1. Within-population variability (chromosome-independent assortment, crossing over, gene conversion)
2. Generation of tandem repeats and conserted evolution (e.g. unequal crossing over, gene conversion)
3. Somatic cell variability (e.g. in asexual plants and *in vitro* conditions)
4. Chromosome evolution (site-specific recombination as a mechanism of chromosome interchanges)

Non-combinatoric effects and F-functions

5. 'Coincidence' with DNA repair functions
6. Role in fertility maintenance in sexuals
7. Role in the reproduction of telomere structure
8. Recombination-dependent rearrangements as regulation events in the control of gene activity
9. Somatic cell variability in the immune system of mammals

Figure 6.1 Some functions (effects) of recombination.

repairing numerous potential premutation changes that accumulate in the germ line during ontogenesis of higher organisms, e.g. in the form of epigenetic DNA damage such as cytosine methylation (Bernstein, 1979, 1983; Walker, 1978; Holliday, 1987, 1988).

The repair explanation is based on a close relationship between recombination, mutagenesis and DNA repair, manifested in the fact that these processes have common molecular mechanisms and genetic regulation systems in a wide diversity of organisms ranging from procaryotes to higher eucaryotes. In particular, many mutagens often prove to be active recombinogens, while mutations for loci controlling meiosis and recombination also tend to exhibit disturbed repair functions. Such a duality of function is exemplified by the *RecA* protein of *E. coli*, which is involved in the early stages of recombination and, simultaneously, in induced *SOS* repair. Nevertheless, the mutation analysis of the *RecA* gene has revealed that these two functions are, in principle, separable. In mammals, DNA repair system was found to be involved in V(D)J recombination of lymphocyte antigen receptor genes (Taccioli *et al.*, 1993). Recombination is also strongly affected by mutations in genes controlling other enzymatic steps in DNA metabolism, e.g. replication (for reviews, see Devoret, 1988; Mahajan, 1988; Modrich, 1991). Genes for DNA repair–recombination exhibit a high degree of sequence, structural or functional similarity across all of biology, including complementation under gene transfer from one species to another (Arrand *et al.*, 1987; Dickstein *et al.*, 1988; Lambert *et al.*, 1988; Banga *et al.*, 1989; Weber *et al.*, 1990; Hoeijmakers, 1993a, b; Ogawa *et al.*, 1993).

A large body of experimental evidence has accumulated indicating a close relationship between repair and recombination in their response to environmental conditions (reviewed by Zhuchenko and Korol, 1985). This is manifested in a substantial increase in recombination (both meiotic and mitotic) in response to most mutagenic factors. The effect can also be regarded as an adaptively significant increase in the role of combinative functions of recombination resulting in a higher efficiency of purifying selection.

Such a close molecular and genetic relationship between the two functions (repair and recombination) undoubtedly points to their common origin. It can probably be supposed that it is the repair functions that played the leading role at early evolutionary stages, having provided opportunities for a sharp increase in the genome size and the transition to diploidy. The latter offered the possibility of recombinational repair of two-strand DNA lesions, which is impossible in haploid systems (Maynard Smith, 1988a). However, the hypothesis that subsequent stages in the evolution of recombination and sex, persistence and further perfection of these mechanisms in higher organisms are associated with repair functions (Bernstein *et al.*, 1985; Bernstein, Hopf and Michod, 1988) seems to be an implausible and disputable one. It is hard to accept the view (Bernstein, Hopf and Michod, 1988) that the combinative effects of recombination exhibited by extant organisms are reduced by selection to a minimum (to be more exact, to such a minimum which is attainable with a given mechanism without a considerable increase in metabolic expenditures).

Apart from population genetic arguments there is some evidence which is difficult to reconcile with the repair hypothesis as the sole explanation:

- 'Justification' of recombination in extant organisms (e.g. *Drosophila*) on the basis of purely repair functions does not appear to be a very plausible explanation, among other things because related species with similar gene orders in their genomes may differ several-fold in the amount of crossing-over (see Charlesworth, 1985).
- If repair is the primary function of meiosis, it is unlikely to be closely connected with crossing-over, otherwise how can one explain the realization of this function in species in which the heterogametic sex exhibits achiasmate meiosis?
- It has been previously found in *Drosophila* that suppression of pairing in chromosome 2 in males heterozygous for a composite inversion results in a higher frequency of radiation-induced lethals compared with that in normal males (Thompson, 1962). But what follows from this is only that repair is likely to be dependent on pairing and not on crossing-over, which is blocked in male meiosis of *Drosophila*.
- The frequency of mutation events occurring in meiotic cells is, probably, higher than that in mitotic ones (Magni, 1964; Koltin, Stamberg and Ronen, 1975; Lindgren, 1975; Auerbach, 1976). This phenomenon, called the 'meiotic effect', is difficult to explain within the framework of the repair hypothesis. As a matter of fact, this effect is the reverse of what might be expected on the hypothesis of repair functions of meiosis.
- It must be emphasized that the 'monopoly' of meiosis over DNA repair is relative: similar systems operate in somatic cells as well. The hypothesis that germ track cells require special (more effective?) systems of DNA repair is not in agreement with the available evidence for the high similarity of meiotic chromosome behavior in animals and plants despite their great differences in differentiation of reproductive system components (notably higher ontogenetic stability of the former and high ecological plasticity and tolerance to genetic imbalance of the latter). The existence of very long-lived woody plants whose mitotic system must ensure a highly faithful transmission of genetic information in the long succession of divisions preceding meiosis should also be viewed from this perspective. One way of improving noise resistance in such systems may be cell selection (Antolin and Strobeck, 1985; Zhuchenko and Korol, 1985; Bernstein, Hopf and Michod, 1988; Hastings, 1991).

All this suggests that DNA repair can hardly be regarded as a primary function of meiosis. Diploidy itself seems to be, in principle, a sufficient precondition for repair of DNA damage (including double-strand breaks) fatal to cells. Therefore, outcrossing, and even meiosis, need not have evolved solely for this purpose. As revealed by an analysis of meiotic and repair mutants in eucaryotes, the majority of loci controlling DNA repair are functional in both meiotic and somatic cells (Baker *et al.*, 1976; Carpenter and Baker, 1982; Jentsch, McGrath and Varshavsky, 1987; Padmore, Cao and Kleckner, 1991; Palladino and Klein, 1992; Shinohara, Ogawa and Ogawa, 1992). If diploidy was but a brief phase and a union of haploid cells was immediately followed by a reduction (as in zygotic meiosis in *Chlamidomonas*), then the evolution of effective repair functions at this diploid stage must have proceeded predominantly within the system of meiosis.

In this context, it is important to take account of the relationship between various factors affecting the occurrence and accumulation of genetic errors (Bell, 1982; Shields, 1982, 1988). If molecular repair is indeed the main factor in maintaining meiosis, then, other things being equal, we should expect lower proportions of ameiotic parthenogens among long-lived species with large genomes. And if selection against deleterious mutations is important for maintaining recombination, then the number of functional loci, rather than the genome size, will be a critical factor. Clearly, the assumption of 'other things being equal' does not correspond to the reality, and in attempting to construct general predictions like these account should be taken of the differences in the efficiency of repair systems of various species. An important factor may be the ecology of the species, in particular fecundity, population size, mode of selection, the effect of environment on mutability, etc.

Recombination and segregation

The simplest 'physiological' explanation of why recombination is maintained in sexuals is its association with chromosome segregation (Maynard Smith, 1978; Hawley, 1988; John, 1990). Darlington remarked that one of the factors maintaining crossing-over in currently existing meiotic organisms may be its role in ensuring regular chromosome segregation. As an objection to such an explanation, examples are usually cited of normal chromosome segregation in species exhibiting sexual dimorphism in recombination. For instance, lack of crossing-over in *Drosophila* males and silkworm females is no obstacle to proper chromosome pairing and segregation.

Certainly, this is a rather strong argument against the physiological explanation. But it does not follow a priori from anything that the solution of the problem of segregation control which has proven suitable for one sex should be effective for the other too. At the same time, analysis of meiotic mutants in which the relationship between recombination and segregation is especially pronounced have revealed situations in which the two phenomena are not coupled. For instance, crossing-over is normally almost completely suppressed in chromosome 4 of *D. melanogaster*. Some *mei* mutants with a reduced general level of exchanges and/or altered exchange distribution within large chromosomes exhibit crossing-over within chromosome 4 and, simultaneously, disturbed segregation of this chromosome (Sandler and Szauter, 1978). Examples of other organisms in which chiasmata and segregation are not coupled have also been found (Maguire, 1992). No less important an argument against the physiological concept of crossing-over maintenance is as follows: none of the known cases of recombination dimorphism violates the Haldane–Huxley rule (if crossing over is fully suppressed in one of the sexes, it is invariably in the heterogametic one) (Burt, Bell and Harvey, 1991). This is very important, since it is in the heterogametic sex that selection tends to reduce crossing over (Nei, 1969). The above is perhaps one of the few general facts which are in perfect agreement with theoretical predictions (for further discussion see section 9.4).

Finally, in connection with the discussion of 'physiological' hypotheses of recombination maintenance, it is also worth mentioning some other non-combinative functions with recombination playing the crucial role (see Figure 6.1 and Hawley and Tartof, 1983; Borst and Greaves, 1987; Golden, Mulligan and Haselkorn, 1987; Hawley and Marcus, 1989; Wang and Zakian, 1990; Haber, 1992; Plasterk, 1992).

6.1.3 Review of modifier models of recombination evolution

The modifier approach

Probably the first attempt to analyze the evolution of recombination without invoking group selection was made by Nei (1967, 1969). He examined changes in the frequency of alleles at a **selectively neutral modifier locus** affecting the rate of recombination between a pair of selected linked loci. It has been shown that, in a constant environment, selection in an infinite random-mating population promotes the establishment of the low crossing-over rate allele (let it be denoted rec^-, with the alternative enhancer allele being rec^+). More rigorous results were later obtained by Feldman (1972) and Altenberg and Feldman (1987). Consequently, the conclusions arrived at in terms of advantages to a population by studying the relation $W(rf)$ coincide with the results from a more justified model of individual selection involving a modifier locus.

The agreement of the results yielded by these approaches has also been demonstrated for some other features of the genetic system (mutations, migration, breeding system) and generalized as 'the mean fitness principle' (Karlin and McGregor, 1974). Unfortunately, it is far from always that such a correspondence is observed (Gillespie, 1981a; Brooks, 1988). Thus, in a number of models of selection in a constant environment, the direction of dynamics of the *rec* modifier depends on its linkage with the selected system (Feldman, Christiansen and Brooks, 1980; sections 7.4, 8.3 and 9.3). This fact is in no way reflected in the form of $W(rf)$. In models for the evolution of recombination under partial selfing, selection can result in the spread of the rec^+ allele reducing mean fitness. Nor is the mean fitness principle in general applicable to varying environments, although Gillespie (1981a) has shown it to be relevant in migration modification if the geometric average of mean fitness (over time) is maximized. It can be suggested that the inapplicability of the mean fitness principle results in all the above cases from violation of the initial assumptions (owing to mutation, varying environment, departure from panmixia).

Since the recombination modifier is assumed to be selectively neutral, selection can have no direct effect on it. Its allele frequencies can only be changed by selection acting on gene combinations whose frequencies in each generation depend on the modifier locus. Such feedback selection (secondary, or induced, selection – Feldman and Balkau, 1973) results in the establishment of the modifier alleles, which ensure an increase in frequencies of favorable combinations of alleles at the selected loci. It is from these combinations that the modifier obtains a hitch.

The selected system of loci (SSL) that are far from being in an equilibrium are those most strongly affected by modifier loci (Wagner and Burger, 1985). Unfortunately, it is equilibrium situations that are considered in the majority of models (particularly analytical ones). Belonging to near-equilibrium situations is also the mean fitness principle. Wagner (1981) has proposed the principle of positive feedback for the 'modifier-selected loci' complexes: the effect of the modifier enhances the intensity of primary selection, resulting in a stronger induced selection on the modifier.

Actually, in discussing modifier models for the evolution of recombination, the neutrality of *rec* alleles can only be regarded as a reasonable assumption for a certain part of the corresponding control system which exhibits a high intrapopulation polymorphism, regional specificity and relatively weak individual effects of loci (sections 6.2 and 6.3). Abundant evidence on meiotic mutants in *Drosophila*, plants and yeast indicates that many *rec* loci involved simultaneously in recombinational and repair DNA metabolism have a dramatic effect on fitness (resistance to stresses, viability, fertility) (for a review, see Baker *et al.*, 1976; Baker and Hall, 1976; Montelone, Prakash and Prakash, 1981; Zhuchenko and Korol, 1985; Jentsch, McGrath and Varshavsky, 1987; Treger, Heichman and McEntee, 1988). The most likely reason for this is that recombination functions seem to have been formed and perfected based on the previously evolved mechanisms for DNA replication and repair (Bell, 1988; Bernstein, Hopf and Michod, 1988; Maynard Smith, 1988a). The restricted applicability of the above assumption of the neutrality of the *rec* loci is also supported by the available phenomenology with respect to the relationship between fitness and crossing-over rate (Tucic, Ayala and Marinkovic, 1981; Zhuchenko and Korol, 1985; Cvetkovic and Tucic, 1986; Zhuchenko *et al.*, 1986).

It should be noted that the 'modifier approach' has also been used in analyzing the evolution of properties other than those of the *R*-system. According to Fisher's (1930) principle of the evolution of dominance, the recessive nature of the majority of harmful mutations results from modifier selection induced by these mutations during the entire past history of the given phylum. In recent years, the problem of the evolution of individual (phenotypic) plasticity has come to the fore (see, for example, Feldman and Krakauer, 1976; Slatkin and Lande, 1976; Via and Lande, 1987; Jain, 1990). Here, too, explanations based on the analysis of changes in the frequency of phenotypic plasticity modifiers have been proposed.

Classification of the models

Given the latest classifications, all the models for the evolution of recombination can be divided into four groups according to two types of factors causing linkage disequilibrium between the selected loci, random (**R**) and deterministic (**D**), and to the effect of recombination within each of the types – conservative (**C**), associated with the elimination of harmful mutations, and prospective (**P**), bringing together useful genes (Maynard Smith, 1988a; Kondrashov, 1988a. For other classifications see also Maynard Smith, 1978; Charlesworth, 1985; Felsenstein, 1988). Classed under the **DP** category is one of the first models with a varying environment pioneered by Sturtevant and Mather (1938) and subsequently elaborated by Charlesworth (1976), Sasaki and Iwasa (1987) and Ronin, Korol and Kirzhner (1993). In this model, the advantage of recombination results from the alternation (in time) of selectively favored allele combinations ($++$ and $--$ in some generations and $+-$, $-+$ in others).

The indicated mode of selection seems to be rather uncommon in nature and has, therefore, been regarded as a very unlikely reason for the maintenance of recombination (Maynard Smith, 1978). It is, nevertheless, assumed that a situation like this may arise under frequency-dependent selection resulting from antagonistic species interactions, for example of the host–parasite type (Jaenike, 1978; Hamilton, 1980; Hutson and Law, 1981; Bronstein and Korol, 1984; Korol, Preygel and Preygel, 1986; Korol and Preygel, 1988b; sections 8.1 and 8.2). Later, a polygenic model with fluctuating optimum has been suggested (Maynard Smith, 1980, 1988a, b; Korol and Preygel, 1986; Chapter 7), in which the advantage of increased recombination arises under more realistic conditions. This provides a good reason to reconsider the role of a varying environment as a factor maintaining recombination (Maynard Smith, 1988a). And again, the same mechanism can produce evolutionary advantages for recombination when fluctuations of the optimum are not due to an abiotic environment but result from antagonistic species interactions or intraspecific competition (Bell and Maynard Smith, 1987; Preygel and Korol, 1990; Korol and Mustyatsa, 1991; Kandel and Korol, 1991; sections 8.3 and 8.4).

Another kind of selection for higher *rf* under a varying environment has been studied in models of subdivided populations (Slatkin, 1975; Maynard Smith, 1977, 1978; Charlesworth and Charlesworth, 1979; Plaetke, Lumme and Koehler, 1988). It has been established that with clinal environmental variation, the *rec*$^+$ allele frequency is only increased under particular relationships between selection intensity, rate of migration and number of subpopulations.

Attempts at carrying out a comparative analysis of two versions of a model for selection in a varying environment – when the environment varies in space and in time – are to be found in Case and Bender (1981) and Bell (1987). Of course, this problem calls for further investigations. Spatial variation in environment is considered as a more effective factor in maintaining genetic polymorphisms than temporal variation (Hedrick, 1986). By contrast, the latter factor seems to be more effective in producing evolutionary advantages for increased recombination.

Under the **RP** (random prospective) category can be classed the models of Strobeck, Maynard Smith and Charlesworth (1976) and Felsenstein and Yokoyama (1976). In the model of Strobeck, Maynard Smith and Charlesworth (1976), induced selection occurs by hitch-hiking, which becomes available to the *rec*$^+$ allele owing to the interaction of a new favorable mutation and a heterotic polymorphic locus maintained at equilibrium. A substantial increase in the *rec*$^+$ frequency can only occur under very tight linkage between these and the modifier. The model of Felsenstein and Yokoyama (1976) is based on the ideas of Fisher and Muller about the accelerating effect of recombination on evolution under mutation pressure. The results of simulations suggest that linkage disequilibrium due to favorable spontaneous mutations at different loci of a finite population (thus, we classify the model as **RP** – random prospective) can be a powerful factor promoting the establishment of the *rec*$^+$ alleles in the population. With unfavorable mutations (model of the **RC**, random conservative, type), an increase in the *rec*$^+$ allele frequency is only possible when it is recessive. Thus, random mutations can, by continually generating linkage disequilibrium in finite populations, prevent the genome from 'congealing'.

Deterministic analogs (**DC**) of the **RC**-type model of Felsenstein and Yokoyama (1976) have also been proposed (Kondrashov, 1982, 1984, 1988a; Charlesworth, 1990). It has been shown that selection against harmful mutations can favor increased recombination in populations of any size when the fitness function is a 'convex' one, i.e. provided selection against mutants is of the truncation type. Similar results have been obtained for recombination in bacteria and RNA viruses (Redfield, 1988; Nee, 1989).

Some authors regard selection against deleterious mutations and selection for beneficial combinations of mutant alleles as one and the same mechanism (Brooks, 1988; Felsenstein, 1988). Actually, however, they differ with respect to biological functions of recombination, conservative versus prospective. The former may, in principle, be independent of sex (outcrossing), whereas the latter is impossible without outcrossing (see also Szathmary and Kover, 1991, and p. 197).

Some conclusions from the work of Felsenstein and Yokoyama (1976) have been analytically confirmed for infinite populations (Feldman, Christiansen and Brooks, 1980). It has been shown that if in the system of selected loci in a stationary state a balance is achieved between mutation and selection against harmful alleles, with the value of D being negative, then the rec^+ allele will increase in frequency provided it is tightly linked to the selected genes. The authors found the rate of changes in recombination frequency due to the interaction of selection and mutation to be very slow, but they do not exclude the possibility of this being increased under a 'coarser' control of recombination involving entire chromosomal segments.

The main idea underlying the 'sib competition' model and some other ecological models is based on the assumption that recombination, by ensuring genetic diversity, reduces the intensity of competition for similar limited resources in a varying environment, thereby allowing a more efficient utilization of the available resources (G.C. Williams, 1975; Maynard Smith, 1978; Young, 1981; Bell, 1982, 1988; Ghiselin, 1988). Numerical analysis (Maynard Smith, 1978) shows that sib competition does confer a substantial selective advantage on higher recombination, even in the absence of linkage between the modifier and the selected system and with moderate linkage between the selected loci. Under multilocus control of resistance to each environmental factor, sib competition will favor higher recombination only when the loci with related functions show tighter linkage than do those with unrelated ones. Then selection will lead to the formation of blocks of functionally related genes and to a higher *rf* between the blocks. A difficulty with this model is the need for reconciling two sometimes opposing requirements: on the one hand, there should be a fairly strong within-family competition, and on the other hand the offspring must find themselves in conditions differing from those of their parents (Maynard Smith, 1978, 1988a).

The assumption itself that competition for limited resources is stronger between sibs than between more distantly related forms is far from always substantiated experimentally (Willson *et al.*, 1987; Martin, Perez-Tome and Toro, 1988). Nonetheless, the idea of sex being capable of reducing the severity of competition for limited resources as well as conferring advantages in interactions with parasites finds confirmation in plants. The experiments of Antonovics' group on grasses have shown a substantial advantage in fitness of sexually produced, i.e. variable, progeny over the genetically uniform, asexually produced progeny (Ellstrand and Antonovics, 1985; Kelley, Antonovics and Schmitt, 1988). Recent tests made by Koella (1993) using published data on nearly 200 plant species also indicate that the sib competition hypothesis could be of interest to explain recombination variation.

External versus internal causes for recombination evolution

At this point, it is appropriate to make two remarks concerning classifications of models for the evolution of recombination.

1. The above classification does not seem to be a comprehensive one since it takes no account of one genetically important difference between the models. In all groups of models, a short-term advantage of recombination results from selection for altered sign of *D*. However, in some cases this occurs because of temporal or spatial variation in the environment, whereas in others it is the result of the finite sample size or of randomness of mutations. It is no less important, in attempting a classification of this kind, to ascertain whether the cause of variation in favored signs of *D* is external or internal (i.e. genetic and, hence, itself subject to evolution). If this approach is taken, the models of linkage disequilibrium due to the flow of mutants (Felsenstein and Yokoyama, 1976; Kondrashov, 1984, 1988a; Charlesworth, 1990) should be referred to the second class and the models based on selection in a varying environment (Charlesworth, 1976; Maynard Smith, 1980, 1988; Korol and Preygel, 1986, 1989) to the first class. Classified under the internal category can also be mechanisms for the evolution of the *rec* system resulting from interaction of recombination and transposable elements (Charlesworth and Langley, 1986), as well as of recombination and repeated DNA sequences, multigene families (Ohta, 1986; Charlesworth, Langley and Stephan, 1986; Stephan, 1986, 1987) and meiotic drive systems (Feldman and Otto, 1989, 1991; Hurst, Hurst and Johnstone, 1992).
2. Different variants of the evolution of *rec* systems by selection of mutations are sometimes classed under opposite categories, with models of selection against deleterious mutations being classified as genetic ones and those of selection for beneficial alleles as environmental ones (see, for example, Kondrashov, 1988a). The asymmetry seems to reflect the neoclassical tradition (discussed by Lewontin, 1974) of classifying the selective values of genes by their effects on fitness, i.e. on an individual's correspondence to its environment. This implies that some new alleles can only be beneficial in changing environments and, conversely, the overwhelming majority of new (mutant) alleles are, to be sure, harmful in constant environments. Moreover, correspondence to the internal genetic environment (an organism's *Bauplan*) is ignored. By contrast, according to the balance concept, the notion of 'a harmful allele' is rather relative, particularly in the case of slightly deleterious genes (see also pp. 154–6). If one takes into account the high genetic variability of 'internal environment' within a population, then, even in constant external conditions, what is actually observed is only a distribution of a given allele's effects on fitness, from deleterious to varying degrees to benefical to varying degrees. The pattern of this distribution is, in its turn, dependent on external conditions. With this approach to allele classification, differences between selection for beneficial mutations and against deleterious ones do not seem to be sufficient reason for classing the corresponding models as definitely environmental and genetic ones.

Understanding co-evolution of the R-system components

The most general conclusion emerging from the above review is that theoretical genetics probably has still a long way to go before solving the key issues of the evolution of recombination. Speaking of the applications of theory to the real world, we seem to be just beginning to formulate relevant questions and requirements for large-scale measurements and estimations. Very little, if anything, is yet known of the evolutionary consequences (for recombination) of varying the values of the remaining parameters of the *R*-system, or of how directional selection on components of the *F*-system will affect various components of the *R*-system under different patterns of environmental change. It is worth mentioning here some works concerned with the evolution of sex and meiosis as an alternative to asexuality (Hamilton, 1980; Moore and Hines, 1981; Manning, 1984; Ellstrand and Antonovics, 1985; Eshel and Weinshall, 1987; Lewis, 1987; Weinshall and Eshel, 1987; Sagan and Margulis, 1987; Bierzychudek, 1987; Edmonds and Rose, 1988; Lloyd, 1988; Kelley, Antonovics and Schmitt, 1988; Kirkpatrick and Jenkins, 1989; Uyenoyama and Bengtsson, 1989; Maguire, 1992), as well as with the evolution of mating systems as part of the sexual process (Lande and Schemske, 1985; Holsinger, 1988; Uyenoyama, 1986, 1988; Campbell and Waser, 1987; Charlesworth, Schemske and Sork, 1987; Charlesworth, Morgan and Charlesworth, 1990; Hoekstra, 1987). On the whole, these issues, while being relevant to the problem of the *R*-system evolution, are beyond the scope of this book.

In terms of the evolution of the system controlling recombination, it is of interest to examine co-evolution of the *rec*-system and mating system. This subject still awaits its researcher, although some steps in this direction have already been made (see, for example, Charlesworth, 1988; Kondrashov, 1988b). Charlesworth examines how linkage between the selected locus concerned with fitness in cyclically varying environment and the modifier gene determining female choice affects the modifier frequency. The next step in the attack on the problem could be the analysis of situations with multilocus control of both fitness components, with the amount of crossing over within each system and between systems also being considered as a genetically controlled polymorphic character rather than a fixed parameter.

The models discussed above are mainly concerned with looking for conditions under which selection favors an increased level of recombination. However, there are also other unsolved problems. A large body of evidence has accumulated on important features of recombination in higher eukaryotes for which theoretical explanations have not yet been provided (sex differences, dependence on environment, dominance–recessiveness relations, interference, etc.). Furthermore, in all of the currently available formal models for the evolution of recombination, the structure of the *rec* system is assumed to be extremely simple. The models usually involve a single *rec* locus active within a single segment or within a few segments but showing equal effects in each of them. However, experimental evidence suggests that some genes can reduce recombination in some genomic regions and sharply increase it in others. On the other hand, the rate of exchanges in a given region may be controlled not by a single gene but by several *rec* genes located in different genomic regions (Simchen and Stamberg, 1969; Chinnici, 1971; Zhuchenko and Korol, 1985; Brooks, 1988). Theoretical models of this kind allowing for complex architecture of the recombination system have not been previously considered.

Attempts at the evolutionary interpretation of features of another extremely important source of variability – the mutation process – are worth mentioning here. First studies along these lines were performed about 50 years ago (Dubinin, 1932; Sturtevant, 1937; Shapiro, 1938; Shapiro and Ignatjev, 1945; Berg, 1948). In subsequent decades this subject has been further developed by various authors, both experimentally (Leigh, 1970, 1973; Chao and Cox, 1983; Kaidanov *et al.*, 1991; Drake, 1989, 1991; Foster and Cairns, 1992) and theoretically (Kimura, 1967; Karlin and McGregor, 1974; Maynard Smith, 1978; Gillespie, 1981b; Holsinger and Feldman, 1983a; Holsinger, Feldman and Altenberg, 1986; Liberman and Feldman, 1986; Nothel, 1987; Ishii *et al.*, 1989; Twomey and Feldman, 1990). One of the main problems in this area is identifying the conditions for various patterns of evolution of mutation rates (relative to the zero or some other 'optimum' level). Holsinger and Feldman (1983a) have shown that a non-zero optimum mutation rate is possible in systems with selfing (complete or partial) in a constant environment. Under panmixia, it is the lowest possible (zero) mutation rate that is the evolutionarily stable one. With a frequently changing environment, selection may favor the spread of a modifier causing non-zero mutation (Gillespie, 1981b; Ishii *et al.*, 1989). Nevertheless, the problem is far from being completely solved. It is particularly important, in our view, to study the **co-evolution** of modifiers of various components of the *R*-system– mutation, recombination, Mendelian segregation, transposable elements, breeding system. This enormous work is practically at its initial stage.

6.2 GENETIC VARIATION IN EXCHANGE FREQUENCY AND DISTRIBUTION

6.2.1 Within-population variation in the *rec* system

Recombination frequency and distribution in the genome are among the most important characteristics of the genetic system of population adaptation. Together with the breeding system and mutation rate and genomic distribution, they determine the dynamics of free variability exposed to selection. The analysis of genetic variation in parameters of the recombination system itself allows a deeper insight into patterns and mechanisms of genome integration and evolution of the *rec* system.

The genetic differences in crossing-over rate seem to have been originally established by Sturtevant, who discovered, in 1913, modifiers sharply reducing crossing over during oogenesis in heterozygous females of *Drosophila*. Sturtevant (1921) supposed these to be inversions. His suggestion was subsequently confirmed on the basis of a cytological analysis of salivary gland chromosomes. Among point mutations affecting recombination, the first to be discovered was the gene *c(3)G*, which almost completely blocks crossing over in *Drosophila* females (Gowen and Gowen, 1922). Unlike inversions, *c(3)G* manifests its effect when in homozygous state, whereas heterozygotes *c(3)G/+* may even exhibit increased values of *rf* (Hinton, 1966).

The abundant recombination data in chromosome 3 of *D. melanogaster* led Gowen (1919) to conclude that crossing-over frequency is one of the most variable biological parameters. Coefficient of variation ranged from 18% to 50% for single exchanges and from 67% to 110% for double exchanges, with relative variation for *rf* increasing with decreasing mean rate of exchanges (both single and double). Serebrovsky (1926) showed these results to reflect not only biological but also purely sampling variation.

Clear evidence for genetic determination of crossing-over frequency in *D. melanogaster* has been obtained in Lawrence's (1958) experiments. He transferred a marked X chromosome (*w–m–B*) into the genetic background of highly inbred strains *Samarkand* and *Oregon*. The effects of autosomes in homo- and heterozygous states on crossing-over rate in the X chromosome were evaluated at temperatures of 15 °C, 20 °C, 25 °C and 28 °C. The strains in question differed substantially in the *rf* value in the distal segment *w–m*, regardless of the temperature regimen (rf_{w-m} in *Oregon* is higher than that in *Samarkand*). In the segment *m–B*, differences of the same sign were observed at temperatures of 15 °C and 20 °C, but these disappeared at temperatures of 25 °C and 28 °C. Thus, the dependence of the response of *rf* to environmental changes on genotype as well as segment specificity of ecological genetic control has been demonstrated. Variation in crossing over in *D. melanogaster* was also studied by Chinnici (1971), Kidwell (1972), Abdullah and Charlesworth (1974), Korol *et al.* (1984), Charlesworth and Charlesworth (1985a, b), Brooks and Marks (1986). These studies have confirmed the general notions about fine, region-specific, control of crossing over and have shown the presence of general (genomic) effects of *rec* modifiers as well. It has also been reported that recombination rate in one region (chromosome) can be dependent on determinants located on other chromosomes.

Recombination studies in *D. ananassae* males with various chromosome substitutions have revealed high variation in crossing-over frequency within marker segments of the second and third chromosomes (Moriwaki, Tobari and Matsuda, 1979). In particular, the *rf* value in the *ml–ru* segment of chromosome 3 varied, depending on genotype, between 2.2% and 42.8%, i.e. 20-fold! One of the reasons for high within-population variation in *rf* in males of this species is polymorphism at enhancers of crossing over (Matsuda and Tobari, 1987). Some of these have been mapped.

Analysis of recombination in higher plants has revealed substantial genotypic differences. As early as in the 1920s to 1930s, results like these were obtained on maize, peas, beans and other organisms. In lion's-mouth, variation in recombination was so wide as to prevent, in a number of cases, the continuation of studies (Baur, 1924; Herzberg-Frankel, 1925).

In Studler's experiments, differences between plants in *rf* within the *sh–su–wx* segment increased as the degree of relatedness decreased and there was correspondence between micro- and macrosporogenesis data. Polymorphisms for B chromosomes and for heterochromatic knobs as well as interactions between these factors have been shown to play a significant role in recombination variation in maize (Chang and Kikudome, 1974; Rhoades, 1978). B chromosomes have been found to affect pairing and crossing over in a number of other organisms, too (reviewed in Zhuchenko and Korol, 1985).

Analysis of genetic variation in *rf* in peas has shown that different hybrids can substantially differ in crossing-over rate between the same marker loci, the hybrids with reduced *rf* for one pair of markers showing closer linkage in other regions as well (Rasmusson, 1927). A similar parallelism was also observed by the author for combinations with loose linkage. For the *le–v* segment (chromosome 4), values of *rf* ranging from 6.6 ± 0.8 to $43 \pm 3.6\%$ have been obtained in various crosses (!).

Genetic variation in recombination between the linked loci has also been established in experiments with sweet peas (Bridges, 1914), wheat (Nillson-Ehle, 1927) and peppers (Lippert, Smith and Bergh, 1966). There is abundant evidence for intraspecific genetic variation in crossing-over rate and/or chiasma frequency in other eucaryotes as well – both lower (*Coprinus*, Day, 1958; *Schizophyllum*, Schaap and Simchen, 1971; *Neurospora*, Stadler and Towe, 1962; Cederberg, 1985) and higher (barley, Gale and Rees, 1970; pearl millet, Manga and Pantulu, 1974; *Crepis*, Whitehouse *et al.*, 1981; jute, Kumar, Paria and Basak, 1981; ryegrass, Karp and Jones, 1982; *Vicia faba*, Fayed, Mandour and Ismail, 1984; rye, Lein and Lelley, 1987; tomato, Korol, 1987, 1988; snail, Price and McBride, 1981; man, Laurie, Palmer and Hulten, 1982; mouse, Siracusa *et al.*, 1989; Ladyghina, Gorlov and Borodin, 1989; rat, Thomson, Westerman and Murray, 1984; lizards, King and Hayman, 1978; grasshoppers, Shaw, 1972, 1974; Fletcher and Hewitt, 1980; Santos, 1993; locusts, Bugrov and Vysotskaya, 1988; silkworm, Ebinuma, 1987).

For several organisms (*Neurospora*, *Drosophila*, maize), cytoplasmic factors have been found to play a role in the control of chromosome pairing and crossing over (Thoday and Boam, 1956; Lawrence, 1963; Moriwaki, Tobari and Mutsuda, 1979; Kruleva *et al.*, 1992). Of course, the best-known example of cytoplasm-dependent effects on recombination is the phenomenon of hybrid dysgenesis (see below).

Rasmusson (1927) holds that crossing-over frequency, like any other quantitative trait, is controlled by a number of genes, variation in this parameter being one of the major properties of hereditary transmission (this, however, does not exclude the possibility of the effects of major *rec* genes and, hence, of Mendelian approach to analyzing genetic variation in *rf*, as suggested in Chapter 3). He advanced, on the basis of extremely high variation in recombination rate in peas usually not found in *Drosophila*, a hypothesis of possible differences in the system of genetic control of recombination between outcrossers (*Drosophila*, maize) and selfers (peas). In the former, the integrated nature of the population gene pool results in the sharing of genes controlling recombination by different individuals and, hence, closeness of their *rf* values to a certain 'standard' level (unless there are sharp deviations resulting from chromosome aberrations). By contrast, selfers rapidly diverge, following rare outcrossing events, into homozygous strains quite different genetically with respect to the *rec*-loci. Subsequent crosses between homozygous strains or cultivars can produce a variety of *rf* values ranging from very small to large ones, depending on the parental stocks used. Discussing the differences in crossing-over rate in autogamous as compared with outcrossing organisms, Rasmusson (1927) believed that the objective of a genetic study is not only mapping factors and achieving an ever-increasing correspondence between genetics and cytology. The phenomenon of variation in linkage merits in itself careful studying, and it would be hardly overoptimistic to hope that these studies will result in new understanding of the problem of heredity in general. His words are of topical interest even today.

Apart from *rec* genes and B chromosomes, differences between individuals in DNA sequences (the 'reacting system' variations), resulting, for example, in heterozygosity for micro- and macroinversions, deletions, etc., may be a factor in within-population variation. It is also possible that a fraction of this variation is due to a polymorphism for the number and distribution of sites of recombination initiation and for location of heterochromatic blocks (for reviews, see Hotta, Chandley and Stern 1977; John and Miklos, 1979; Hotta and Stern, 1984; Zhuchenko and Korol, 1985; Stern, 1986; Treco and Arnheim, 1986; Brooks, 1988; John, 1990; Smirnov, 1991; Pearlman, Tsao and Moens, 1992).

6.2.2 Interstrain variation in crossover rate and distribution in *Drosophila*

Consider first the data on variability of recombination parameters in a near-centromeric region of chromosome 2 of the F_1 heterozygous females. These were obtained (by the authors of the present book in collaboration with Dr L.P. Kovtyukh) by crossing 12 wild-type strains with a marker line, *b cn vg* (Korol *et al.*, 1984). The estimation was performed based on the first four 3-day broods. Also studied in the course of the experiment were genotypic differences in the amount of temperature-induced crossing over in the *b–cn–vg* region. To this end, heterozygous females were exposed to 30 °C for 12 h. The results of the experiment suggest that there are considerable differences between the hybrids with respect to both the frequency and distribution of exchanges in the *b–cn–vg* region and response of these parameters to thermal treatment (some examples are given in Tables 6.1 and 6.2).

Differences between genotypes reach 300% and more with respect to *rf* in the *b–cn* segment and 250% with respect to *rf* in *cn–vg*. The distribution of crossovers within two segments, *b–cn* and *cn–vg*, is described by the ratio $\alpha = rf_{b-cn} / (rf_{b-cn} + rf_{cn-vg})$, whose maximum and minimum values differ more than sixfold. Differences between certain hybrids were so large that an individual analysis (without pooling) of separate broods of individual F_1 females revealed no overlapping of the distributions of their *rf* values. Therefore, for some of the comparisons made, the combined effects of non-genetic variation in *rf* cannot, even in part, mask the genetic differences. The observed variation between hybrids with respect to rf_{b-cn} and rf_{cn-vg} is enhanced following the thermal treatment. Nevertheless, the correlation between the *rf* values prior and subsequent to exposure is highly significant, being 0.74 ± 0.21 for the *b–cn* region and 0.75 ± 0.20 for *cn–vg*.

Table 6.1 Genetic variation in the exchange frequency and distribution in the *b–cn–vg* region of chromosome 2 among some *D. melanogaster* hybrids (data of the first four 3-day broods)

F_1 *between b cn vg and the strain*	N	*rf* (%) within segment *b–cn*(1)	*cn–vg*(2)	*c*	$\frac{1}{1+2}$ (%)
Domodedovo32	998	2.89 ± 0.53	5.58 ± 0.73	3.07 ± 1.20	34.1
Domodedovo18	1796	6.93 ± 0.60	9.68 ± 0.70	0.74 ± 0.23	41.7
Florida	1165	12.18 ± 0.96	11.50 ± 0.94	0.96 ± 0.21	51.4
Boa esperance	1524	1.77 ± 0.27	14.47 ± 0.88	1.75 ± 0.43	7.6
Algeria	1441	6.83 ± 0.67	10.61 ± 0.81	1.13 ± 0.29	39.2
GK80	1221	6.93 ± 0.73	10.89 ± 0.89	0.86 ± 0.29	38.9
net	1470	9.33 ± 0.76	13.20 ± 0.88	0.82 ± 0.19	41.4

Table 6.2 Genotype-dependent changes in *rf* in heterozygous *Drosophila* females exposed to thermal treatment (averaged over three replicates)

F_1 *between b cn vg and the strain*	N	Δ *rf* (%) within segment *b–cn*	*cn–vg*
Domodedovo32	1585	0.75 ± 0.70	-1.42 ± 0.88
Domodedovo18	2214	2.22 ± 0.86	1.45 ± 0.97
Florida	1932	0.07 ± 1.21	-0.93 ± 1.37
Boa esperance	1931	1.36 ± 0.45*	-2.10 ± 1.15
Algeria	2534	-1.02 ± 0.82	-3.51 ± 0.99*
GK80	1799	8.05 ± 1.11**	2.28 ± 1.20
net	1606	10.15 ± 1.25***	0.12 ± 1.22

The effect of treatment is significant at *$P < 0.05$, **$P < 0.01$ and ***$P < 0.001$ (based on Student's test with d.f. = 4).

The coefficient of coincidence (c) for the considered region of chromosome 2 also shows rather large between-genotype variation. The value of c varies from 0.74 (weak positive interference) to 3.07 (rather strong negative interference – excess of double exchanges). It was assumed that negative interference is mainly characteristic of short chromosome segments and is associated with intragenic recombination (Kushev, 1974). Nevertheless, cases are known of significant excess of double crossovers in adjacent regions observed in large chromosome segments. Thus, in *D. melanogaster*, within a segment 4 cM long accounting for about 25% of the cytological length of chromosome 3 and spanning the centromere, a significant excess of multiple exchanges has been found (Sinclair, 1975). Similar results have been obtained by other workers (Green, 1975; Dennell and Keppy, 1979). Dennell and Keppy suggest that negative chromosome interference is characteristic of all regions exhibiting a very low average density of exchanges per unit physical length.

By our data, in strains showing negative interference, recombination in both segments, *b–cn* and *cn–vg*, of chromosome 2 is relatively low (e.g. Table 6.1). This is in agreement with the hypothesis of Dennell and Keppy. The conclusion about strong negative interference in chromosome 2 of the *Domodedovo-32* strain has been also confirmed in our other two independent experiments: in experiment I, $c = 2.41 \pm 0.33$, sample size being $n = 11,413$; in experiment II, $c = 3.40 \pm 0.41$, $n = 6223$. Note that within each experiment c tended to depend on the brood (age of females) and sex of flies in the progeny of heterozygotes: the estimated value of c for daughters is higher than that for sons (in experiment I, $c_{♀} = 2.79 \pm 0.46$, $n_{♀} = 5876$; $c_{♂} = 1.92 \pm 0.46$, $n_{♂} = 5537$; in experiment II, $c_{♀} = 4.18 \pm 0.62$, $n_{♀} = 3061$; $c_{♂} = 2.73 \pm 0.54$; $n_{♂} = 3162$). This may be due to selective elimination of double recombinants, which seems to be more intense among males.

6.2.3 A *P* element-induced increase in variation of crossing-over rate in *Drosophila*

It is known that some crosses in *Drosophila* result in a set of genetic abnormalities termed hybrid dysgenesis and involving male recombination (MR), distorted segregation, sterility, chromosome breakage and increased mutation rate (Kidwell, Kidwell and Sved, 1977; Bregliano *et al.*, 1980). The effect takes place upon crossing females from laboratory strains with males from natural populations. It has been found that this type of cross can, in certain cases, induce changes in the location of transposable genetic elements (TE). A precondition for a transposition event is the interaction of the so-called *M* cytotype with a genomic *P* element carrying a structural gene for the enzyme transposase and having long repeats at both its ends.

One of the known genetic sources of *P* elements is a strain with the *T-007* second chromosome. Females heterozygous for marker genes and derived from an *M* × *T-007* cross have been shown to exhibit, in addition to the above-mentioned dysgenic abnormalities, substantial changes in crossover distribution patterns. In particular, the value of *rf* in a distal region of chromosome 2 is nearly halved compared with the control level, whereas in short proximal regions it is increased more than three- to fourfold (Slatko, 1978). For the segments studied, Slatko also observed a reduction in the amount of interference resulting from the MR system effects. In general, heterozygosity for the *T-007* chromosome causes substantial redistributions of exchanges similar to the effects of *mei* mutations while having no serious effects on the total crossing-over rate.

Higher mutability due to transpositions of TEs can be theoretically expected to increase the genetic variability of a population and to enhance the response to selection. This has been confirmed by Mackay: selection for higher bristle numbers in *Drosophila* is increased, as is the variance of this character, in dysgenic crosses (Mackay, 1984, 1985, 1987; see also Torkamanzehi, Moran and Nicholas, 1992). Changes in fitness components during selection are also attributed to transpositions of TEs (Kaidanov *et al.*, 1991).

Variability in crossing-over rate is also known to be quantitative in nature. Therefore, it is worth testing the possibility of increasing variability in *rf* by introducing TEs into a population. This has been done on the basis of crosses with the *T-007/Cy* strain (Zhuchenko, Vereschagina and Korol, 1988; Vereschagina and Korol, 1991). The *T-007* males were introduced into a synthetic hybrid population derived from crossing 21 wild-type strains with subsequent panmixia over five generations; the *T-007* males were also introduced at generations 15 and 20 (in all three cases the proportion of *T-007* flies was 5%). To estimate the levels of recombination in the experimental and control populations, a previously synthesized four-marker chromosome 3 (*th ri red ss*) was transferred, by crossing with balancer lines, into the genetic background of the populations. The resultant tester analogs

$$\frac{g_1}{g_1}\frac{g_2}{g_2}\frac{th\ ri\ red\ ss}{th\ ri\ red\ ss} \quad \text{and} \quad \frac{g_1^*}{g_1^*}\frac{g_2^*}{g_2^*}\frac{th\ ri\ red\ ss}{th\ ri\ red\ ss}$$

(here g_1 and g_2 are chromosomes 1 and 2 in the control and g_1^* and g_2^* in the experimental populations) were individually crossed with females of compared populations in which *rf* was estimated in the indicated regions of chromosome 3. The direction of crosses is such that the resulting heterozygotes can exhibit no hybrid dysgenesis even if there is an MR factor in the population by the time of estimation. It should be noted that the region chosen (*th–ss*) is characterized by substantial suppression of exchanges. As expected, the introduction of *T-007* into the population resulted in a wider variation in crossing-over frequency with the rise in the σ^2_{rf} value in individual chromosome segments being extremely large (Figure 6.2). The mean crossing-over rate was also considerably increased in all the chromosome 3 segments studied, but the pattern of exchange distribution remained unchanged.

The results obtained suggest that introducing TEs into a population causes

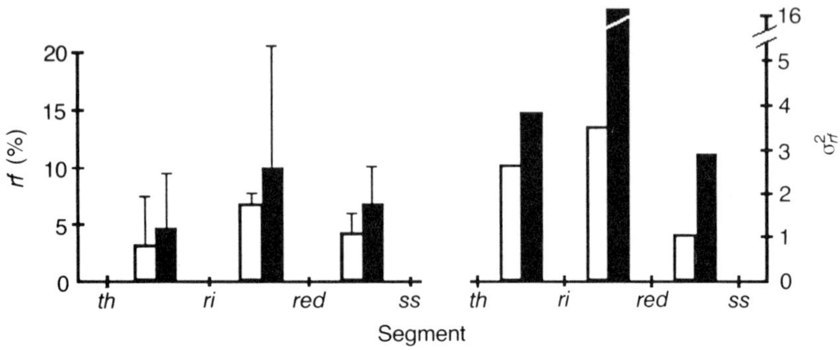

Figure 6.2 Effect of chromosome *T-007* on variability of recombination in the chromosome 3 of *D. melanogaster.* Vertical bars indicate the maximum values of *rf* among the tested 27 females from the experimental population (■) and 30 from the control one (□) (based on data from Zhuchenko, Vereschagina and Korol, 1988).

changes in recombination. The above effects may, in principle, be due to the following reasons:

- locally altered recombination properties of the chromosome (Yurchenko, Zakharov and Golubovsky, 1985; Sved *et al.*, 1991);
- hybrid dysgenesis is accompanied by the formation of crossovers in those chromosome regions which are normally 'excluded' from the crossing-over process (Slatko, 1978), resulting in the formation of new allelic combinations at loci modifying recombination;
- insertion of TEs at sites adjacent to the *rec* genes may alter the latter's activity, thereby affecting crossing over in segments controlled by them.

There seems to be a strong relationship between structural features of the genome in terms of the number and locations of TEs and its functional and recombinational organization (Charlesworth and Langley, 1986). Experimental studies of the relationship between recombination and TEs (McClintock, 1965; Slatko, 1978; Greenblatt, 1981; Yurchenko, Zakharov and Golubovsky, 1985; Dooner and Kermicle, 1986; Sved *et al.*, 1991) should, in our opinion, be complemented by similar studies at the population level. The above assessment of the effect of *T-007* chromosome on within-population variance for *rf* in chromosome 3 of *Drosophila* seems to be one of the first studies of this kind (see also Montgomery, Charlesworth and Langley, 1987).

6.2.4 Crossing-over variation following distant crosses

Differences between interspecific hybrids

Altered exchange distribution within chromosomes upon crossing different species and varieties has been reported in various organisms, such as mice, rye, maize, cotton, tomato, etc. In tomato, the relevant studies have mainly been performed by Rick (1969, 1972) and concern variation in *rf* in crosses between marker strains of red-fruited cultivated *L. esculentum* and green-fruited species. The taxonomically well-represented genus *Lycopersicon* (having species both closely related to and, to varying degrees, distant from the cultivated tomato) enables an assessment of the relationship between the degree of genetic divergence of the parental stocks and variation in crossover distribution in meiosis of the corresponding hybrids. This problem is, undoubtedly, of great significance to the breeder in that it provides insights into possible restrictions in using some or other forms as donors of genes. Phylogenetic and molecular genetic studies of tomato have revealed that speciation within this genus is primarily due to the accumulation of allelic substitutions at different loci and, to a lesser extent, to structural changes in chromosomes (Tanksley, 1987; Zamir and Tanksley, 1988). If that is the case, then changes in the recombinational behavior of chromosomes of hybrids should be regarded, in the first place, as reflecting changes in the system of *rec* genes controlling frequency and distribution of exchanges. Nevertheless, a certain role of minor (cryptic) structural changes as well as of possible changes in heterochromatin block location cannot be excluded (Stebbins, 1945; Stephens, 1950; John and Miklos, 1979; Miklos, 1982).

Korol and Bocharnikova (1988) have made an attempt to identify tomato genomic regions in which changes in crossing-over occur upon crosses between different *Lycopersicon* species. The studies involved hybrids between marker strains *aw d* (chromosome 2) and *m2 c* (chromosome 6); *Me wv* (chromosome 2); *wv d* (chromosome 2); *sy sf bls* (chromosome 3); *ful e* (chromosome 4); and *hl a* (chromosome 11) and wild tomato forms. Hybrids between multimarker mutants and the cultivar *Nistru* (*L. esculentum*) were used as the control.

It has been established in the course of this study (Figure 6.3) that recombination rate in individual chromosome segments of the hybrids examined may be reduced by several-fold compared with the level characteristic of an intraspecific hybrid, whereas in other regions an increase in *rf*, possibly a compensatory one, occurs. It is natural to expect the amount of reduction in *rf* to be inversely proportional to the degree of taxonomic relatedness between the parental forms. Then, the smallest reductions should be observed in hybrids of *esculentum* with *racemigerum* and *pimpinellifolium*, greater reduction with *minutum* and *hirsutum*, and the greatest ones with *pennellii*. The actual evidence points, however, to the lack of direct relationship between the extent of exchange inhibition in F_1 and the taxonomic distance between the parental strains. In particular, in some regions, recombination is considerably lower in a hybrid of *esculentum* with *pimpinellifolium* than with the more distantly related forms *minutum* and *pennellii*.

The data obtained point to the necessity of directly evaluating the recombination behavior of each form involved in the breeding process. A considerable reduction in the rate of exchanges within individual segments (e.g. *hl–a*)

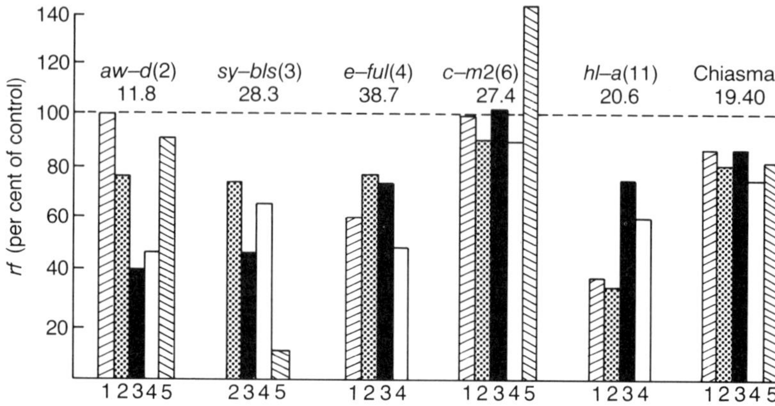

Figure 6.3 Crossing-over rate in crosses of *L. esculentum* with: 1, *racemigerum*; 2, *pimpinellifolium*; 3, *minutum*; 4, *hirsutum* (var. *glabratum*); 5, *L. pennellii* (100% – the level corresponding to crosses within the *L. esculentum* species).

upon crossing the cultivated tomato with the closely related forms *pimpinellifolium* and *racemigerum* suggests that it is probably allelic changes in *rec* genes, rather than structural differences among homologs, that play the decisive role in exchange frequency variations in tomato. Indeed, as a result of 15-year-long experiments, Stubbe (quoted by Zhuchenko, 1973) suggested that the formation of the cultivated tomato had proceeded by selection of gene mutations rather than chromosome aberrations. In other words, structural distinctions of the tomato chromosomes from their homologs in *racemigerum* and *pimpinellifolium* are insignificant and, probably, cannot be responsible for the two- to threefold reduction in *rf*. On the other hand, in *Lolium*, normal chromosome pairing is possible in an interspecific hybrid even with the 40% difference in DNA content between the parents (Rees *et al.*, 1982). Note that a drastic reduction in rf_{hl-a} upon crossing *esculentum* × *pimpinellifolium* has been observed by other authors as well (Griffing and Langridge, 1963).

A good argument in favor of the assumption of the genic (rather than structural) nature of the reduction in *rf* is the following fact. Hybrids of the strain *hl a* (markers on the chromosome 11) with the *racemigerum* and *pimpinellifolium* forms show a greater reduction in rf_{hl-a} than do the hybrids with the *hirsutum* species. If the reported reduction in *rf* relative to the normal level is assumed to be due to structural incompatibility of chromosomes of the parental stocks, then just the reverse would be expected to occur. Indeed, the green-fruited species *L. hirsutum* is taxonomically much more distant from the cultivated tomato than are the red-fruited forms *pimpinellifolium* and *racemigerum*. Moreover, it is for chromosome 11 (along with 9) that some differences with respect to the chromomere organization of chromatin at pachytene are observed between the green-fruited and red-fruited species (reviewed by Zhuchenko, 1973). Another example of restricted recombination between closely related species is provided by the data of Doebley and Stec (1991). They observed a two- to sevenfold reduction in map distances for a number of segments in hybrids of maize × teosinte relative to the values in maize × maize hybrids, although taxonomically maize and teosinte are rather close.

One more indirect argument in favor of the genic nature of changes associated with the formation of the cultivated tomato is the data on the copy number of cloned structural genes (Bernatzky and Tanksley, 1986; Zamir and Tanksley, 1988). For the majority of cDNA clones, the corresponding genes are present in the genome as single loci or two unlinked loci. In potato, whose genome bears a close organizational resemblance to the tomato's genome, the proportion of single-copy probes (among 239 examined) was nearly 80%. Also in this plant, a substantial reduction in recombination was observed in interspecific hybrids relative to the intraspecific level (see Bonierbale, Plaisted and Tanksley, 1988; Gebhardt *et al.*, 1991; DeVicente and Tanksley, 1991).

Variation in rf within F_2 populations of interspecific hybrids

Rick (1972) has shown that the extent of crossing-over inhibition in marker regions of the tomato in advanced backcross generations may be increased relative to the level in F_1. Similar results have been obtained with interspecific hybrids of cotton (Saunders, 1969). Experiments with the *esculentum* × *racemigerum* and *esculentum* × *hirsutum* hybrids conducted by Korol and Bocharnikova (1988) have revealed a high variation between heterozygotes F_2 in crossing-over rate in a pericentric region, *hl–a*, of chromosome 11. Moreover, a hybrid between taxonomically more distant forms (*esculentum* × *hirsutum*) was also notable for the fact that its progeny showed a wider variation in rf_{hl-a} (Table 6.4): a threefold variation (from 8.3% to 27.2%) has been obtained with as few as nine F_2 genotypes heterozygous for both markers (based on their F_3 progeny examination). This suggests that selection on *rf* among hybrid progeny has good prospects. The above conclusion about a higher crossing-over rate in the F_1 hybrid *hla* × *hirsutum* compared with the F_1 (*hla* × *racemigerum*) hybrid has been confirmed in F_2: rf_{F_2} (*hla* × *hirsutum*) > rf_{F_2} (*hla* × *racemigerum*). From comparison of the rf_{hl-a} values in F_1 and F_2 genotypes of each of the hybrids (see Table 6.4), it follows that recombination restriction may be more severe in F_1 than in F_2. Formally, this is equivalent to dominance of *rec* genes reducing crossing over.

Thus, the results of analyzing the behavior of *rf* in F_2 and F_1 enable one to question the generality of findings that in advanced generations the rate of exchanges should necessarily be lower than that in F_1 (e.g. Rick, 1972). For the two hybrids considered here, the average rate of exchanges in F_2 is higher than that in F_1, the highest *rf* values among the F_2 genotypes exceeding the corresponding values for F_1 by more than twofold.

In general, recombination properties of interspecific hybrids compared with

Table 6.3 Crossing-over frequency (rf_{hl-a}) in F1 and F_2 of interspecific tomato hybrids

	hla × *L. hirsutum*		*hla* × *L. racemigerum*	
Hybrid	N	rf(%)	N	rf(%)
F_1	566	12.60 ± 1.95	1078	7.96 ± 1.05
F_2	163	21.90 ± 3.75	870	11.10 ± 1.14
	651	21.16 ± 1.89	552	18.74 ± 1.88
	159	8.68 ± 2.36	572	11.20 ± 1.41
	133	19.55 ± 3.91	325	10.19 ± 1.79
	243	27.18 ± 3.45	496	14.86 ± 1.76
	147	10.44 ± 2.69	362	10.31 ± 1.70
	228	10.01 ± 2.12	412	15.79 ± 1.99
	188	8.30 ± 2.12	576	21.45 ± 3.94
	205	25.58 ± 3.63	364	15.46 ± 2.09
Characteristics of F_2				
	Mean	15.19 ± 2.52	Mean	12.96 ± 1.23
	σ^2	57.2	σ^2	13.6
	CV(%)	49.8	CV(%)	28.3
	χ^2_8	65.6 ($P < 0.001$)	χ^2_8	26.2 ($P < 0.01$)

those of intraspecific ones can be regarded as a good model for studying a wide range of problems pertaining to the nature of species differentiation, genetic control of recombination, dependence of exchanges on the degree of DNA homology between parental stocks and the influence of recombination properties of hybrids on possibilities of introgressing useful genes and gene blocks into cultivars.

6.3 ARTIFICIAL SELECTION FOR ALTERED RECOMBINATION FREQUENCY

This topic has received growing attention in recent years, mainly because of the interest in the problem of the evolution of recombination. Following the unsuccessful attempt of Gowen (1919), selection for altered crossing over in *Drosophila* and genetic analysis of its consequences have been carried out by Detlefsen and co-workers (Detlefsen and Roberts, 1921; Detlefsen and Clementi, 1923). Selection for lower *rf* in the *w–m* region (X chromosome) proved to be ineffective during the first 5–6 generations. However, starting with generation 7, recombination began to decrease rapidly, reaching 0–1% by generations 10–15 in the first replication (prior to selection, *rf* was 27–28%). Following the termination of selection, the *rf* values ranged, for the next 20 generations, from 4% to 12%, fluctuating about the mean value of 6%.

Some authors believe that these data cannot be interpreted in terms of selection of modifiers, since response to selection in the experiment just considered might have been caused not by the accumulation of *rec* genes with relatively small effects, but by the formation of inversions. This assumption provides, as it does, an explanation for the first replication results, but it ignores the data from the second replication in which reduction in *rf* was gradual, and even after the values of $rf \approx 4.6\%$ has been reached fluctuations of *rf* up to 10–12% were observed in some generations.

Selection for increased crossing-over rate in the experiment by Detlefsen and Roberts (1921) proved to be unsuccessful. The authors attribute this result to increased frequency of double exchanges in the *w–m* region. In general, Detlefsen and Roberts viewed these findings as evidence for the multifactorial control of crossing over. To evaluate the effect of selection for rf_{w-m} on crossing over in other regions, the selected strain was crossed with the stock *sc cv ct m f* with a marked X chromosome (notice that locus *w* is located in the segments *sc–cv*). It turned out that, on average, F_2 genotypes satisfied the inequality $\Delta rf_{sc-cv} > \Delta rf_{cv-ct} > \Delta rf_{ct-m} > \Delta rf_{m-f}$, i.e. recombination reduction in response to selection for lower rf_{w-m} was the largest in the distal region and diminished as the distance from the centromere became shorter (Detlefsen and Clementi, 1923).

It was not until 35 years later that subsequent experiments on selection for altered recombination rate in *Drosophila* were carried out. In 1958, Parsons reported a positive response after two generations of selection for increased recombination in the pericentromeric region, *b–pr*. The value of *rf* was, however, rather small. Acton (1961) has made an attempt to alter recombination frequency in approximately the same region of chromosome 2. Using a technique similar to that of Detlefsen and Roberts (1921), he performed both upward and downward selection for rf_{cn-vg}. In both cases, after 11 cycles of selection (23 generations) the crossing-over frequency remained at the original level. The population exposed to selection does not seem to have exhibited a sufficient amount of genetic variation.

As already noted, the species *Drosophila ananassae* exhibits a high level of spontaneous male recombination and considerable genotypic variation in *rf*. With this in mind, Mukherjee (1961) carried out up- and down-selection for altered *rf* in the *px–pc* region of chromosome 3. In the start population, the value of $rf_{♂}$ ranged from 0 to 25%, the mean being 11.4% ($rf_{♀} = 23\%$); after eight cycles of selection, a strain was obtained with $rf_{♂}$ values in the range of 0–5% (1.5% mean). Selection for increased *rf* proved to be ineffective. Nevertheless, by generations 6–7, the lower value of *rf* approached the mean of the foundation population. Therefore, we may assume that continuation of the upward selection would succeed. It will be remembered that in the experiment of Detlefsen and Roberts, too, selection for reduced recombination rate was ineffective during the first 5–6 generations. Approximately the same result was obtained in another experiment on selection for altered *rf* in *D. ananassae* males (Kale, 1968). Selection for 15 generations was more effective in the downward direction, although it cannot be ruled out that response to selection in this case is due to an inversion rather than to polygene modifiers.

It follows that upward selection for altered *rf* usually proves to be ineffective. This can be easily explained if we assume that selection tends to increase the frequency of double exchanges in sufficiently long segments (Detlefsen and Roberts, 1921). If that is the case, then what is responsible for the lack of the effect in some of the above experiments on selection for lower *rf*? The following are what appear to be the most likely causes.

The majority of the above experiments are characterized by a narrow genetic basis of the start population used and/or close inbreeding in the course of selection sharply reducing variability. As reported by Valentin (1973), the additive genetic variance for *rf* in the segments examined by him in a specially constructed population of *Drosophila* accounted for a small part of σ^2_{rf}. Therefore, for a strong effect to be obtained, selection should go on for a relatively long time or it should be performed on a fairly large material. It is known that in selection experiments one can observe the so-called accelerated response to selection when selection is continued beyond the point at which a plateau was reached (Mather and Harrison, 1949). Some of the negative results seem to be due to a premature termination of selection.

An attempt to bring any trait, including recombination rate, to a certain limit by artificial selection can be opposed by increasingly strong natural selection. The effect of natural selection will be enhanced if artificial selection for the altered trait (e.g. recombination rate) brings about a correlated change in other traits as a result of pleiotropy and linkage (if *rec* genes are located in chromosomal regions of substantial adaptive significance). This may be manifested in lower fertility and viability of certain genotypes, in altered physiological reactions, etc.

The drawback of experiments on selection for altered *rf* resulting from limited variability at the start and/or from inbreeding in the course of selection has been overcome in studies carried out over the last two decades (Chinnici, 1971; Abdullah and Charlesworth, 1974; Dewees, 1975). Analyzed in these works are the nature of genetic changes accumulated in certain regions during selection on *rf* and their effects on other genomic regions.

By alternating individual and family estimations, Chinnici (1971) altered, in 33 generations of two-way selection, crossing over in the *sc–cv* region of *D. melanogaster* X chromosome from 15.4% to 22.1% and 8.5% respectively, both changes (Δrf^+ and Δrf^-) being highly significant. The author believes that the gradual variation in *rf* from generation to generation can be taken as an indication of the polygenic nature of the observed response to selection. Chinnici considered simultaneously the segments *cv–sn3–m*, but these, unlike the target region *sc–cv*, were subjected only to individual selection (without family selection). By generation 33, *rf* in these regions significantly decreased in both strains relative to the original level. Genetic analysis revealed strong interaction between the chromosomes of the selected lines in terms of recombination.

In a similar experiment with dominant markers, *Gl*, *Sb*, on chromosome 3, an increase in *rf* was achieved in 12 generations from 14.8% to 30%; an attempt to alter downward was unsuccessful (Kidwell, 1972). To produce a start population having an extensive genetic basis, a marker strain was crossed to four wild-type strains. Then, F_1 heterozygotes were backcrossed twice to the same strains. The resulting groups were intercrossed according to the full diallele scheme to form, after two generations, a start population. After 20 generations of selection, the reduction in fitness proved so large as to force Kidwell to discontinue the selection. The selected forms were maintained over the course of subsequent generations by mass intrastrain crosses; in 20 generations, the value of *rf* dropped to 20.5% in the 'plus' strain and to 9.6% in the 'minus' strain. It is to be admitted, in view of these results, that Dewees (1975) was right in pointing out the need for maintaining control populations in parallel with selected strains. Surprisingly, before Dewees this was practised by none of the experimenters performing selection for altered *rf*.

By way of simulating natural selection for tighter linkage between interacting loci, the following experiment has been carried out (Abdullah and Charlesworth, 1974). Selection for reduced *rf* in chromosome 3 of *D. melanogaster* was based on dominant lethal markers, *Sb* and H^2, with double non-crossover heterozygotes in repulsion ($Sb + / + H^2$) being preserved in each generation for the population maintenance. As in the experiments by Chinnici and Kidwell, special steps were taken to detect and eliminate chromosome rearrangements. That the changes in *rf* observed by Abdullah and Charlesworth are due to selection on modifier genes is also suggested by a gradual decline in *rf* in the two selected strains.

In attempting to extrapolate these and other data on artificial selection for altered *rf* to natural populations, it is however necessary to answer the question of how representative the start populations used are. Therefore, the procedure adopted in another experiment with *Drosophila* appears to be a very logical one (Charlesworth and Charlesworth, 1985a,b). The following has been done: marker genes *Gl*, *Sb* and *Sb*, *Ser* of chromosome 3 were transferred into the genetic background of flies from a natural population; then, a two-way selection for *rf* was performed over the course of 18 generations. In only one strain out of 16 did the authors achieve a substantial increase in *rf* within the *Gl*–*Sb* region (from 13% to 22%), and in none of the variants was there a marked response to selection for reduced *rf*. A genetic analysis using whole-chromosome substitutions showed the response to selection in the 'plus' strain to be due to chromosomes 2 and 3, to a lesser extent to X, as well as to epistatic interactions between these.

Interesting results have been obtained by Dewees (1975) in an experiment involving markers *rb* and *j* of the flour beetle, *Tribolium castaneum*. Since here recombination occurs in both sexes, it was necessary to choose the sex in which to perform selection. According to the author's data, $rf_{♀rb-j} = 21\%$ and $rf_{♂rb-j} = 33\%$. The former value is suitable for a two-way selection, whereas the latter is suitable for downward selection only, since the consequences of selection for higher *rf* would be difficult to record here. After 20 generations (15 of selection and five of mass breeding), mean recombination frequency in heterozygous females *rbj*/+ + from the 'plus' strain was 39.2% and 41.5%, and in those from the 'minus' strain 24.8 and 21.0%, i.e. selection for increased *rf* proved to be highly effective, but the population showed no response to the downward selection. Importantly, there was good agreement between the results from replications. Test of progeny of reciprocal crosses *rbj*/*rbj*(♀) × *rbj*/+ +(♂) have revealed similar asymmetry of response to selection, although the differences in Δrf^+ and Δrf^- are not as large. A genetic analysis of differences between the 'plus' and 'minus' strains showed the dominant alleles reducing recombination to be responsible for the asymmetry or response to selection.

It follows that direct selection for altered *rf* in females results in a correlated change in $rf_♂$, the pattern of correlated response depending on the direction of selection and phase of markers (*cis*- or *trans*-combination in the heterozygote in question). The observed difference in the effects of *rec* alleles in heterozygous males and females seems to be due to the influence of sex (as a condition) on the degree of dominance at the corresponding heterozygous *rec* loci. Differential behavior of *cis*- and *trans*-heterozygotes may reflect the fact that a proportion of heterozygous *rec* loci are linked to the region in question (*rb-j*).

Having completed five cycles of selection for higher and lower chiasma frequency in the locust *Schistocerca gregaria*, Shaw (1974) obtained a difference of four chiasmata (20% of the mean value) between the 'plus' and 'minus' strains. Analysis of the progeny of crosses between the selected strains has revealed a large additive component in the *rec* system responsible for the differences accumulated in the course of selection.

Selection for altered *rf* proved to be most effective in the silkworm males. In as few as ten generations, Hasimoto (1960, cited by Turner, 1979) succeeded in enhancing *rf* in the *p–y* region of chromosome 2 from 25.6% to 37–39% and in lowering it down to 5–7%, i.e. the difference between the upward and downward selection was 32%. A subsequent genetic analysis by other workers showed the differences resulting from selection to be due to both recessive and dominant modifiers (some of these exhibiting appreciable effects, up to 8–11%). Nevertheless, the system of multifactor control shows, in general, no prevailing dominance of enhancer or reducer alleles (Turner, 1979; Ebinuma and Yoshitake, 1981; Ebinuma, 1987).

Although genetic variation in *rf* is extremely high in a number of species of higher plants, we know of only one attempt to alter this parameter by means of selection (Allard, 1963). The author used three hybrid combinations of lima beans heterozygous for segments *rw–l* (*rf* = 38%), *s–l* (26.9%), and *wl–d* (2.1%). The selection procedure used was as follows. In each combination, exchange frequency was estimated for five randomly chosen heterozygous F_2 plants based on the F_3 progeny. Then from each family, with the lowest and the highest *rf* values, three F_3 genotypes were chosen at random, their F_4 progeny tested, and so on. In all the segments studied, recombination rate during upward selection in F_5 (as estimated from the F_6 family data) substantially increased relative to the original value; downward selection proved to be almost ineffective. On the whole, the difference $rf^+ - rf^-$ between the selected strains was 10% for the *rw–l* segment, 20% for *s–l* and 7.5% for *wl–d*.

Allard attributes such a strong asymmetry of response to selection to the peculiarities of the organism used. Employed in the experiment were cultivated forms of lima beans, and since their growing conditions place a high selective premium on genetic uniformity, the preceding selection reduced *rf* to a minimum. A similar idea was put forward by Jain and Gupta (1971) in discussing evidence for higher chiasma frequencies in a group of primitive varieties of maize as compared with contemporary ones. It may be, however, that this general argument, valid under certain conditions, has no direct bearing on the experiment in question. Indeed, we have already seen that selection for reduced recombination in different populations of the same species (*D. melanogaster*) can produce entirely different results under approximately equal environmental conditions. Selection effectiveness depends on the presence of a sufficient amount of genetic variability for *rf* in the start population, on the selection scheme, duration and intensity and on the genomic regions studied. The limited genetic basis, close inbreeding, low intensity and short duration of selection (only four cycles, F_2–F_5) are all factors that could also have been responsible for the lack of changes in *rf* in the downward direction in Allard's experiment. Moreover, comparative cytological evidence indicates that in mammals, domestication (and the associated) increase in the uniformity of environmental conditions) has resulted not in a reduction but, conversely, in a considerable increase in recombination (Burt and Bell, 1987).

A direct two-way selection for altered *rf* has been performed by Landner (1971) in *Neurospora*. After three cycles of selection, he identified, among the high- and low-recombination strains, a number of isolates showing significant differences in *rf* in the region in question. This material also enabled the effect of selection for recombination in other regions to be assessed. Selection for high and low recombination was also performed with *Schizophyllum commune* (Simchen, 1967). The resultant selections varied in *rf* in different segments.

In conclusion, natural and laboratory populations of a number of organisms have been experimentally found to contain a considerable amount of genetic variability in recombination, available for selection to act upon. The effectiveness of directional selection for altered *rf* has been shown to depend on the segment under study, the size, structure and origin of the start population, the breeding system, the estimation procedure and the intensity and duration of selection. These experiments provided the necessary material for an in-depth study of the problem of the evolution of recombination. The data obtained suggest that the exchange frequency and distribution in the eukaryotic genome are under a complex control of polymorphic *rec*-systems, the rate of recombination in a given region depending on both linked loci and genes located on other chromosomes.

6.4 TWO PROBLEMS IN EXPLAINING THE *rec* SYSTEM EVOLUTION

6.4.1 Effects and function: the 'relay race' principle

Section 6.1 provides a review of population genetic models seeking to solve one of the most difficult problems of modern theory of evolution, namely to account for the wide occurrence of sexual reproduction and maintenance of non-zero recombination in living organisms. As we have already seen, natural populations exhibit substantial variability in *rf* and, unless one assumes this variability to be absolutely neutral, it is necessary to explain the reasons for maintaining it. The first impulse that one may feel on acquainting oneself with the diversity of mechanisms proposed is to take the very diversity of proposals as indicating the absence of a genuine explanation which one would like to be straightforward and generally applicable. It is not impossible that such a conclusion would be justified if the history of life on the earth could be likened to a one-act play with a single scenario. The reality, however, seems to be better approximated by a situation with sequential–parallel succession of scenarios, making attempts to provide a functional (adaptationist) explanation much more difficult.

Elucidation of the nature of difficulties posed by the functional approach was facilitated by the discussion of this issue in the general conclusion to the *Proceedings of the Symposium of Molecular Evolution* (Maynard Smith, 1982). The point that was made was that if some structure or mechanism affects a certain process, the fact itself is an insufficient reason for asserting that this is the function of the structure in question. For instance, from the fact that satellite DNA affects recombination rate it does not follow that it has been formed for this reason. Although mobile genetic elements enhance genetic variation, one cannot say that this effect was the main cause of their appearance, evolution and persistence in their present form.

With respect to the reasons for maintaining non-zero recombination, two functional explanations have been proposed so far, one of which proceeds from the assumption of recombination origin and evolution due to its repair (conservative) functions (Bernstein, 1979, 1983; Holliday, 1987, 1988; Bernstein, Hopf and Michod, 1988; Shields, 1988). The other, more conventional and, in our opinion, more justifiable approach is based on the apparent combinative effects of recombination (Fisher, 1930; G.C. Williams, 1975;

Maynard Smith, 1978, 1988a; Brooks, 1988). Clearly, models of Muller and of his followers (Muller, 1932, 1964; Felsenstein and Yokoyama, 1976; Manning, 1982, 1984; Kondrashov, 1984, 1988a; Charlesworth, 1990), which 'outwardly' seem to be repair ones should also be referred to the class of combinative explanations.

Which of the two effects of recombination then should be considered a primary or true 'function' determining the pattern of macroevolution of the phenomenon? While being very interesting and important, this question does not appear to be a crucial one since an answer to it can only be expected at the verbal rather than the operational level. By contrast, the situation is basically different with respect to problems concerning the evolution of *rec* systems and it is quite realistic to formulate questions and obtain answers regarding the factors responsible for recombination rate decrease or increase, advantage of non-zero recombination, *rec* gene polymorphism, and so on (Charlesworth, 1985). Now we have perhaps too many answers to questions like these and do not know how to choose the correct one. But that is not so bad after all, since one cannot rule out the possibility that all the factors considered by population genetics (section 6.1, Michod and Levin, 1988) are important in the evolution of *rec* systems. Of course, their relative roles depend on particular situations (pattern of ontogenetic cycle, genome size and organization, mutation rate, breeding system, pattern of environmental change in space and time, etc.) and vary with the circumstances. Consequently, each particular evolving system, including the *rec* system, can have its own limiting factors determining the pattern of its dynamics. A change in circumstances accompanied by changes in limiting factors results in altered relative roles of mechanisms responsible for the evolution of both the entire system and its *rec* properties. To put it in another way, one can speak of a 'relay race' principle in evolution (Ratner, 1990).

Assuming the relay principle allows one to think more freely in terms of **effect** and **function**: what can only be regarded as an **effect** at one stage becomes a function at another, thereafter evolving as a **function**. This by no means rules out the preservation of older functions, including those of repair (Williams, 1988). Since the *R*-system properties (mutation rate, breeding system, recombination and linkage) are themselves affecting the evolutionary process, such a relay race can be viewed as an evolution of factors of evolution (Darlington, 1939; Malinowsky, 1941; Schmalhausen, 1968; Zavadsky and Kolchinsky, 1977). In general, the pluralistic approach to the problem of the evolution of sex and recombination has been increasingly accepted in recent years (Michod and Levin, 1988).

Even if DNA repair had been the primary function of recombination at early stages of evolution with combinative variability being its by-product, the 'relay baton' could have passed to combinative functions of recombination following the transition to diploidy and development of sufficiently effective systems of genetic repair. Therefore, we are justified in regarding the effect of recombination on genetic variation observed in the currently existing organisms as a key factor (or at least one of the key factors) of the *rec* system evolution. It appears impossible to decide at present which of the combinative explanations of the evolution of recombination is the main one in each particular situation or group of situations (Maynard Smith, 1978, 1988a; Felsenstein, 1985; Crow, 1988; Fletcher, 1988; Seger and Hamilton, 1988).

6.4.2 Parametric versus multiple modifier models

A careful study of each of the theoretical models is a necessary step in elucidating their plausibility, though even with this rather modest formulation of the problem we are still far from its solution. Actually, in currently available models the *rec* locus dynamics depends on two groups of **fixed parameters**: one specifying the genetic system (*R*-parameters) and the other the individual's adaptation to environmental conditions, abiotic or biotic (*F*-parameters). It is known, for example, that the rate of selfing may determine not only the extent but also the direction of the rec^+ allele frequency change within a population (Charlesworth, Charlesworth and Strobeck, 1979; Holsinger and Feldman, 1983b). Mutation models of the evolution of recombination have mutation rate as one of their parameters. A variety of models for the evolution of multigene families and other repeated DNA sequences reveal a significant effect of crossing-over level or gene conversion rate (depending on the mechanism in question) on the system's behavior. The recombination rate can affect: (a) changes in the frequency of mobile genetic elements within a population (Charlesworth and Langley, 1986); (b) the fate of modifiers of meiotic drive (Eshel, 1985; Lessard, 1985); (c) the fate of *rec* alleles (Feldman, Christiansen and Brooks, 1980; sections 7.4 and 9.3); and (d) the conditions for the spread of selfing modifiers within a population (Karlin and McGregor, 1974).

We believe that the models under study will be more realistic if the fixed values of mutation rate, of selfing rate, recombination level, ratio of reciprocal to unequal exchanges, etc. (here referred to as 'parametric models') are replaced by their modificatory analogs in which all the relevant properties are determined by polymorphic loci rather than being fixed. Of course, analysis of the dynamics and equilibria of systems like these, with several modifier loci, is much more complex than that of parametric ones, but it permits a more adequate study of the problems related to the evolutionary role of recombination and to the evolution of recombination itself. By analogy with models for the evolution of selected loci, it is probably worthwhile to provide the simulated systems with an ability for 'self-adjustment' of the repertoire of allelic effects of modifier loci.

This approach to the *R*-system modeling is not only warranted but even indispensable. Indeed, the evolution of *R*-properties due to selection for *F*-properties is only possible via a feedback mechanism, i.e. selection for genes and combinations of genes of the selected system brings about associated changes in frequency of the *R*-system genes that have brought these combinations into being. Analyzing the behavior of parametric systems with fixed values of such parameters as mutation rate, rate of selfing, etc., actually implies studying a situation with an open feedback loop, i.e. the model fails to fit the phenomenon under study. One of the first examples of a model of this type was suggested by Thomson and Feldman (1974), who considered simultaneous dynamics of meiotic drive modifier and *rec* modifier, the latter controlling the linkage of the former to the driven locus. Bengtsson and Uyenoyama (1990) proposed a two-modifier model of segregation evolution including gene conversion and meiotic drive. Wiener and Feldman (1991) analyze co-evolution of modifiers of migration and mating system (although the latter modifier in this model is not a neutral one and acts as a selectable locus).

It is also important to take into account the possible non-stationarity of *F*- and *R*-system co-evolution processes. Evolutionary changes in some of the *R*-system characteristics can produce a shift in the range of effects of *F*-loci, thereby creating new opportunities for further evolution of *R*-properties. A model of this type has been suggested for analyzing the co-evolution of selfing rate (an *R*-property) and inbreeding depression (Charlesworth, Morgan and Charlesworth, 1990, 1991).

6.4.3 Many properties of recombination wait for evolutionary explanations

In the review of models of the *rec* system dynamics (section 6.1), we have largely focused on the evolution of **recombination rate**. The reason for this is that most of the theoretical work in the field has been devoted to studying this problem (see also the results discussed in Chapters 7 and 8). However, there also exists another group of problems, concerned with a number of properties of recombination, that are no less important for understanding the role of recombination in the population dynamics of multilocus systems and in the evolution of recombination itself and of other *R*-system components. In other words, the basic properties of recombination are valid for evolutionary explanations and the functional significance of recombination features should be examined. In view of differences in the meaning of the terms 'effect' and 'function', such a treatment, of course, will only imply ascertaining that the explanations proposed are possible in principle and logically consistent with the population genetic theory.

Among recombination properties requiring an evolutionary (functional) interpretation, the following seem to be worth mentioning: the dependence of *rf* on the environment (a fact established almost 80 years ago); non-uniform genomic distribution of recombination events (both between and within chromosomes) and interference of exchanges (known since 1914); inter-specific variation of the conversion/crossover ratio; sexual differences in recombination rate and distribution, especially difficult to account for in hermaphroditic species; interchromosomal effect of rearrangements on recombination; allelic relations (dominance–recessiveness) at *rec* loci; etc.

Important evidence for the evolutionary determination of some of the above properties is that they are altered in meiotic mutants, sometimes dramatically. In particular, it has been shown, in *Drosophila* and some plant species, that mutations at individual *mei* loci reduce interference and result in a redistribution of exchanges, in favor of segments where crossing over normally occurs at lower frequencies than would be expected based on physical lengths (Carpenter and Sandler, 1974; Jones, 1967; Moens, 1969; Sandler and Szauter, 1978; Carpenter and Baker, 1982; Zhuchenko and Korol, 1985; Smirnov, 1991). In other words, non-randomness of recombination is due not only to peculiarities of primary DNA structure, but also to normal *rec* alleles.

While a considerable amount of literature has been accumulated on the problems of forces maintaining non-zero recombination (the problem of 'recombination level'), the corresponding work on the 'recombination phenomenology' has scarcely begun. An exception are separate works on the evolution of non-uniform distribution of exchanges (Charlesworth, Langley and Stephan, 1986) and Maguire's (1980) verbal model concerned with the role of interference. Actually, an adaptationist explanation is also needed (and can be provided) for such cardinal properties of the *R*-system of diploid sexual organisms as the Mendelian segregation ratio of 1 : 1 and random assortment of non-homologous chromosomes (Gershenson, 1928; Darlington, 1939, 1979; Malinowski, 1941; Lloyd, 1984; Eshel, 1985; Williams, 1988; Feldman and Otto, 1991; Lyttle, 1991, 1993; Silver, 1993).

It seems reasonable to view normal Mendelian segregation as a product of evolutionary co-adaptation, of adjustment of genomic components within a species. Indirect evidence for this is provided by the fact that segregation distortions frequently occur in the progeny of interspecific hybrids and are similar in manifestation to meiotic drive systems (Bonierbale, Plaisted and Tanksley, 1988; Paterson *et al.*, 1988; Seldin, Howard and D'Eustachio, 1989; Doebley and Stec, 1991, 1993; Gebhardt *et al.*, 1991; Eshed *et al.*, 1992; also reviewed in Zhuchenko and Korol, 1985). Nature provides clear though comparatively rare examples of possible segregation distortions. It is necessary to understand how and why 'normal' Mendelian segregation has come to be the general norm (Hurst, Hurst and Johnstone, 1992).

It is of interest to note that interspecific hybridization can lead to still more dramatic disorders in the sexual system, right up to its replacement by parthenogenesis (Vrijenhoek, 1990). Why a switch like this is confined to a comparatively small group of taxa and what features of the genetic system make the very change possible remains unclear.

In concluding this chapter, we would like to recall the essential difference between the problems of explaining the evolutionary reasons for the maintenance of sex and recombination. The tendency of many authors to view these tasks as identical may result in a substantial bias in comparative assessment of potentials of various hypotheses and models. In particular, selection against harmful mutations should probably be considered to be one of the most important mechanisms ensuring the evolutionary (selective) advantage of non-zero recombination. Within this class of models, the interaction of selection and recombination is a conservative factor maintaining the attained level of organization. It appears that recombination here might have evolved within ameiotic asexual systems, as, for example, in immune response systems of mammals, or in the form considered by Szathmary and Kover (1991). A contrasting situation is encountered in dealing with selection for new favorable mutations, or with adaptation to envrionmental conditions (physical or biotic) varying in time or space. In this case, recombination cannot in principle be effective without being associated with the sexual process (amphimixis). Its role here is that of generating the **novel** rather than that of protecting the **old**. The difference between these two explanations of recombination functions reflects two approaches to the concept of variability: the balanced one and the neoclassical one (see Lewontin, 1974). On the other hand, it can be hypothesized that the advantages of sex (**segregation**) are in no way associated with **recombination** (Moore and Hines, 1981; Eshel and Weinshall 1987; Kirkpatrick and Jenkins, 1989). But how then could one explain the fact that in eukaryotes sexual reproduction and segregation always coincide with meiotic crossing over (at least in one of the sexes)?

7 Selection for increased recombination due to environmental fluctuations

7.1 THE PROBLEM

The existence in nature of substantial within-population variance for recombination rate as well as obvious deleterious effects of recombination (break-up of co-adapted gene blocks, a dissipative tendency to return any deviation to the population mean, twofold disadvantage of sexual reproduction as compared with parthenogenesis) raise the question of evolutionary mechanisms responsible for the origin and persistence of recombination and sex. As already noted, there are several classes of models accounting for selection in favor of recombination. These assure the advantage of high recombination by mechanisms, deterministic or stochastic, generating linkage disequilibrium.

One of the most popular explanations of the deterministic type is provided by the polygenic model pioneered by Maynard Smith (1980, 1988b). He has shown that directional selection as well as a combination of stabilizing and directional selection can result in a rapid increase in the frequency of the rec^+ allele. He examined a selected system composed of six equidistant loci with equal additive effects and a *rec* modifier. According to Maynard Smith, the observed increase in *rf* under directional selection is the result of the formation of negative linkage disequilibrium in early generations (excess of combinations of the *101010* type) which subsequently, being disrupted by crossing over, produces selectively advantageous gametes of the *111111* or *000000* type (depending on the direction of selection). An increase in the proportion of recombinant genotypes in the population results in a higher frequency of the rec^+ allele. He also examined the situation with a cyclically varying optimum (fluctuating selection). In all cases, there was a loss of variability (see also Lande, 1976). To maintain polymorphism, Maynard Smith introduced into the selected system a mutation process with a very high mutation rate (10%).

The role of fluctuating (in space and/or time) selection as a factor maintaining genetic polymorphisms is one of the central problems in population genetics and has a direct bearing on the general controversy about the causes of the variability observed in nature. Its solution is beset with difficulties even in the one-locus case, the more so when attempting to interpret the experimental evidence (Felsenstein, 1976; Gillespie, 1976; Hedrick, Ginevan and Ewing, 1976; Hedrick, 1986).

Subsequent analysis has shown that Maynard Smith's model permits a number of quite acceptable parameter combinations resulting in polymorphism persistence for the selected system of loci (SSL) for an extended period of time even in the complete absence of mutation (Korol and Preygel, 1986, 1989; Maynard Smith, 1988b). Actually, a considerable reduction in variability due to allele frequencies in the SSL tending to deviate from the initial level of 0.5 can occur very soon, but the progress to fixation, if any, usually takes many hundreds of generations to complete (a fact also mentioned by Lewontin, 1974; and Vol and Koval, 1985). It turned out that this time interval is usually sufficient for dramatic changes to occur at the locus modifying recombination (e.g. fixation of the rec^+ allele even if its initial frequency is 1–5%). Considering situations with a small (1–2%) selective advantage of heterozygotes removes the problem of fixation. Moreover, for a number of fitness function types, the maintenance of a polymorphism can also be achieved with purely additive effects of genes on the selected trait (under unequal gene effects), or with partial dominance of either allele (no matter which) at one or several loci (Korol, Kirzhner and Ronin, 1993). All this enables one to question the established opinion that viewing fluctuating environments as a universal factor in the evolution of recombination holds no promise because such fluctuations are unlikely to prevent gene fixation.

7.2 DESCRIPTION OF THE MODEL

We have considered the *rec* system dynamics in multilocus infinite random-mating diploid populations in variable environments (Korol and Preygel, 1986, 1989). The number of selected loci (k) (with two alleles, *0* and *1*, at each locus) can be set in a broad range (say from 2 to 6). Selection may be defined by a function $w(x, z) = w(|x - z|)$ like

$$w(x, z) = \exp[-(x - z)^2/2s^2], \tag{7.1}$$

where z is an optimum value of the trait in a given generation, s is a coefficient determining the intensity of selection against genotypes deviating from the optimum value and x is the genotypic value of the trait. We considered the case of loci with equal additive effects on the selected trait (x is equal to the number of *1* alleles) as well as that of partial or complete dominance. In the complete dominance case, x is equal to the number of loci with at least one *1* allele. We have provided for the possibility of specifying heterozygous advantage as an additional factor maintaining polymorphisms. The overall fitness of genotypes with character x, heterozygous for k loci, is given here by $w_k(x) = w(x)(1 + \delta)^k$, where δ is the magnitude of the selective effect of single-locus heterozygosity.

In addition to formulating the $w(x, z)$ function as in equation 7.1, we have considered a number of other ways of specifying stabilizing selection (Figure 7.1). In all these cases, possibilities are provided for considering either equal or unequal additive effects on the selected trait, partial or complete dominance, as well as heterosis for fitness. In the case of two loci (A/a and B/b), the value of x is specified by the matrix

$$\begin{array}{cccc} & aa & Aa & AA \\ bb & m - d_a - d_b & m + h_a - d_b & m + d_a + d_b \\ Bb & m - d_a + h_b & m + h_a + h_b & m + d_a + d_b \\ BB & m - d_a + d_b & m + h_a + d_b & m + d_a + d_b \end{array}$$

where d_a and d_b are additive effects, h_a and h_b are dominant effects and m is the overall mean.

For simplicity, it was assumed that only single and double exchanges are possible within the SSL. Let r denote the exchange rate over the entire segment containing the SSL. Then, assuming no chiasma interference, $f_0 = (1 - r)^2$, $f_1 = 2r(1 - r)$, $f_2 = r^2$. Provided that p_s is the recombination rate between the adjacent equidistantly located selected loci, then

$$p_s = 2r(1 - r)/(k - 1) + r^2(k - 2)/C^2_{k-1} = 2r/(k - 1).$$

The difference from Maynard Smith's (1980) scheme is only in the way of

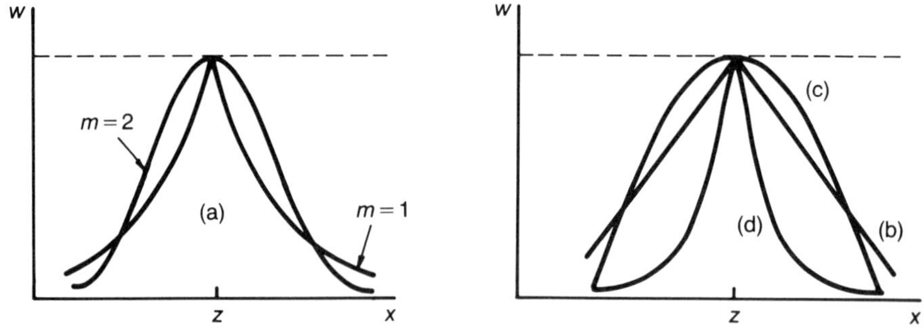

Figure 7.1 Types of fitness function $w(x, z)$ used in numerical analysis of the *rec* system dynamics. The shape of $w(x, z)$ is defined as follows: (a) $w = \exp[-(|x - z|/s\sqrt{2})^m]$; (b) $w = 1 - s|x - z|$ for $|x - z| < 1/s$ and $w = 0$ in other cases; (c) $w = 1 - s(x - z)^2$ for $|x - z| < 1/\sqrt{s}$ and $w = 0$ in other cases; (d) $w = 1/[1 + (x - z)^2/s^2]$.

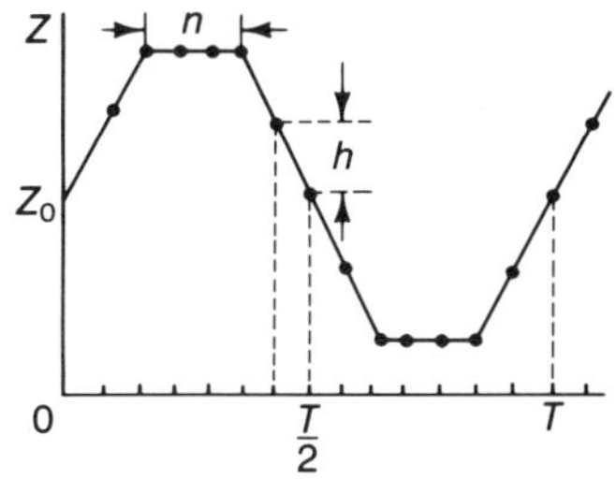

Figure 7.2 Trapeziform variation of the optimum with time in computer simulations.

specifying recombination between the *rec* modifier and the SSL, the latter being defined by a separate parameter that is independent of r.

As an initial point, a vector of gamete frequencies was usually chosen with linkage equilibrium between all the selected loci as well as between these and the *rec* locus. It was assumed that selection occurs with a fluctuating optimum whose temporal variation is, in the general case, trapeziform (Figure 7.2).

The model discussed in this section describes situations of the genotype–environment interaction since it is assumed that for each genotype there is an environmental state in which its fitness is maximum, and when the environmental state is changed there is a change in the ranking of selectively favored genotypes (Glotov and Tarakanov, 1985; Bell, 1988). Contrary to the older notions, it has recently been found that environmental fluctuations under genotype–environment interactions and, even, additive effects of fitness genes can promote the maintenance of genotypic variability (Gillespie and Turelli, 1989). Via and Lande (1987) have shown that the genotype–environment interaction plays an important role in the dynamics of genetic variation under disruptive selection in a spatially heterogeneous environment. Also, there is experimental evidence that spatial and temporal environmental heterogeneity may lead to an increase in genetic variance in a population or to stabilization of variability (Allard and Kahler, 1974; Mackay, 1981; Barker, Starmer and Vacek, 1987; Verdonck, 1987; Mukai, 1988; Altukhov, 1989; Mitchell-Olds, 1992; Nevo *et al.*, 1993).

Variations in gametic frequencies from generation to generation were calculated in our model by the method of 'genetic operators' (Lewontin, 1964). In essence it is as follows: deterministic equations for the dynamics are not written explicitly but are formulated and simultaneously solved with the aid of computer-implemented logical rules. Let g_n and g_{n+1} be gametic frequencies in generations n and $n+1$. Formally, this can be written as $g_{n+1} = \Omega g_n$, where Ω is an operator transforming gametic frequencies. The Ω operator is a sequence of several operations:

$$\longrightarrow \underbrace{[g_n] \xrightarrow{M} [z_n]}_{\text{generation } n} \xrightarrow{S} [z_s] \xrightarrow{R} \underbrace{[g_s] \xrightarrow{W} [g_{n+1}]}_{\text{generation } n+1} \longrightarrow$$

where M is a crossing operation generating a matrix of zygotic frequencies in generation n, $z_n = g'_n\ g_n$, and S is a selection operation consisting in multiplying each element, z_{ij}, of matrix z_n by corresponding fitness values, w_{ij}. The matrix of frequencies of the selected zygotes, z_s, is used to generate a vector of the selected gametes of the next generation, g_s, with the aid of recombination operator, R, which consists in multiplying each z_s element by the probabilities of various crossover events and adding the result of each multiplication to the corresponding g_s element. Finally, g_s is transformed, by a normalization procedure, into g_{n+1}. The normalization coefficient, $w = \Sigma g_{si}$, is the mean fitness of the population in generation n.

7.3 RESULTS

7.3.1 Feedback selection for increased recombination with a cyclically varying optimum

Maintenance of polymorphism

The possibility of maintaining a polymorphism within the SSL depends on the shape of the fitness function $w(x, z)$, the mode of action of genes on the selected trait x, the relative contributions of stabilizing and directional selection and *rf*. The following are some of the results obtained for two-locus selected systems in the absence of heterozygous advantage (Korol, Kirzhner and Ronin, 1993). The types of $w(x, z)$ examined (a–d) are shown in Figure 7.1 (a–d). Numerical examples which appear in Table 7.1 correspond to a trapeziform environment with a period $T(1, n, 1, n)$, where states with intermediate values of z last for one generation each and those with extreme values for n generations. The value of z is x_{AaBb} in the intermediate state, while $z_{\min} = x_{aabb}$ and $z_{\max} = x_{AABB}$ in the extreme states. The volume of the attracting domain of stable polymorphic trajectories was calculated using random initial points (haplotype frequencies) evenly distributed in the four-dimensional space.

(a) For exponential functions of the type shown in Figure 7.1 (a), the polymorphism is unstable for any k and s when $d_a = d_b$ and $h_a = h_b = 0$. However, with unequal effects of additive genes or a departure (even a small one) from additivity, stable polymorphic trajectories are possible unless *rf* exceeds some critical level, *rf**. In all cases (under both stable and unstable polymorphisms), linkage disequilibrium D remains negative within the cycle (see also Maynard Smith, 1980, 1988a, b). Further analysis shows that the maintenance of stable sign of D within the period critically depends on the sign of the 'integral fitness disequilibrium coefficient' (Kirzhner, Korol and Ronin, 1993). The critical role of unequal effects of additive genes has also been shown for the maintenance of polymorphisms under selection in constant environments (Hastings and Hom, 1990).

(b) When $w(x, z)$ changes in a linear manner with deviation of x from z, the

Table 7.1 Polymorphism under two-locus stabilizing selection with cyclically changing optimum

Shape of $w(x, z)$	Parameter values	Period $(n, 1, n, 1)$	$x(A/a, B/b)$ d_a	d_b	h_a	h_b	Range of rf[†]	Range of v[‡]
$\exp\text{-}(\lvert x\text{-}z\rvert/s\sqrt{2})^m$	$m=1, s=4$	1414	2	1	0	0	0–0.5	1–0.38
	$m=2, s=3$	1414	2	1	0	0	0–0.05	1–0
	$m=1, s=1.5$	1212	1	1	0.1	0	0–0.017	1–0
	$m=1, s=4$	1212	2	1	0	0.1	0–0.06	1–0
$1-s\lvert x-z\rvert$	$s=0.1$	1414	2	1	0	0	0–0.026	1–0
	$s=0.1$	1212	2	1	0	0	0–0.04	1–0
	$s=0.1$	1414	1.5	1	0	0	0–0.01	1–0
	$s=0.2$	1414	1	1	0.2	0.2	0–0.013	1–0
$1-s(x-z)^2$	$s=0.05$	1414	1.5	1	0	0	0–0.026	1–0
$1/\{1+[(x-z)/s]^2\}$	$s=2.0$	1414	1	1	0.2	0.2	0–0.014	1–0
	$s=2.0$	1414	2	1	0	0	0–0.021	1–0
	$s=1.0$	1414	2	1	0	0	0–0.5	≈ 1

[†] The range of rf values comapatible with polymorphism for selected loci is presented; for each of the tested values of rf the volume (v) of the attracting domain of stable polymorphic trajectories was calculated using 10^5 random initial points.

[‡] The range of the v-values corresponding to the rf range is presented. In most of the cases a critical value $rf = rf^*$ exists such that $v = 1$ if $rf < rf^*$ and $v = 0$ if $rf > rf^*$.

additivity of the effects of genes on the selected trait leads (at certain parameter values) to the additivity of their effects on fitness. As in case (a), a polymorphism is possible here when $d_a \neq d_b$ or $h_a \neq 0$ and $h_b \neq 0$. With increasing *rf*, the area of polymorphism attraction diminishes. Within the cycle, $D < 0$.

(c) For $w(x, z)$ of the type in Figure 7.1 (c), a stable polymorphism is only possible when $d_a \neq d_b$. With increasing selection intensity and decreasing *rf*, the polymorphism attraction area expands. Within the cycle $D < 0$.

(d) Provided selection is sufficiently intense, a stable polymorphism is possible in this case even for $d_a = d_b$ and $h_a = h_b = 0$, such that, within the cycle, $D > 0$ under tight linkage and changes sign when *rf* is sufficiently large. With less intense selection, a polymorphism is possible either under unequal gene effects or when h_a and h_b are non-zero.

Thus, contrary to the general belief based on asymptotic results (Lande, 1976), stabilizing selection with a fluctuating optimum can promote maintenance of polymorphism for additive genes (see also Korol, Kirzhner and Ronin, 1993). It is worth mentioning that limiting behavior of even simple two-locus systems under cyclical haploid selection could be rather complex, with long supercycles, T-cycles and chaotic-like dynamics as examples (Kirzhner *et al.*, 1993).

Changes in the rec system

Most of the results presented below pertain to Gaussian selection (Figure 7.1 (a)). For all the systems studied – from the two- to six-locus one – it is possible to find, for various types of gene action (additive and dominant, with and without heterozygous advantage), a number of parameter sets for which recombination is selectively advantageous. Examples of the rec^+ allele dynamics in an environment with a fluctuating optimum in the absence of linkage between the modifier and the selected system are given in Table 7.2. Under purely stabilizing selection or low variation of the optimum, there is a decrease in the frequency of the rec^+ allele. Selection on the *rec* locus induced by adaptation to varying environments proves to be effective for both large and small differences in *rf* between the rec^+ and rec^- alleles. The rec^+ allele increases in frequency both when dominant over the rec^- allele and when recessive. As is seen from the table, rec^+ can spread through the population regardless of the number of loci in the selected system.

One important fact is worth noting. In Maynard Smith's (1979, 1980) model, the modifier is linked to the selected loci. Our computations show that a rapid change of the genic composition at the *rec* locus in favor of the high recombination allele also occurs when the locus is unlinked to the SSL. The explanation based on the hypothesis of 'hitch-hiking due to linkage' is, therefore, inapplicable to this case, and one can only speak of **co-selection** of the *rec* locus and the selected system.

Table 7.2 The effect of stabilizing selection with a cyclically fluctuating optimum on the rec^+ allele frequency $[w(x,z) = \exp - (|x - z|/s\sqrt{2})^2]$

					rf value in genotype		*No. of generations*	*Frequency of rec⁺*	
k	δ	s	h	T	rec^-	rec^+	(n)	p_0	p_n
2	0	0.75	0.40	16	0.01	0.2	100	0.5	0.73
3	0	0.5	0.50	16	0.01	0.05†	200	0.01	0.66
3	0	0.7	0.50	16	0.01	0.05	500	0.05	0.91
4	0	1.0	1.50	16	0.01	0.02†	150	0.01	0.72
4‡	0	1.0	1.00	16	0.01	0.1	200	0.05	0.75
4	0.05	1.0	1.00	8	0.01	0.05	300	0.5	0.95
5‡	0	1.0	0.75	16	0.01	0.1	300	0.05	0.74
5	0	1.0	3.50	8	0.01	0.02	150	0.10	0.98
6	0	1.4	2.00	8	0.01	0.05	35	0.5	0.74
6	0	1.4	3.00	8	0.01	0.1	35	0.5	0.99

† rec^+ allele is dominant; in the other variants it is recessive.
‡ Variants with a dominant *1* allele at the selected loci, the effects of alleles on the selected trait being additive in the other cases.
k, number of selected loci; δ, heterozygous advantage per locus; h, change of the optimum per generation; T, the period of changes.

Apart from the question of the evolutionarily stable recombination rate between the selected loci, of considerable interest is the analysis of the rec^+ allele dynamics in circumstances favoring increased recombination. In most of the cases studied, these dynamics are in the form of pulsations whose 'depth' depends on the intensity of linkage between the modifier and the SSL. An increase in the rec^+ allele frequency occurs in the 'unfavorable' portion of the period (when the population mean shows maximum deviation from the current optimum), while in the 'favorable' portion *rf* rises less rapidly or even declines (Figure 7.3).

Intuitively, this behavior is quite understandable. It is in the unfavorable portion of the period that the gene combinations selected for are in the minority so that increased crossing over has a positive net effect and produces recombinants that confer a selective advantage on the rec^+ allele. This advantage can be realized with a greater or smaller amount of lag, depending on specific parameters of the model: the maximum in variation of the p_{rec^+} value is somewhat shifted relative to the minimum in the population mean fitness. Conversely, in the portion of the period with high mean fitness, recombination will tend to break up favorable combinations constituting the majority of the gametes, thereby resulting in lower fitness in the next generation. A rise in the rec^+ frequency is therefore slowed down, or even followed by a temporary decrease. But since the environment varies in time, the population reaches a point at which its genetic make-up is maladapted to the environment, recombinants are needed again, and they increase in frequency, resulting in a higher frequency of the rec^+ allele.

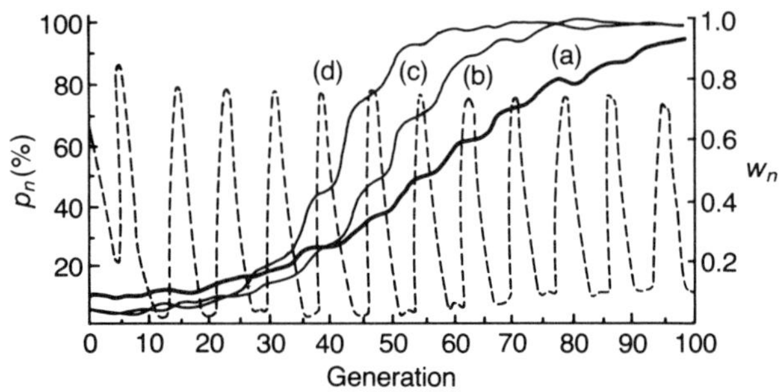

Figure 7.3 The effect of selection in a cyclically varying environment (period $T = 16$) on the frequency (p_n) of the rec^+ allele: (a) $k = 3$, $h = 1$; (b) $k = 4$, $h = 1.5$; (c) $k = 5$, $h = 1.75$; (d) dynamics of the mean fitness for case (a). In all the variants, the fitness functions used is $w = \exp[-(|x - z|/s\sqrt{2})^2]$, selected loci are additive, $\delta = 0$, $s = 1$ and the rec^- allele corresponds to $rf = 0.01$ and rec^+ to $rf = 0.1$, the rec^- allele being dominant.

Fluctuating selection with an environmental state, optimal for the entire population

Following Maynard Smith, we proceeded in our simulations from the assumption that corresponding to each environmental state is a certain optimum value of the quantitative trait and, conversely, for each genotype with its own trait value there is a corresponding optimum environmental state. Biologically, the assumption that there exists an environmental state, optimal for the entire population, for which selective differences between individuals are minimum seems to be more plausible. When the environment deviates in either direction from this optimum value, fitness is decreased such that genotypes with larger (relative to the optimum) trait values show smaller fitness reductions with the environmental deviation to the right and those with smaller trait values with the environmental deviation to the left of the optimum state. Numerical analysis confirmed the previous general conclusion: temporal variation in environment can place a selective premium on non-zero *rf* over zero *rf* and on a high rate of exchange over a low one (Korol and Preygel, 1986, 1989).

7.3.2 Is a linked *rec* modifier always better?

The effect of linkage of the *rec* modifier with the SSL on the effectivenss of selection for recombination has previously been considered in a variety of models (Nei, 1969; Charlesworth, 1976; Strobeck, Maynard Smith and Charlesworth, 1976). According to the general conclusion, close linkage promotes a more rapid spread of the allele, enhancing recombination. In a model with a varying environment, Charlesworth (1976) observed an increased rate of rec^+ allele spread when it was linked to the SSL. He has also established that for rather long periods of environmental fluctuation it is possible to obtain situations in which not only the rate but also the direction of the modifier dynamics is dependent on the intensity of linkage of the modifier with the SSL: if the modifier is unlinked to the selected loci then the rec^+ allele is eliminated from the population, and if it is linked then it becomes established. The same effect was observed in a model with a balance between selection and mutation (Feldman, Christiansen and Brooks, 1980). An increase in the evolutionary stable recombination value with a closer linkage of the *rec* locus with the SSL has also been reported for a haploid model with a varying environment (Sasaki and Iwasa, 1987).

Other workers discussing the microevolution of recombination in fluctuating environments (e.g. in fluctuating optimum models, Maynard Smith, 1980, 1988b; or in systems of the host–parasite type, Hutson and Law, 1981; Bell and Maynard Smith, 1987) did not consider the effects of linkage of the modifier with its target region on selection for changed recombination. It is generally assumed, probably under the influence of the above results, that linkage enhances the intensity of selection for increased recombination (Brooks, 1988). This has led some authors to conclude that *rec* modifiers affect the rate of exchanges only in segments adjacent to them.

Speaking of the effectiveness of selection for *rf*, we will mean, in the first place, the direction of selection and the rate of change in the frequency of *rec* alleles rather than the equilibrium value of *rf*. The conclusion of selection in favor of the rec^+ allele being enhanced with tighter linkage between the modifier and the SSL is not a universal one. It only holds for an environment in which moments (or periods) of change in the direction of selection alternate with rather long periods of constant conditions. If periods of constant conditions are lacking (or short) then selection for recombination will be most effective when the *rec* locus is unlinked (or loosely linked) to the SSL. Thus, the dependence of the effectiveness of selection at the *rec* locus on the linkage of the modifier with the selected genes is not a monotonic one but depends, to a considerable extent, on environmental characteristics. Corresponding to each type of environment is a certain 'optimum' (i.e. determining the strongest selection in favor of the rec^+ allele) degree of linkage of the modifier with its target region (Korol, Preygel and Preygel, 1990). This conclusion is illustrated below by some numerical results.

1. *Charlesworth's model*

Consider a random-mating diploid population with two selected loci, A/a and B/b, and a *rec* locus (M/m) controlling crossing over within the SSL, such that allele M increases the exchange rate and m decreases it. The recombination fraction, R, between A/a and B/b is equal to one of the three values, R_{MM}, R_{Mm}, or R_{mm}. Let r denote the exchange rate between the *rec* modifier and the SSL. Let also the environment be a periodically changing one (with period T), and let the population stay in each of the two possible environmental states for an equal number of generations, allelic combinations *AB* and *ab* being favored by selection in one state and *Ab* and *aB* in the other. Our simulation experiments utilized the same matrices as those used by Charlesworth (1976).

Figure 7.4 (a) shows the dynamics of the *M* allele for increased recombination with tight linkage ($r = 0.05$) and free recombination between the *rec* modifier and the SSL ($r = 0.5$) in the environment with fluctuation period $T = 4$. As is seen from the figure, the unlinked modifier spreads faster. For comparison, Figure 7.4 (b) depicts changes in the frequency of the linked and unlinked

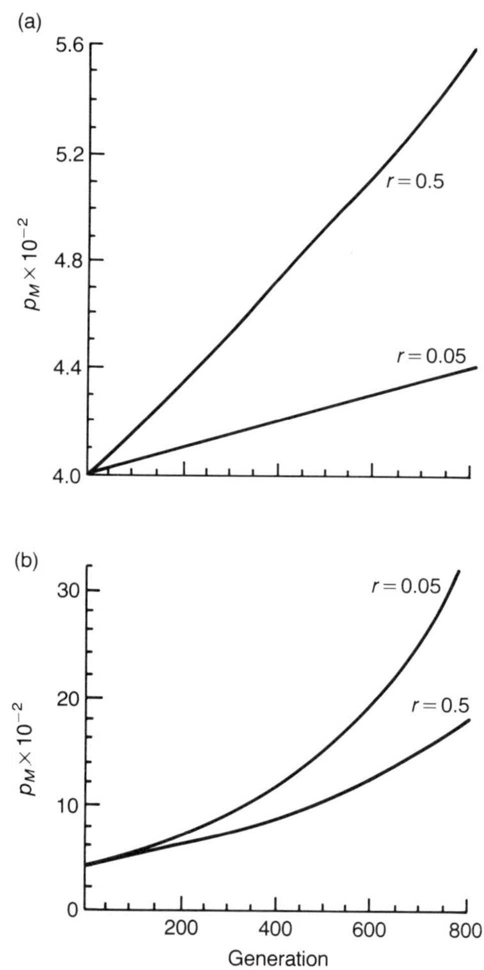

Figure 7.4 Dependence of the *rec* modifier frequency changes on its linkage to the selected system (based on Charlesworth's model): (a) $T = 4\ (2 + 2)$; (b) $T = 10\ (5 + 5)$.

M allele in a less variable environment ($T = 10$). In an environment like this, the rec^+ allele linked to the SSL is the one which spreads faster.

How can one explain these results? When there is a change in the environmental state, the allele M for high rf must have the opportunity for hitching a ride, i.e. forming associations with allele combinations favored by selection in the new environment. Clearly, a hitch-hiking event like this will take the shortest time to occur when the modifier is unlinked. It is therefore the unlinked modifier which is optimum (in terms of the effectiveness of selection for increased recombination) at the moment of change in the environmental state. At the same time, in the region of constant environment, allele M spreads faster the tighter the *rec* locus is linked to the SSL. Indeed, in a constant environment selection, following an unsteady-state process, causes the allele reducing rf to form associations with gene combinations favored in a given environment, thereby decreasing the chances of their being broken up by recombination. Understandably, the tighter the linkage of the modifier with the SSL, the more difficult it is for such associations to form and, hence, the longer the hitch obtained by allele M (rec^+) from favorable combinations of alleles at loci A/a and B/b.

Therefore, an optimum (in terms of selection for increased recombination) degree of linkage between the *rec* modifier and the SSL is a point of balance between two opposite trends: selective advantage of the loosest linkage in a varying environment and of the tightest linkage in a constant one. If the environmental fluctuation period is large, i.e. the population stays long enough in constant conditions, then selection for increased rf, if any, will be most effective with a linked modifier (see Figure 7.4). Results like these have been reported by Charlesworth (1976) and Sasaki and Iwasa (1987). If, on the other hand, changes in environmental states occur rather frequently, an opposite situation is possible in which the allele enhancing recombination spreads through the population faster when the modifier is loosely linked than when it is tightly linked (see Figure 7.4 (a)).

2. Maynard Smith's model with a fluctuating optimum

The advantage of an unlinked modifier can be expected to be manifested to the full in an environment of the type in which selective ranks of genotypes change from generation to generation. The polygenic model suggested by Maynard Smith (1980) is a useful system for elucidating the role of linkage of the modifier with the SSL when the problem is stated like this. Comparing selection for recombination in different environments, Maynard Smith draws the conclusion that, for an advantage of increased rf, periods of varying selection in different directions should alternate with periods of constant selection. It will be shown that Maynard Smith's conclusion is only valid when linkage between the modifier and the SSL is sufficiently tight.

Temporal variation of the optimum was assumed to be, in the general case, 'trapeziform' (see Figure 7.2). If periods of constant conditions are lacking, one can speak of a 'serrated' environment. In the case of $T = 2n$, i.e. when there are only two states, the change from one state to the other occurring suddenly, we can speak of a 'rectangular' environment.

Figure 7.5 shows the dynamics of a recessive allele M for various values of parameter n in a serrated environment with short (Figure 7.5 (a)) and long (Figure 7.5 (b)) fluctuation periods and in a trapeziform environment (Figure 7.5 (c)). As expected, in a serrated environment with a short period, selection for increased *rf* is most effective when the *rec* locus is unlinked or loosely linked to the selected system. Tighter linkage **alters the direction of change**, resulting in the allele reducing recombination being fixed in the population. With a longer oscillation period (and a shorter range of variation of the optimum), i.e. with a more stable environment, it is an intermediate linkage that is optimal in terms of selection for increased *rf*. Finally, in a still more stable environment with long periods of constant conditions, selection for recombination is most effective under tight linkage.

Thus, we have shown, even if numerically, that the effect of linkage of the *rec* modifier on the effectiveness of selection for recombination in a fluctuating environment is more complex than previously suspected and depends considerably on the pattern of environmental variation. Corresponding to each type of environment is a certain 'optimal' intensity of linkage of the modifier with the selected system, i.e. assuring the strongest selection for recombination. In a stable environment with a relatively long period of fluctuation and stretches of constant selection, tight linkage of the *rec* locus with the selected

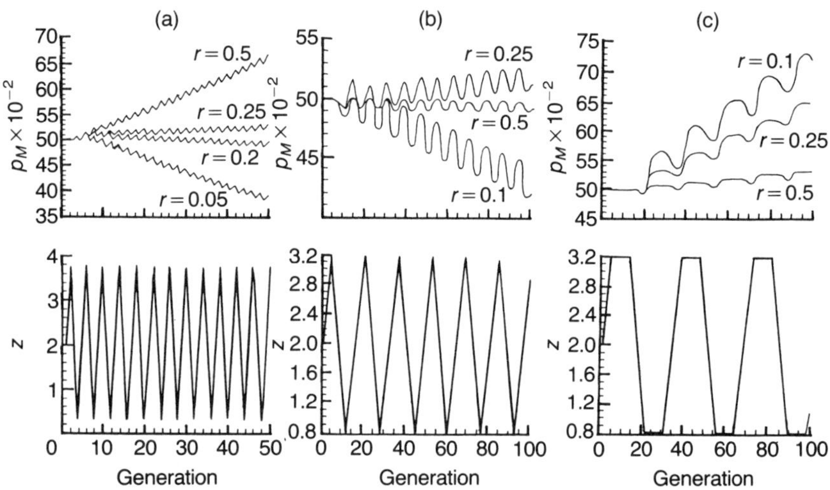

Figure 7.5 Dependence of the rec^+ allele dynamics in a varying environment on its linkage to the selected system (based on Maynard Smith's model; a two-locus case with $s = 0.75$): (a) $T = 4$, $n = 0$, $h = 1.75$; (b) $T = 16$, $n = 0$, $h = 0.3$; (c) $T = 32$, $n = 8$, $h = 0.3$.

system not only accelerates the spread of the rec^+ allele through the population but, in many cases, is a necessary condition of such a spread. If the environment is a serrated one, i.e. selective ranks of genotypes change between one generation and the next, then the situation is qualitatively different from that observed by Charlesworth (1976). Here selection in favor of the rec^+ allele is most effective when the modifier is unlinked or loosely linked to the SSL, whereas tighter linkage may result in a loss of rec^+. Situations are also possible (generally, in a serrated environment with a long period) in which the rec^+ allele has a selective advantage at intermediate linkage intensities only.

The effect of the linkage of a *rec* modifier with the SSL is also important when the selected loci undergo mutations (Feldman, Christiansen and Brooks, 1980). In particular, the existence of a critical recombination level, R^*, such that corresponding to the values of $R < R^*$ and $R > R^*$ are the opposite directions of the *rec* modifier dynamics has been established. A similar effect of linkage has been found for mutation rate modifiers (Twomey and Feldman, 1990).

Since the linkage of the *rec* modifier with its target region can have a dramatic effect on the rate of the *rec* allele frequency changes, it would be of interest to use linkage intensity as a variable parameter as well. In other words, it would be worth introducing, in addition to the ordinary *rec* modifier, another ('second order') modifier determining the intensity of linkage of the *rec* locus to its target, the SSL. A simple model of this type will be described in Chapter 9. We will show that selection can, in principle, modify the degree of linkage between the modifier and the SSL, i.e. one does seem to be justified in speaking of the evolution of their relative positions.

Recently, a number of attempts have been made to model experimentally, using laboratory populations, the microevolution of recombination by selection in a varying environment (Flexon and Rodell, 1982; Zhuchenko, Korol and Kovtyukh, 1985; Zhuchenko, Korol and Gorodetsky, 1988; Wolf, Wohrmann and Tomiuk, 1987; Gorodetsky, Zhuchenko and Korol, 1990; section 7.6). These are, undoubtedly, but the first steps in this direction. Studies of this type are a logical continuation of theoretical works that were largely performed during the 1970s to 1980s. Of what importance are the results reported in this section in designing experimental studies? It is known that the rate of crossing over in any region of the genome may be the product of combined actions of recombination regulators, both linked and unlinked to the target region (Simchen and Stamberg, 1969; Chinnici, 1971; Charlesworth and Charlesworth, 1985a,b; Brooks and Marks, 1986; Ebinuma, 1987; Brooks, 1988). Their actual relative effects in a start population are usually unknown. We believe, therefore, that in order to bring under selection as many of these modifiers as possible (and to try to compare their relative roles in determining the rate of recombination in the regions in question), it would be worth simultaneously conducting several experiments differing in the pattern of environmental change – high- and low-frequency, serrated and those with long periods of stability. Although an approach like this is much more labor-intensive, it sharply reduces the chances of obtaining a negative result stemming not from the essence of the problem but from a wrong choice of the experimental parameters.

In general, the results suggest that temporal variation in environment may be a universal factor maintaining comparatively high rates of recombination in natural populations as well as the cause of rapid rearrangement of the *rec* system at times of drastic changes in environment. Our estimates indicate that only a few tens of generations may be sufficient for substantial (experimentally detectable) changes in *rf* to occur within a population. A short-term advantage of increased recombination can also be obtained under directional selection with a constant or fluctuating optimum (Maynard Smith, 1980, 1988b). This effect cannot persist for a long period of time and terminates upon fixation of chromosomes carrying the selected gene complexes (Bergman and Feldman, 1990). Of course, other things being equal, the rate of fixation is the lower the greater is the number of loci controlling the selected trait. However, even long before fixation in the selected loci higher recombination may become disadvantageous. The outcome depends on the net effect of recombination which produces new allele combinations involved in the response to selection and at the same time disrupts adaptive gene complexes formed in previous generations. This contradiction is relevant also to the role of recombination in breeding programs (e.g. Zhuchenko and Korol, 1985; Pfeiffer, 1993).

7.4 EXPERIMENTAL MODELING OF THE *rec* SYSTEM MICROEVOLUTION

7.4.1 Review of some published results

As already noted, there is a voluminous literature concerned with the evolution of sex and recombination. Several groups of models have been proposed in which various forms of selection on fitness loci induce changes at the *rec* locus. But despite the keen interest in the problem displayed by theoreticians, it has been poorly studied experimentally, and what little evidence has accumulated so far (see below) is almost unknown to them. In two recent books on the evolution of sex and recombination (Stearns, 1987a; Michod and Levin, 1988), the experimental studies are not even mentioned. In fact, attention has largely been given to indirect (with respect to the evolution of recombination) evidence of the effects of artificial selection on crossing-over rate coming from a variety of organisms (section 6.3).

Few of the experiments related to the problem of maintenance of sex involve testing initial model assumptions (e.g. of a twofold advantage of asexual reproduction and the conditions under which it is reduced) rather than evaluation of the effect of various forms of selection on recombination and/or rate of sexual reproduction. Unfortunately, no significant progress has been made with this approach (see Bierzychudek (1987) for a review).

Results from experiments with artificial selection for altered *rf* values can be viewed as strong evidence for genetic variation at the *rec* loci. But can **induced selection** affect this variation and produce directional changes in recombination? We know of only seven experiments and one series of observations whose results have a direct bearing upon the question raised.

1. In experiments of Palenzona and co-workers, selection for altered wing size in *D. melanogaster* resulted in changes in *rf* between marker genes *b*, *cn* and *vg* (Palenzona and Zattoni, 1970; Mochi, Guerra and Palenzona, 1978). In one experiment with selection for a reduced wing size, the value of *rf* remained unchanged, while it was lowered when the direction of selection was reversed. Such a correlated response in *rf* was observed in segment *b–cn* only. In each cycle of selection, heterozygous females non-recombinant with respect to the *b–cn–vg* region were crossed with recessive males from the same population. Such a mating scheme can result in a selective pressure for reduced *rf* in the marker region of chromosome 2. The net recombination effect in this region is, therefore, the product of combined actions of (a) induced selection for *rf* in an upward and/or downward direction; (b) selection for wing size (resulting from pleiotropic effects of the selected alleles on *rf* or their linkage to the *rec* modifiers); (c) negative direct selection against the rec^+ alleles. Whatever the reasons for the observed changes in *rf*, it emerges from the data that selection on a phenotypic trait is capable of changing the *rec* system.
2. Clegg, Horch and Kidwell (1979) have performed an experiment involving four isozyme loci on chromosome 3 of *D. melanogaster*. The experiment was started on a large cage population with a maximum possible linkage disequilibrium between all pairs of loci. About every five generations, *rf* was estimated by testing progenies of heterozygous females. The authors expected selection in such a population maintained under constant environmental conditions to result in a reduction in *rf*. The experiment, however, revealed no general tendency for *rf* to vary between generations.
3. A synthetic population established by mixing 28 strains of *D. melanogaster* was subjected to selection for resistance to DDT for more than 300 days (Flexon and Rodell, 1982). The selection proved to be effective: by the time of estimation, resistance increased 10- to 100-fold relative to the control level. The authors compared the control and treatment populations with respect to crossing-over rate in all the three large chromosomes. It has been established that direct selection for DDT resistance increased the frequency of estimation, resistance increased 10- to 100-fold relative to the control 3 (*Gl*, *Sb*) but had no effect on *rf* in the sex chromosome (*w*, *m*, *f*). Analysis of individual chromosomes has revealed a significant positive correlation

between the chromosome contribution to resistance and the extent of change in *rf* relative to the control value.

4. Selection of *D. melanogaster* flies for altered locomotor activity and rate of synapse inactivation during ether narcotization yielded two strains sharply differing in the characters indicated. Subsequent estimation has shown that these strains also differ considerably in crossing-over rate in the *b–cn–vg* region of chromosome 2: corresponding to high activity are higher values of *rf* (Lobashev *et al.*, 1973).
5. Mustard (*Brassica campestris*) was exposed to disruptive selection for flowering time for five generations in a population composed of self-compatible and self-incompatible types (Harinarayana and Murty, 1971). The authors observed a substantial increase in chiasma frequency during microsporogenesis (as revealed by meiocyte examination at diakinesis) under two-way selection in self-incompatible populations. In self-compatible types, an increase in chiasma frequency only occurred in 'plus' strains. The greater increase in the rate of exchanges resulted from selection for late flowering.
6. Studied in the yeast has been the effect of temporal variation in nutrient medium on the amount of recombinational variation or, to put it more exactly, on the rate of sporulation (sexual reproduction) (Wolf, Wohrmann and Tomiuk, 1987). The data obtained indicate that an increase in environmental heterogeneity is accompanied by a higher rate of sexual reproduction, although the result depends on the duration of the experiment and on the rate of environmental change over generations. No direct estimation of changes in the rate of crossing over was made.
7. In experiments with mice Gorlov *et al.* (1992) have established that selection for higher body weight results in increased chiasma frequency in male meiosis. Lowered exchange frequency was found after selection for increased litter size.
8. Studies of the relationship between male chiasma frequency and features of reproductive strategy, longevity, body size, etc. of more than 30 mammalian species have shown that domesticated species exhibit a much higher rate of exchanges than might be expected on the basis of a general interrelationship among these parameters (Burt and Bell, 1987). The authors attribute this result to the advantage of increased recombination, which provides opportunities for overcoming unwanted correlations between characters in the course of selection.

7.4.2 Selection for increased recombination in *Drosophila* resulting from adaptation to adverse temperature conditions

An attempt has been made in our laboratory to test experimentally the possibility of a rapid readjustment of the *rec* system in large heterogeneous populations through selection induced by changes in abiotic environment.

The experiments were performed on four *Drosophila* cage populations derived from crosses between

(a) strains *b cn vg* and *Domodedovo-18* (population 1);
(b) strains *b cn vg bw* and *Domodedovo-18* (population 2);
(c) strain *b cn vg bw* and the population established as a result of interbreeding four strains – *Domodedovo-18*, *Domodedovo-32*, *Algeria* and *Canton S* (population 3);
(d) strain *b cn vg bw* and each of the strains *Domodedovo-18*, *Domodedovo-32*, *Algeria*, *Canton S* with subsequent interbreeding of the F_1, F_2, etc. hybrids with one another (population 4).

The experiments were carried out with populations subdivided into two variants – control (C) and treatment (T) – and maintained in cages each with 30 vials; the generations were non-overlapping. The populations were divided into subpopulations (C and T) only after several (usually 5–10) generations of random mating under optimum conditions. The C variants were maintained at $25 \pm 1\,°C$, the T variants in a climate chamber under daily fluctuating temperature with fluctuation amplitude increasing with generations (Figure 7.6). For each population, the number of individuals was varied from 700–1500 to 7000–10 000.

Consider first the results for population 1. After 15 generations of dynamics in contrasting conditions both subpopulations were assessed for recombination rate in the *b–cn–vg* region. To this end, phenotypically normal virgin females were taken from both subpopulations. From among these, heterozygotes were selected by evaluating the progeny of a test cross with *b cn vg* males. For each of the selected heterozygous females, *rf* was estimated in marker regions. Since temperature conditions may directly affect crossing over, one-half of the T variants were kept at optimum temperature (denoted as TC), while the others were kept in the same conditions as the entire T population (TT); the controls (CC) were kept in standard conditions. For comparison, the same estimation was made for the F_1 flies. A repeat estimation of the T and C variants of population 1 was carried out after an additional 13 generations. At that time the proportion of flies heterozygous for markers *b* and

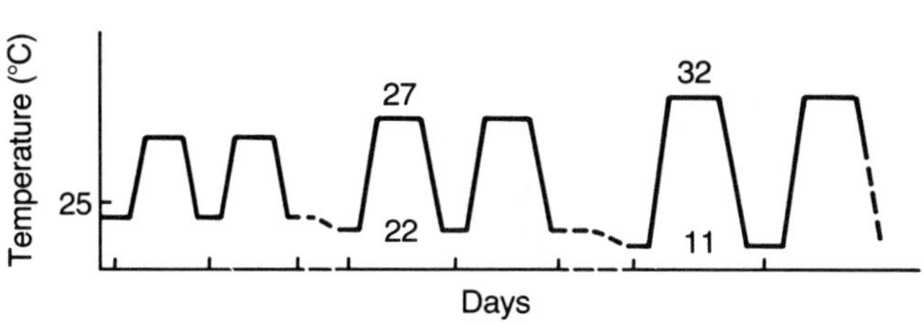

Figure 7.6 The mode of daily variation of temperature with increasing amplitude in the treatment (T) variant.

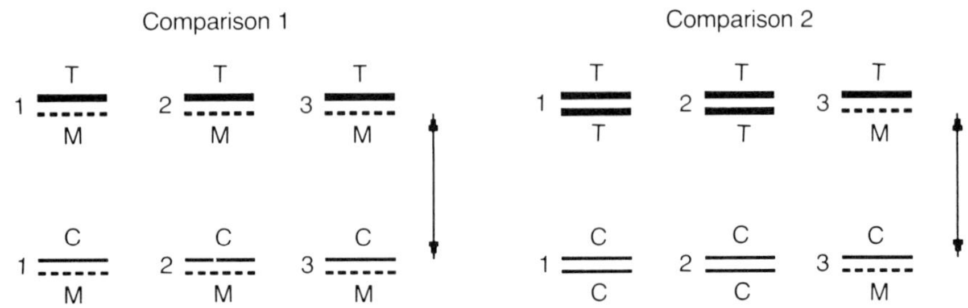

Figure 7.7 Two types of comparison of experimental and control populations with respect to *rf* in the third chromosome: *T*, *C* and *M* are, respectively, chromosomes of experimental and control populations and of the marker strain *rucuca*.

cn had already substantially decreased. Therefore, instead of comparing estimates obtained in the genetic background of each population (let us call it comparison 2), heterozygotes were produced by crossing flies from subpopulations T and C with the marker strain. The obtained heterozygous females (T and C) were compared for *rf* in the segment in question on the basis of test cross progeny identification (let it be comparison 1). Clearly, under comparison 1, between-variant differences will only be due to the *rec* genes dominant or semidominant with respect to the corresponding alleles in the marker strain (Figure 7.7).

In population 1, crossing-over frequency in both regions (*b–cn* and *cn–vg*) as well as throughout the interval *b–vg* turned out to be higher in the experimental population than in the control one. This is equally true for the TT and TC samples, i.e. an excess over the control level is genetically determined and is not due to the *rf* estimation conditions (Table 7.3). Assessment of heterozygotes after 28 generations of thermoadaptation (employing comparison 1) shows that the *rec* modifiers which have been accumulated in the T population through selection are not recessive with respect to their counterparts in the tester (and, probably, in the C population).

Consider now the results from experiments with populations 2–4 (Zhuchenko, Korol and Gorodetsky, 1988; Gorodetsky, Zhuchenko and Korol, 1990; Korol, Preygel and Preygel, 1990). In estimating *rf* in the *b–cn–vg* region of population 2, measurements were performed on individual females; in the rest of the cases they were made on replicates. The number of heterozygous females per replicate and the number of replicates and broods in producing the test cross flies varied from experiment to experiment, but in all cases these parameters were identical (or very close) for variants T and C (Table 7.4). Conditions under which the heterozygous females tested for *rf* were maintained were normal ($t = 25$ °C).

For population 2, nearly significant differences between the T and C variants are observed for the *b–cn* region as soon as after 10 generations. After only 30

Table 7.3 Variation in *rf* (%) within chromosome 2 of *Drosophila* owing to selection for resistance to fluctuating temperature (from Zhuchenko *et al.*, 1983)

	Segment								
	b–cn			*cn–vg*			*b–vg*		
Variant	*n*	*N*	*rf*	*n*	*N*	*rf*	*n*	*N*	*rf*
Hybrid F_1									
C	6	4235	11.4 ± 0.5	6	4235	10.5 ± 0.5	6	4235	19.0 ± 0.6
T	8	3741	11.9 ± 0.5	8	3741	11.8 ± 0.5	8	3741	21.4 ± 0.7
Generation 15									
C	3	2154	11.8 ± 2.0	7	4019	9.5 ± 0.8	3	1744	16.9 ± 1.6
TC	8	6059	15.1 ± 1.5	8	6059	12.2 ± 0.9*	8	6059	22.9 ± 1.1*
TT	8	4270	14.3 ± 1.8	7	3466	13.0 ± 1.6	7	3466	25.0 ± 2.4*
TT + TC	16	10329	14.9 ± 1.2	15	9525	12.5 ± 0.8*	15	9525	23.2 ± 1.2**
Generation 28									
C	7	3942	9.8 ± 1.1	7	3942	9.1 ± 0.6	7	3942	17.6 ± 1.2
T	3	1559	11.4 ± 1.2	4	2100	14.1 ± 2.0*	3	1559	22.2 ± 1.2*
Generations 15 + 28									
C	10	5096	10.5 ± 1.2	14	7961	9.3 ± 0.6	10	5686	17.4 ± 1.0
T	19	11888	14.4 ± 1.4*	19	11625	12.8 ± 0.9*	18	11084	23.1 ± 1.1***

n, number of families; *N*, number of flies.
Deviations from the control are significant at $^*P < 0.05$, $^{**}P < 0.01$, $^{***}P < 0.001$ respectively.

Table 7.4 The effect of selection for adaptation to temperature fluctuations on recombination within chromosome 2 of *Drosophila*

				rf (%) *for*		
Population	*Generation*	*Type*†	*Variant*	*b–cn*	*cn–vg*	*vg–bw*
2	10	2	C	10.48 ± 0.51	10.84 ± 0.40	32.56 ± 1.05
			T	14.42 ± 0.91	11.44 ± 0.51	34.27 ± 0.78
	30	2	C	7.65 ± 0.65	9.97 ± 0.62	29.42 ± 1.79
			T	19.78 ± 1.87	16.11 ± 1.46	35.72 ± 1.36
3	40	2	C	6.57 ± 0.97	11.15 ± 1.02	32.37 ± 1.83
			T	12.51 ± 1.12	10.94 ± 0.90	35.37 ± 1.40
	70	1	C	4.85 ± 0.60	9.28 ± 0.66	32.96 ± 1.36
			T	13.86 ± 2.32	12.62 ± 2.40	34.56 ± 1.56
4	72	1	C	3.55 ± 0.47	8.07 ± 0.84	34.85 ± 1.44
			T	10.37 ± 0.54	9.13 ± 0.63	35.03 ± 1.77

†Here and in Tables 7.6 and 7.7 the type of heterozygotes used for comparison of the T and C variants is as shown in Figure 7.7.

generations, the experimental and control populations differ to such an extent that values of rf_{b-cn} for individual females show no overlapping: even rf_{max} in the control is lower than rf_{min} in the treatment. It is interesting that the effect is due to a simultaneous reduction in *rf* in the control population and increase in *rf* in the experimental one. It should be noted that higher *rf* values in the T variant relative to those in the C variant were also observed for regions *cn–vg* and *vg–bw*, the relative excess in this case being, however, much lower.

A significant excess of rf_{b-cn} in the experimental variant over the control level has also been established for populations 3 and 4. A comparison of the recombination estimates from population 3 confirms the above conclusion that the differences between the T and C values of *rf* are the result of the crossing-over frequency being simultaneously increased in the treatment and decreased in the control variant. Moreover, at generation 70 an analysis based on comparison 1 yielded qualitatively the same results as those obtained from comparison 2 at generation 40. Two additional conclusions from the obtained data are possible:

- between generations 40 and 70, there was a process of divergence (with respect to the rate of exchanges) going on (at least in the *b–cn* region) in the populations compared;
- both the reduction in *rf* in the control and its increase in the treatment are due to genetic factors which are dominant, or at least semidominant, with respect to the corresponding allele in the tester strain *b cn vg bw*.

Judging by the data from population 4, significant differences between the T and C variants are also possible in the absence of any increase in *rf* in the former. That is, in optimal conditions selection will tend to reduce crossing over, whereas in extreme conditions the initial rate of exchanges is maintained. It should be mentioned here that, at the start, population 4 was the most variable among the four populations studied. Rough estimations which have been performed in Professor Zhivotovsky's laboratory (personal communication) show that its variability level for quantitative (morphological) traits is similar to that observed in natural populations.

It seems quite reasonable to assume that the higher the initial variation of the population, the lower is the selection for increased recombination in unfavorable conditions and, moreover, the higher is the selection for decreased crossing over in optimum stable environments. The results suggest that we have been the first to verify experimentally the hypothesis, advanced by Fisher (1930) 60 years ago, that in a constant environment selection favors tighter linkage between genes. They also indicate, albeit indirectly, that selection against hamful mutations alone seems in itself to be insufficient for maintaining the normal level of recombination, otherwise such a decrease in *rf* in control populations (e.g. nos. 3 and 4) would have been impossible. Using the same data, a possible increase in *rf* induced by selection for fitness has been first demonstrated clearly and reproducibly on large populations. Both effects – decreased and increased *rf* – proved to be segment specific.

With this (segment specificity of responses to induced selection) in mind, further assessment of the effect of selection on recombination within the second chromosome of population 2 was made on the basis of comparison 1 at generation 65 using the marker strain *al b pr c px sp* (Table 7.5).

The material presented here supports the above data on high rate of response of the chromosome 2 near-centromeric region to induced selection: for region *b–pr*, the rf_T/rf_C ratio is $16.8/6.1 = 2.6$, as it is for the segment *b–cn* from the same population (containing this region as its subinterval) but under comparison 2 in generation 30 ($19.78/7.65 \approx 2.6$). Confirmed for this region has also been the conclusion of non-overlapping ranges of variation in *rf* in the control and experimental populations. In generation 30, this was based on *rf* measurements in individual females, while in generation 65 it was based on measurements of replicates comprising progeny of several females. In addition to *b–pr*, a tendency toward a higher *rf* (albeit a weaker one) was observed in other regions as well, even becoming reversed in the *pr–c* region. It must be emphasized that the distributions of summed-up (along the chromosome) values from the T and C variants show practically no overlapping.

We also compared the T and C variants of population 2 with respect to the amount of interference. Significant differences in the values of coefficients of coincidence have only been found for four pairs of adjacent and non-adjacent regions:

Intervals		Coincidence	
		C	T
al–b,	*px–sp*	1.72 ± 0.11	1.42 ± 0.08
b–pr,	*px–sp*	0.26 ± 0.13	1.08 ± 0.05
b–c,	*c–px*	0.67 ± 0.05	1.03 ± 0.05
al–c,	*c–sp*	0.69 ± 0.04	0.92 ± 0.03

Clearly, the T variant shows a weaker interference of exchanges compared with the C variant.

Table 7.5 Segment-specific variation in *rf*(%) within the second chromosome of population 2 of *D. melanogaster* under induced selection for thermoadaptation

	Range of rf (min and max) in population		*Mean crossing-over rate*	
Region	*C*	*T*	*C*	*T*
al–b	24.7–31.6	30.1– 39.2	28.1 ± 0.9	33.4 ± 1.0
b–pr	4.3– 8.4	11.0– 19.0	6.1 ± 0.5	16.1 ± 0.9
pr–c	13.7–26.5	13.7– 20.4	18.3 ± 0.9	16.8 ± 0.8
c–px	6.7–21.5	11.8– 25.4	17.3 ± 0.9	19.6 ± 0.8
px–sp	0.8– 6.8	2.7– 11.5	5.3 ± 0.6	8.0 ± 0.8
al–sp (Σ)	66.5–84.7	83.2–101.2	74.8 ± 1.3	93.4 ± 1.4

For chromosome 3, the effect of selection on *rf* was assessed for population 2 in generation 45 on the basis of comparisons 1 and 2. The conclusions from the analysis of variation in *rf* within chromosome 2 hold for chromosome 3 as well (Table 7.6), especially when evaluating the consequences of selection on the basis of comparison 2, i.e. upon introducing the marker chromosome in the genetic background of populations C and T (with the aid of a balancer strain). Thus, the effect of segment-specific variation in *rf* is a pronounced one, but in chromosome 3, unlike chromosome 2, it was not the near-centromeric region that was most responsive, but the left distal segment *ru–h*. The distributions of individual *rf* values within replicates for the most responsive region, *ru–h*, and sums of all intervals of chromosome 3 show no overlapping.

In view of the above results, the following question is of interest: what is the genetic nature of differences in recombination between the control and experimental populations and are these differences due to the recombination modifiers accumulated in the course of selection or do the observed effects result from changes in the recombination properties of the genomic regions? In order to exclude possible effects of chromosome rearrangements on the results obtained, a special test of the material has been undertaken. Salivary gland chromosome preparations of heterozygotes between flies of populations T and C, on the one hand, and of the marker strains, on the other, were examined. In none of the cases were rearrangements found.

Comparisons of the results from analysis for recombination within chromosome 3 population 2 after 45 generations of selection (Table 7.6) led to the

Table 7.6 The effect of selection for resistance to temperature fluctuations on crossing-over rate (%) in chromosome 3 of *D. melanogaster*

Region	Type of test	*Range of rf (min and max) in population*		*Mean crossing-over rate*	
		C	*T*	*C*	*T*
ru–h	1	6.4 – 21.1	13.5 – 26.5	13.6 ± 1.2	20.1 ± 1.2
	2	4.6 – 10.5	18.0 – 26.3	8.1 ± 0.8	21.1 ± 0.9
h–th	1	7.6 – 20.6	10.6 – 27.3	16.2 ± 1.6	19.5 ± 1.8
	2	10.0 – 19.0	14.8 – 23.5	15.0 ± 0.9	17.4 ± 1.4
th–st	1	0.0 – 0.9	0.0 – 5.6	0.3 ± 0.6	1.1 ± 0.8
	2	0.0 – 0.4	0.0 – 0.5	0.1 ± 0.3	0.3 ± 0.1
st–cu	1	0.9 – 11.8	1.3 – 15.2	5.4 ± 1.0	6.3 ± 1.1
	2	1.5 – 6.6	1.8 – 6.2	3.9 ± 1.3	4.3 ± 0.5
cu–sr	1	4.4 – 14.5	6.0 – 18.4	11.2 ± 1.6	11.7 ± 0.9
	2	2.1 – 9.0	8.1 – 16.4	7.1 ± 1.1	11.4 ± 0.8
sr–e	1	4.4 – 13.6	6.3 – 19.5	9.5 ± 1.1	13.0 ± 1.4
	2	2.8 – 9.1	6.7 – 15.6	7.3 ± 0.7	10.4 ± 0.6
e–ca	1	30.6 – 44.5	28.8 – 45.2	39.9 ± 1.1	36.8 ± 1.1
	2	29.9 – 39.6	28.1 – 36.6	35.6 ± 0.7	33.2 ± 0.5
ru–ca (Σ)	1	68.3 – 104.8	87.7 – 119.8	86.9 ± 2.2	106.1 ± 2.5
	2	65.4 – 82.7	88.0 – 109.0	74.7 ± 1.0	98.9 ± 1.0

conclusion that at least some of the differences between the T and C variants result from changes in the system of recombination-modifying genes. Indeed, genotypes from the T variants in comparisons 1 and 2 are identical with respect to the marker chromosome (Figure 7.7); the same is true for the C variants. Nevertheless, comparisons of the T and C variants with respect to *rf* in the two types of heterozygotes do not yield completely identical results. From this, it can readily be inferred that a proportion of genetic factors responsible for the observed differences are located outside the chromosome in question, i.e. at least some of the effects are due to changes in the controlling system, presumably to *rec* modifiers that are unlinked to their target regions. Earlier (sections 7.2 and 7.3), we showed that selection acting on multilocus systems can favor increased frequency of the rec^+ modifier even if the latter is **unlinked to the selected system**. In the absence of linkage, we can only speak of 'co-selection', experimental evidence showing this to be a real mechanism for the evolution of the recombination system. Another indirect conclusion emerging from the data concerns the expression of genetic modifiers in heterozygous condition. The *rec* alleles on chromosomes 1 and 2 increasing *rf* in the *ru–h* region of chromosome 3 (which have been accumulated during selection in extreme temperatures) can be assumed to be dominant over the corresponding alleles in the *ru cu ca* strain, while those reducing $rf_{ru\text{–}h}$ in the control population are more likely to be recessive. Thus, our result from computer simulations, suggesting the possibility of altering dominance relations at the *rec* loci under induced selection, seems to find experimental support (see section 9.2 for further discussion).

Similar arguments lead to the conclusions about the effect of selection on interference (Table 7.7). In both types of comparison, a weaker interference is observed in the T variant compared with the C variant for the majority of segment pairs where $c < 1$, the effect being manifested both within the chromosome arm (*ru–h, h–th; cu–sr, e–ca; sr–e, e–ca*) and across the centromere (*ru–h, sr–e; h–th, e–ca*). It will be remembered that a tendency toward a weaker interference in the T variant has also been noted for chromosome 2.

Table 7.7 Some examples of the effect of selection for thermoadaptivity in *Drosophila* on interference within chromosome 3

		Coincidence in heterozygotes of type			
		1		2	
Segment pair		*C*	*T*	*C*	*T*
ru–h,	*st–cu*	1.13 ± 0.25	0.94 ± 0.16	0.25 ± 0.12	0.53 ± 0.08
ru–h,	*sr–e*	0.70 ± 0.15	0.90 ± 0.10	0.30 ± 0.10	0.82 ± 0.06
h–th,	*sr–e*	1.64 ± 0.19	1.24 ± 0.11	1.06 ± 0.12	0.93 ± 0.07
st–cu,	*sr–e*	0.38 ± 0.19	0.77 ± 0.18	0.14 ± 0.10	0.50 ± 0.11
cu–sr,	*sr–e*	1.25 ± 0.23	2.25 ± 0.17	0.64 ± 0.15	0.85 ± 0.09
sr–e,	*e–ca*	0.33 ± 0.07	0.89 ± 0.07	0.23 ± 0.04	0.41 ± 0.04

With negative interference ($c > 1$), the effect of selection is less definitive. Collation of the results from the two types of comparison also leads to the conclusion that both the degree and sign of interference in chromosome 3 may be determined by genetic factors located on other chromosomes since the corresponding variants in comparisons 1 and 2 have identical genetic constitutions with respect to chromosome 3.

Thus, we have obtained indirect evidence for genetic differences in interference being determined by modifiers and demonstrated, for the first time, the possibility of induced selection to act on interference modifiers. Given the relevant arguments concerning dominance–recessiveness relations between alleles of recombination-modifying loci, one can conclude that allelic differences between populations T and C with respect to genes modifying interference are also characterized by dominance–recessiveness relations. Of course, it may be that loci affecting crossover rate are, at the same time, altering interference.

The discussion of our data on the *Drosophila rec* system variation under the influence of population adaptation to daily temperature fluctuations can probably be concluded by the following remark. Although selection on modifiers of crossing over seems to be the most plausible explanation for the results obtained, other possible mechanisms cannot, in principle, be ruled out. In particular, the effect of stress on the genetically heterogeneous start population could have resulted in transpositions of mobile elements which in some cases affect recombination (Greenblatt, 1981; Yurchenko, Zakharov and Golubovsky, 1985; Zhuchenko, Vereschagina and Korol, 1988; Sved *et al.*, 1991; Vereschagina and Korol, 1991). As mentioned earlier, the observed changes in recombination are unlikely to be caused by interchromosomal effects of heterozygosity for large chromosome rearrangements. A possible effect of a small, cytologically undetectable, rearrangement can also be excluded since estimation of crossing-over rate (practically throughout the genome) has revealed no regions exhibiting a substantial reduction in *rf* in the T variants.

7.4.3 Increased recombination in *Drosophila* resulting from selection for positive and negative geotaxis

Most studies concerned with the *rec* system evolution have employed one-way selection for fitness traits. An exception is the aforementioned experiment with mustard (*Brassica campestris*) (Harinarayana and Murty, 1971), in which a considerable increase in chiasma frequency in microsporocytes has been established following selection for both early and late flowering. The limitation of one-way selection schemes is the danger of a change in recombination being a trivial result owing to initial linkage disequilibria between selected genes and *rec* modifiers. In the above-described experiments with *Drosophila* thermoadaptation, this danger was reduced by parallel studies on different populations.

With this in mind, we have also studied the effect of selection of *D.*

melanogaster flies for positive and negative geotaxis on recombination (Korol and Iliadi, 1994). The experiment was conducted on a large cage population established previously by mixing 21 lines (the control population from collaborative experiments with Dr N.M. Vereschagina; p. 181) and maintained in the laboratory for more than 50 generations. Selection for geotaxis was carried out using a vertical labyrinth. Flies (150 males and 150 virgin females) were estimated at each stage of selection, males and females being tested separately. To establish the next generation, extreme flies (10%) were used for each direction of selection; the control material was chosen randomly. A rapid response to selection for geotaxis was observed from the first stages of selection. Then the rate of response slowed down and a selection plateau was achieved. The estimates of geotactic activity for generations 33 and 49 (in units corresponding to the labyrinth design used in the experiment) are given below:

	geo^+	geo^-	*Control*
F_{33}	7.06 ± 0.24	3.11 ± 0.39	4.60 ± 0.41
F_{49}	7.31 ± 0.25	2.31 ± 0.33	4.26 ± 0.62

The recombination frequency was estimated for chromosomes 2, 3 and X in generations 36, 40 and 44 respectively. Females of the compared strains were crossed with marker stock males *y cv ct v car*, *al b cn vg bw* and *ru h cu sr e* (markers on chromosomes X, 2 and 3 respectively). Then the test cross progeny resulting from crosses between hybrid females and marker males were analyzed (in eight replications).

From the results obtained (Figure 7.8), it is clear that selection for altered geotaxis **in both directions** brought about a dramatic increase in crossing-over rate in different portions of the genome. The total increment in recombination for the genome portion considered is 78 cM for geo^+ and 66 cM for geo^-. Selection for negative geotaxis did not result in *rf* changes in chromosome 2, while selection in the opposite direction caused a fourfold increase in *rf* in the *b–cn* segment and a significant, albeit smaller, increase in the adjacent regions, *al–b* and *cn–vg*. As we have seen, such a marked response in the *b–cn* segment (in terms of *rf* changes) as this was also observed in the above experiments with selection for thermoadaptation in *Drosophila*.

How can these results be interpreted? We attribute the observed unidirectional change in *rf* caused by two-way selection for geotaxis to the advantage conferred by selection on recombinants. Moreover, the data as a whole indicate that changes in *rec* loci are responsible for the effect. An important question regarding these data is the possibility of random effect caused by the low number of selected progeny and the absence of parallel replicates. The main arguments against such an interpretation and in favor of the one presented here are:

- The unidirectional mode of changes in *rf* in both geo^+ and geo^- (which are more than simply two independent replications).

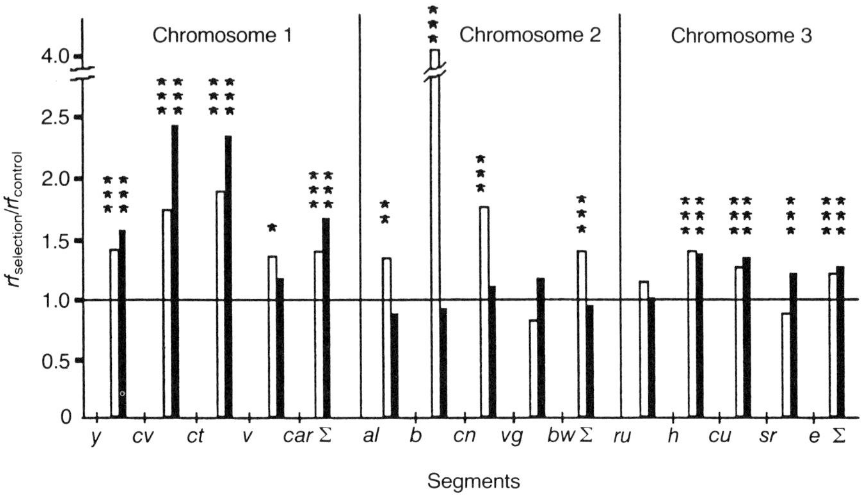

Figure 7.8 The effect of two-way selection for geotaxis on recombination across the genome of *D. melanogaster* (□, geo^+; ■, geo^-). The symbol Σ denotes the results for the sums of *rf* values for each of the chromosome. Cases in which $rf_{selection}/rf_{control}$ deviates from unity are marked by one, two and three asterisks (corresponding to $p < 0.05$, $p < 0.01$ and $p < 0.001$ respectively).

- Reaction of different genome regions in the same direction.
- Very high rates of *rf* reaction are not very common, even in experiments with direct selection for recombination. In fact, the observed differences between opposite selection lines rarely exceed the two- to threefold level (Chinnici, 1971; Kidwell, 1972; Dewees, 1975; Charlesworth and Charlesworth, 1985a,b).

Based on our and other experimental evidence (see pp. 213–4), it can be concluded that directional selection for any trait favoring new gene combinations for many generations can play a role as a factor in the *rec* system evolution. Of course, the selection regime is of paramount importance for such evolution, for its direction and rate (Maynard Smith 1988b; Korol and Preygel, 1986, 1989; Bergman and Feldman, 1990; Charlesworth, 1993).

In general, the obtained data experimentally substantiate the theoretical conclusions of possible rapid changes in the *R*-system during selection on the *F*-system. These changes affect the control of both the frequency and genomic distribution of crossovers. Changes in the dominance–recessiveness relations of the *rec* modifiers under selection have also been demonstrated experimentally. The results indicate that there is a need for an evolutionary explanation of recombination phenomenology. They also suggest that experimental testing of the proposed models for *rec* system evolution is a feasible and promising approach. It seems reasonable to assume that no further significant progress in understanding the evolution of recombination is possible without combining theoretical analysis with direct large-scale modeling on convenient experimental organisms and with observations in nature, contrary to the occasionally voiced pessimism over such a possibility.

8 Species interactions as a factor in the evolution of recombination

8.1 INTRODUCTION

The previously discussed models for the evolution of recombination can be divided into two classes on the basis of factors determining changes in the frequency of the *rec* modifier (internal or external). A classification like this will place mutation models (the finite population model of Felsenstein and Yokoyama and the infinite population model of Kondrashov) in the first class. As a rule, models of the second class assumed a very special (and probably rarely occurring in nature) pattern of environmental change, such that the sign of D for selectively favored allelic associations at the interacting loci varied from generation to generation (Charlesworth, 1976) or between locations (Slatkin, 1975). Therefore, fluctuating environment models have until quite recently been considered unattractive as a tool for explaining the reasons for maintaining non-zero recombination in natural populations (see Maynard Smith, 1978). They began, however, to draw increasing attention in connection with the reported effectiveness of selection on multilocus systems in a fluctuating environment as a factor favoring recombination and sexual reproduction in general (Maynard Smith, 1980, 1988b; Moore and Hines, 1981; Manning and Dickson, 1986; Korol and Preygel, 1986, 1989; Charlesworth, 1993). It has been shown (sections 7.2–7.3) that the model suggested by Maynard Smith (1979, 1980, 1988b) actually has a much larger potential for explaining high recombination than other models of this class.

Thus, in each of the two classes of models in question there exist relatively satisfactory formal explanation for induced selection in favor of selectively neutral *rec* modifiers responsible for non-zero recombination. At the same time, a number of authors have advanced the idea that it is the heterogeneity of biological environment of a given species rather than that of its abiotic environment that is important to the evolution of sex in general and recombination in particular.

As early as 1957 Stebbins pointed out that in plants recombination is higher in central than in marginal populations. Plants in marginal habitats exhibit higher rates of selfing or apomixis than plants in tropical environments in spite of the physical environment in tropical regions being much more stable (Levin, 1975). However, as pointed out by Meijen (1987), it is this stability or optimality of the abiotic environment which, by reducing selective pressure against sharply deviating genotypes, enhances the probability of new macroevolutionary taxa being formed in equatorial regions (at least for higher plants). One reason for apomicts being mostly found in marginal conditions is, undoubtedly, the independence of their multiplication from the population density, the latter being a critical factor for outcrossers. This may also coincide with high vulnerability of meiosis and gametogenesis to unfavorable conditions resulting in lowered fertility of sexuals in general. Finally, it cannot be ruled out that apomicts are less sensitive (in their vegetative characteristics) to environmental fluctuations (Bierzychudek, 1989).

Levin (1975) explains the predominance of sexual forms in tropical communities by their higher complexity and diversity and by the abundance of pests and herbivores, which are continuously evolving new abilities to attack plants. In order to survive, a plant species must, in turn, constantly evolve new resistance mechanisms (one of the formulations of the Red Queen hypothesis; Stearns, 1987a). Undoubtedly, recombination as a main mechanism for rapid release of variability is the basis for such evolutionary improvement. Of great interest therefore are models of genetic interactions between antagonistic species (Jaenike, 1978; Glesener, 1979; Hamilton, 1980, 1982; Hutson and Law, 1981; May and Anderson, 1983; Bronstein and Korol, 1984; Bremermann, 1987; Bell and Maynard Smith, 1987; Korol and Preygel, 1988b).

The idea of the evolutionary advantages of sex and of recombination being the result of biotic interactions has also been formulated as the Tangled Bank hypothesis (see Bell, 1982, 1987; Ghiselin, 1988). This is based on the assumption that intraspecific competitive relations resulting from the genotype–environment interactions in saturated conditions may account for the widespread occurrence of recombination. The sib competition model can be regarded as a version of the tangled bank. Bell (1987) holds the view that difference between the Red Queen and the Tangled Bank is in the mode of the genotype–environment interaction, that is whether the changes in selective ranks of genotypes occur in time or in space respectively.

Another explanation for the increased frequency of occurrence of parthenogenetic and self-fertilizing forms in marginal habitats may be their higher colonizing capacity (Vrijenhoek, 1990). Incidentally, the notion itself of sex predominating in tropical communities as opposed to mesic ones (with their unstable abiotic environments) does not seem to be as unquestionable as generally believed. Thus, according to Nevo's observations (personal communication), the Amazonian tropical jungles in general exhibit a much higher incidence of asexual reproduction compared, for example, with those in Europe or Israel.

The existing formalizations of the *rec* system evolution due to interspecific contradictions are based on the following arguments. Consider two species, host and parasite. The most common genotype in the host population causes selection for increased frequency of that type in the parasite population which shows the highest performance in infecting this genotype. This will result, after a period of time, in a reduction in the frequency of the host genotypic class in question. This implies a reduction in fitness of the parasite type that has spread as a result of the above cycle. In general, interactions within antagonistic systems such as 'host–parasite' or 'predator–prey' can result, under certain co-selection regimes, in both components becoming polymorphic (Levin *et al.*, 1982).

If the host resistance or immunity retards the development of the corresponding parasite genotype, then co-selection will be frequency dependent: the fitness of a given genotypic class in the host depends on the frequency of the virulent strain of the parasite whose frequency, in turn, depends on that of the host class in question in the previous generation. In general, one may speak of the fitness of a genotypic class being inversely related to its frequency (Antonovics and Ellstrand, 1984; Burdon, 1987; Jones and Harvey, 1987). The resulting selection regime is capable of inducing undamped oscillations of the genetic constitution of both populations. Here, also, we can have a version of the evolution of recombination due to varying biotic environmental conditions, but it is the population itself that plays an active role in determining the pattern of selective pressure of these conditions. This can be regarded as a sort of 'hybrid' version of the evolution due to interactions between 'extrinsic' and 'intrinsic' factors. Furthermore, this class of models can be considered to be a 'hybrid' one also with respect to another dichotomous classification associated with a deterministic or a stochastic pattern of linkage disequilibrium formation (Seger and Hamilton, 1988).

Suppose that the host longevity is k times that of the parasite. Then, during one cycle of development on each host genotype infected by a mixture of parasite genotypes, the parasite will undergo microevolutionary changes increasing its fitness through 'local specialization'. If that is the case, then the more different the infected progeny genotype is from that of its parents, the smaller is the chance of its becoming a suitable food resource for the parasite type adapted to the parental genotype. Under asexual reproduction, the parental and progeny genotypes are, with rare exceptions, identical. Sexual reproduction and recombination substantially weaken the parent–offspring correlation, thereby reducing the probability of the offspring being infested by a virulent strain of the parasite. Lloyd (1980) has formulated the idea of the advantage of sexuality as a hypothesis of recruitment of protective genes.

Thus, antagonistic relations of the host–parasite type, universal in animate nature, may be a significant factor placing a high selective premium on sex, genetic variation and, in particular, recombination (Jaenike, 1978; Hamilton, 1980; Lloyd, 1980; Tooby, 1982; Rice, 1983a,b; Bronstein and Korol, 1984; Bell, 1988; Bierzychudek, 1987; Bremermann, 1987; Seger and Hamilton, 1988; Hamilton, Axelrod and Tanese, 1990; Korol, Preygel and Preygel, 1990). The indicated relations can also play a significant role in the evolution of another important component of the sexual reproduction system associated with regulation of variability–mate choice (sexual selection) (Kirkpatrick, 1986; Read, 1987; Zuk, 1988).

It should be noted that many vegetatively reproducing plant species acquire resistance to pathogens not as a result of genotypic variation but through the effects of morphological and/or biochemical resistance mechanisms (Parker, 1988b), i.e. ontogenetic adaptation. It is, however, higher vertebrates that have evolved the most advanced mechanisms for resistance to pathogens: during ontogenesis of each individual, the genetic diversity of specialized cells of the immune system assuring adequate responses to a wide variety of antigens is generated within the multigene family of immunoglobulins based on a high frequency of mitotic recombination and the hypermutation process. That is, resistance is assured here not only on the basis of population genetic diversity (e.g. because of the HLA system) but also because of **intraorganismal genetic variation**. The latter is, to a great extent, non-random (reviewed by Rodin and Rzhetsky, 1987).

Let us, however, return to the main models for the evolution of recombination with interspecific interactions. In the model of Hamilton (1980), coefficients of fitness of genotypes of a two-locus haploid host are defined as functions of their frequencies: the effect of the parasite is defined implicitly and manifests itself in that with increasing frequency of the host genotype its fitness declines. Let X_1, X_2, X_3, and X_4 be the frequencies of four possible genotypes, AB, Ab, aB and ab, of the host in a given generation. Then, in the next generation

$$X_i' = (X_i - \alpha_i rf_{ab} D)w_i/w, \quad i = \overline{1,4},$$

where rf_{ab} is recombination fraction between loci A/a and B/b, $\alpha_i = +1$ for $i = 1, 4$ and $\alpha = -1$ for $i = 2, 3$; w_i is fitness of the ith genotype, w the mean fitness of a population. Hamilton specifies $w_i = w(X_i)$ as $w_i = \exp[\delta(1 - 4X_i)/2]$.

Provided selection intensity, as measured by parameter δ, is large enough ($\delta > 4 - 2rf$), then stable genotype frequency fluctuations of varying complexity are set up in the system, such that $X_2 = X_3$ and $X_1 = X_4$ (i.e. what is actually oscillating is D). With some selection intensities, stochastic fluctuations are also possible. Hamilton has also shown that in a mixed host population composed of asexual and sexual types, the latter are at an advantage **even when their fecundity is half** that of the former. Asexual types are either eliminated from the population or maintained at low frequencies. The advantage of sex becomes greater with increasing recombination frequency. If the population is composed of sexual types only, then, with sufficiently strong selection, the mean fitness, w, rises with increasing rf. This phenomenological approach has been further elaborated allowing for epidemiological features of host–parasite interactions including not only frequency- but also density-dependent effects (May and Anderson, 1983).

Hamilton (1980) has shown, in terms of mean fitness, that interspecific interactions inducing frequency-dependent selection may be a factor promoting increased recombination. As already noted, models for the evolution of recombination based on the analysis of changes in the frequency of modifier genes are genetically more attractive. Hutson and Law (1981) were the first to attempt such an analysis for a system of the host–parasite type. Their model differs from the one described above in two important aspects: the haploid host species is represented by two selected loci concerned with resistance to the parasite and a neutral modifier controls recombination frequency between the two; as in the model of Hamilton, the parasite species is not specified explicitly and its effect is only manifested in that coefficients of fitness of the host depend on the frequency of its haplotypes, and frequency-dependent selection is assumed to operate with time delay

$$w_i^{(n)} = f(X_{i1}^{(n-k)} + X_{i2}^{(n-k)}), \quad i = \overline{1,4},$$

where $f(X)$ is a continuous, positive monotonically decreasing function and X_{i1} and X_{i2} are frequencies of genotypes *AB*, *Ab*, *aB* and *ab* associated with alleles rec^- and rec^+ respectively.

The authors have analytically studied the population behavior in the vicinity of point $X^o\{X_{11} = X_{21} = X_{31} = X_{41} = 1/4,\ X_{12} = X_{22} = X_{32} = X_{42} = 0\}$. When selection is strong enough or delay k is large, X^o is unstable; non-linear oscillations of D_{ab} arise in the population, which may result in increased frequencies of X_{i2}, i.e. in a higher frequency of the rec^+ allele. This conclusion has been substantiated numerically. It has been shown that on introducing into a population without recombination a dominant mutant allele rec^+ (at an initial frequency of 0.1%) responsible for free recombination between loci A/a and B/b, its frequency is increased several ten- or hundred-fold depending on the coefficients specifying the regime of frequency-dependent selection – its intensity and time delay.

The model suggested by Hutson and Law can be regarded as a valuable (even if very crude) approximation. Some of its limitations have been pointed out by the authors themselves:

- selection is assumed to act on the host haplophase only – a diploid model was not examined;
- the fitness of a genotype in a given generation depends on genotypic frequencies occurring k generations earlier, though frequencies from other generations also seem to be important;

- the rec^+ allele is assumed to be dominant;
- only the initial behavior of the modifier (the stability of the initial point) was studied.

So far, in none of the works has the behavior of a system with a diploid host been studied analytically, but some important results can also be obtained with computer simulations. Highly objectionable is the implicit way of formalizing genetic relations between antagonistic species, reducing their interactions to specifying frequency-dependent coefficients of the host fitness (with or without time delay): in explicit form, parasite is actually absent from constructions of this type. Models such as these might just as well be referred to any system with frequency-dependent effects, for instance to the situation when genotypes 'deteriorate' the environment, rendering it unsuitable only for those individuals that resemble them. The problem of the role of frequency-dependent selection in the evolution of sex and recombination is in itself very important (Clarke, 1979; Ellstrand and Antonovics, 1985; Kelley, Antonovics and Schmitt, 1988). But in this case such a generality should rather be regarded as a limitation of the model since, with an approach like this, the biological peculiarity of a particular situation (interaction of two species, say host and parasite) is to a considerable extent lost. This may pose difficulties in attempting to test the hypotheses experimentally (see Bierzychudek, 1987, for a critical review).

8.2 A MODEL OF CO-EVOLUTION WITH SELECTION ON GENE COMBINATIONS

8.2.1 Description of the model

We have proposed a more realistic model for the co-evolution of host and parasite populations (Bronstein and Korol, 1984; Korol, Preygel and Preygel, 1986; Korol and Preygel 1988b). Unlike previous models, this one reflects the genetic constitution of both populations and their interactions (see also Bell and Maynard Smith, 1987). This enables the host and parasite genotype frequency changes to be studied, and the pattern of relations between species to be specified naturally and in a legitimate way. The host resistance and the parasite virulence are assumed to be controlled by systems of loci, such that parasite genotypes differ in their performance in exploiting a given host genotype. Conversely, the degree of resistance to a particular parasite genotype depends on the host gene combination. An approach like this also results in frequency-dependent selection.

In the model suggested (Figure 8.1), both populations are infinite, random mating and diploid. That both components are diploid is a fairly reasonable assumption for pairwise antagonistic interactions of the predator–prey or herbivore–plant type. For the host–parasite systems, the requirement for the

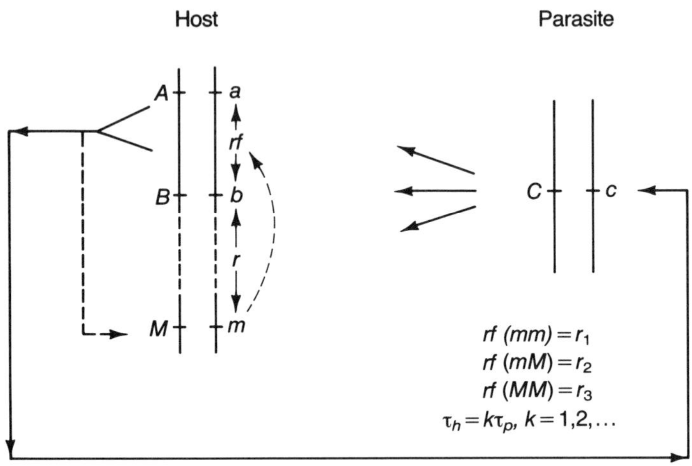

Figure 8.1 Structure of a model for the evolution of recombination under host–parasite interactions. Solid arrows indicate interactions of the selected loci, broken arrows represent effects due to the modifier locus; τ_h and τ_p are life cycle lengths of the host and parasite respectively.

parasite to be haploid is frequently more feasible. Clearly, the proposed model can readily be restated to include this case. The host resistance is determined by two diallelic loci, A/a and B/b; the rate of recombination, rf_{ab}, is controlled by a neutral *rec* locus with alleles M and m. A single locus, C/c, is concerned with the parasite virulence. The host longevity is assumed to be k times that of the pathogen (in a particular case, conceivably, $k = 1$). The relations between host and parasite populations are such that homozygotes CC infect genotypes AB/AB and ab/ab, homozygotes cc genotypes Ab/Ab and aB/aB. The overall fitness of each of the indicated genotypic classes of the host is the product of the abiotic and biotic components, abiotic selection being specified by the matrix:

$$W = \begin{array}{c} \\ BB \\ Bb \\ bb \end{array} \begin{array}{c} \begin{array}{ccc} AA & Aa & aa \end{array} \\ \begin{bmatrix} \alpha_0 & \beta & \alpha_1 \\ \gamma & 1 & \gamma \\ \alpha_1 & \beta & \alpha_0 \end{bmatrix} \end{array}$$

The overall fitness of genotypes AB/AB and ab/ab is a function of the parasite allele C frequency: $w_{11} = w_{33} = \alpha_0\varphi(p_C)$. Similarly, for Ab/Ab and aB/aB, $w_{13} = w_{31} = \alpha_1\varphi(p_c) = \alpha_1(1 - p_C)$, where $\varphi(p)$ is a continuous, monotonically decreasing function defined on the interval $[0, 1]$. The overall fitness of other genotypes of the host is assumed to be equal to their abiotic fitness. Provided abiotic fitnesses of the parasite genotypes CC, Cc and cc are specified by coefficients $\widetilde{W}_C$, 1 and $\widetilde{W}_c$, then their overall fitnesses may be defined as $W_c = \widetilde{W}_c\psi(X_{AB} + X_{ab})$ and $W_C = \widetilde{W}_C\psi(X_{Ab} + X_{aB})$, where $\psi(x)$ is a continuous, monotonically increasing (on the interval $[0, 1]$) function.

Changes in the frequency of the parasite alleles are defined by the ordinary equation for a one-locus random-mating population with selection, such that k generations of the parasite succeed one another in a single generation of the host. The equations describing the behavior of the host three-locus system (two selected loci and a modifier of recombination) are taken from Feldman (1972). If $X_1(n), \ldots, X_8(n)$ are frequencies of haplotypes $ABM, \ldots, abm$ in generation n, and p_c is the frequency of allele c, then transition equations for numerical iterations may be written as

$$X_{ABM}(n+1) = X_1(n+1) = \mathrm{F}_1\{X_1(n), \ldots, X_8(n); r, r_1, r_2, r_3; [w_{ij}(n)]_{i,j=1}^{3}\}$$
$$X_{AbM}(n+1) = X_2(n+1) = \mathrm{F}_2\{X_1(n), \ldots, X_8(n); r, r_1, r_2, r_3; [w_{ij}(n)]_{i,j=1}^{3}\}$$
$$\cdots\cdots\cdots\cdots\cdots\cdots\cdots\cdots\cdots\cdots\cdots\cdots$$
$$X_{abm}(n+1) = X_8(n+1) = \mathrm{F}_8\{X_1(n), \ldots, X_8(n); r, r_1, r_2, r_3; [w_{ij}(n)]_{i,j=1}^{3}\}$$
$$p_c(n+1) = f[p_c(n), W_c, W_C].$$

Computations have been carried out for various values of k with different combinations of the rf_{ab} values in genotypes MM, Mm and mm. The initial point was chosen such that $D_{ab} = D_{am} = D_{bm} = 0$, $p_c \approx 0.5$, $p_a = p_b = 0.5$, $p_{rec^+} = p_M = 0.04$. Most of the computations performed were based on a symmetrical fitness matrix with $\alpha_0 = \alpha_1 = 1$, $\beta = \gamma = 0.5$. Figure 8.2 shows the general form of the functions φ and ψ used in various combinations.

In our model, in contrast to that of Hutson and Law, the dependence of fitness coefficients of genotypes on their own frequencies and the time delay are not specified a priori but arise as a result of species interactions (Bronstein and Korol, 1984, Korol, Preygel and Preygel, 1986, 1990; Korol and Preygel, 1988a).

8.2.2 Behavior of the model

Computer simulations have shown that, regardless of the form of functions φ and ψ chosen and with a given selection intensity and no polymorphism at locus M/m, one can find a range of values of $rf \in [rf_1, rf_2]$ for which the system exhibits non-linear oscillations of D_{ab} and p_c of the limit cycle type. Over the range indicated, a reduction in rf results in increased amplitude and period of oscillations in a steady-state regime. When $rf < rf_1$, divergent oscillations with subsequent quenching of oscillations occur in the unsteady-state process. If $rf > rf_2$, then, following a perturbation, D_{ab} converges to an equilibrium value of D monotonically or in the form of damped oscillations. Provided selection is not too strong, then conceivably $rf_1 = 0$, i.e. oscillations will be stable even with complete linkage between resistance genes of the host, which is equivalent to the case of one-locus four-allele stability. On the other hand, when selection is rather strong, oscillations are also possible for $rf = 0.5$. The stability of the fixed point $\pi(p_c = 0.5, D = 0)$ depends on the form of the functions φ and ψ; but even if π is locally stable, with sufficient intensity of selection and

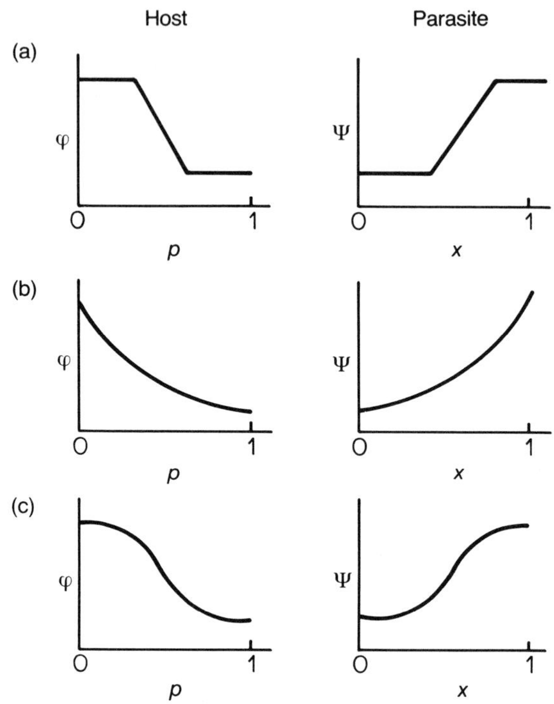

Figure 8.2 Types of frequency dependences used in computer simulations: (a) $\varphi(p)$ and $\psi(x)$ are linear functions; (b) $\varphi(p) = \exp(-\alpha p)$, $\psi(X) = \exp(\beta X)$; (c) $\varphi(p) = \alpha_1 - \beta_1 \tan^{-1}(2p-1)$, $\psi(X) = \alpha_2 - \beta_2|2X-1|^{\gamma}$, for $X < 0.5$ and $\psi(X) = \alpha_2 + \beta_2|2X-1|^{\gamma}$ for $X > 0.5$ (which will be referred to as $\psi(X) = d_2 \pm \beta_2|2X-1|^{\gamma}$).

Table 8.1 The effect of rf_{ab} on the amplitude and period of oscillations of D_{ab} under host–parasite interactions

Selection in				*D_{ab} oscillation regimes*		
Host (p)	*Parasite* (X)	k	rf	$D^{\dagger}_{ab}$	$\|D\|_{max}$	*Period* (T)
$1-p$	$1 \pm 0.9\|2x-1\|^{0.5}$	5	0.05	O	0.06	4
			0.0		0.22	10
$1.5-(3/\pi)\tan^{-1}(2p-1)$	$1+(2/\pi)\tan^{-1}(2X-1)$	5	0.1	O	0.22	12
			0.2		0.16	12
$1-p$	$4X$	5	0.5	A	0.07	—
			0.15	O	0.17	24
		1	0.2	O	0.14	24
			0.0	A	0.24	—
			0.05	O	0.22	27
$1 \pm 0.9\|2p-1\|^{0.5}$	$1 \pm 0.9\|2X-1\|^{0.5}$	5	0.00	Q	0.21	10
			0.15	O	0.06	5

†Regimes of D_{ab} behavior are as follows: O, oscillatory; A, aperiodic regime; Q, quenching of oscillations.

a proper value of *rf*, small deviations of p_c from 0.5 will, anyhow, produce unattenuated non-linear oscillations. The results just described are shown in Table 8.1 and in Figures 8.3 and 8.4.

Studies concerned with species interactions – both genetic (when analyzing gene and genotype frequencies) and ecological (in analyzing population sizes of the system components) – show that, for certain parameter combinations, the key variables defining the system may exhibit irregular behavior like that of stochastic nature (May, 1974, 1987; Hamilton, 1980, 1982; May and Anderson, 1983). This pattern of behavior of deterministic objects resulting from non-linear interactions of their subsystems is termed 'chaotic'. An intermediate type of behavior corresponds to aperiodic regimes of oscillations, in particular *T*-cycles. In the model described, we observed all these types of behavior (see Figures 8.4 to 8.6). Sets of parameters have been ascertained corresponding to 'beats' during the unsteady-state process. Thus, frequency-dependent selection induced by interaction of two antagonistic (diploid) species can result in stable oscillations of the coefficient of linkage disequili-

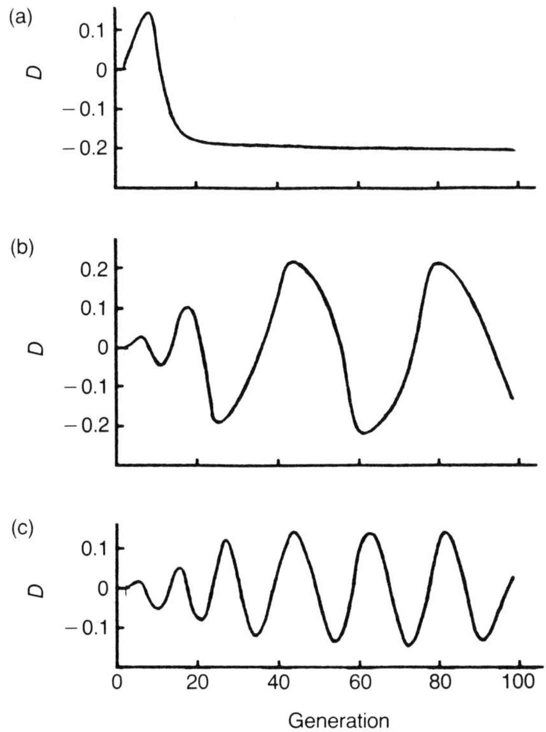

Figure 8.3 Examples of behavior of D_{ab} under various selection regimes and *rf* values: (a) $\varphi = 1 - p$, $\psi = 4X$, $k = 5$, $rf = 0.1$; (b) $\varphi = 1.5 - (3/\pi)\tan^{-1}(2p - 1)$, $\psi = 1 + (2/\pi)\tan^{-1}(2X - 1)$, $k = 5$, $rf = 0.1$; (c) same as in (b) except that $rf = 0.2$.

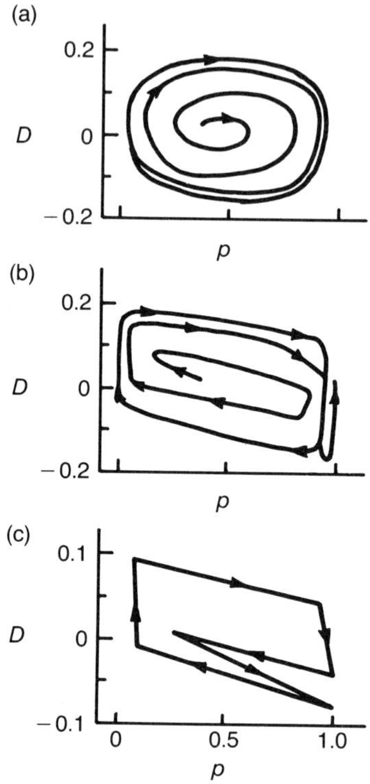

Figure 8.4 Oscillation patterns for the host–parasite system under various selection regimes. (a) Oscillations: $\varphi = 1 - p$, $\psi = 1 + (4/\pi)\tan^{-1}(2X - 1)$, $k = 1$, $rf = 0.05$; (b) Quenching of oscillations: $\varphi = 2(1 - p)$, $\psi = 2 \pm 1.9|2X - 1|^{0.2}$, $k = 5$, $rf = 0.2$. (c) Six-point cycle: $\varphi = 3(1 - p)$, $\psi = 1 \pm 0.9|2X - 1|^{0.1}$, $k = 8$, $rf = 0.01$.

brium between genes of one of the species, both in degree and in sign. The amplitude and period of oscillations depend on selection intensity and crossing-over rate.

Can these oscillations of D be a factor in the evolution of recombination? As already mentioned earlier, in models with a fluctuating environment, it is changes in the sign of the equilibrium value of D that are considered to be a necessary condition for an increase in the rec^+ modifier frequency (e.g. Charlesworth, 1976). For the case of host–parasite interactions, an answer to the question is partially provided, but for a haploid host only, in the work of Hutson and Law (1981). The effect of parasite has been defined implicitly in the form of host genotype fitness function as a decreasing function of parasite frequency with some time delay. Before proceeding to analyze the behavior of the *rec* modifier alleles, consider how the mean fitness of the host (as expressed by the geometric mean) depends on crossing-over rate between the selected loci.

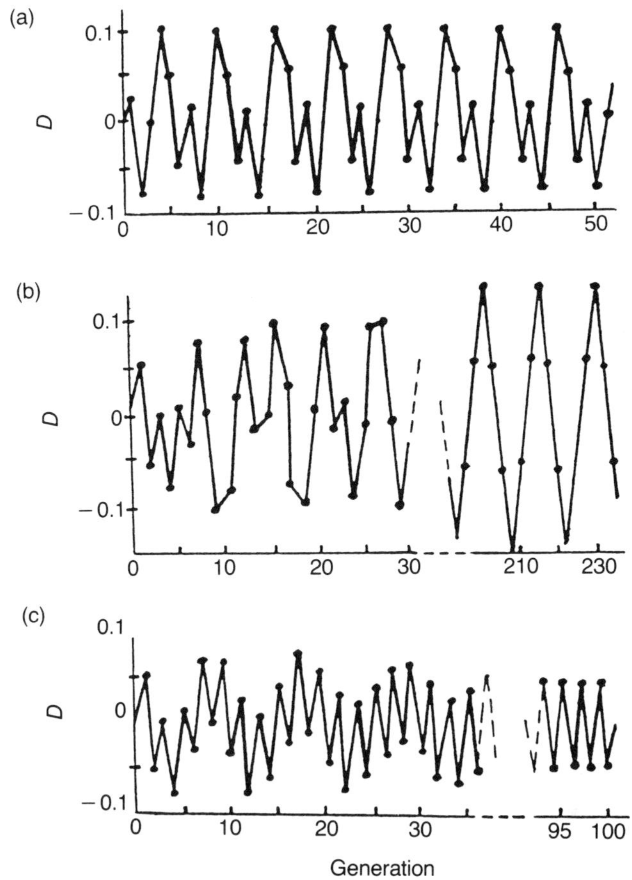

Figure 8.5 Examples of aperiodic regimes of D_{ab} oscillation. (a) Six-point cycle: $\varphi = 3(1-p)$, $\psi = 1 \pm 0.9|2X-1|^{0.1}$, $k=8$, $rf=0.01$. (b) 'Chaotic' behavior in the unsteady-state process: $\varphi = 3(1-p)$, $\psi = 1.9 \pm 1.4|2x-1|^{0.1}$, $k=5$, $rf=0.2$. (c) 'Beats' in the unsteady-state process: regime same as in (b) except that $rf=0.22$.

As already noted (see section 6.1), for a constant environment the correspondence between the two criteria for comparing *rec* strategies – altered frequency of modifier M and 'the magnitude of mean fitness' – is established by the Karlin–McGregor principle. Although, in general, the system in question (host–parasite) evolves in a constant environment, each of its components creates a varying environment for the other one. On the whole, one can hardly expect the correspondence between the indicated two criteria to persist in a situation like this. Nevertheless, it is advantageous to ascertain how the mean fitness of the host population depends on rf_{ab}. The results from several selection regimes shown in Figure 8.7 suggest that in estimations based on mean fitness increased recombination between loci for resistance to parasite

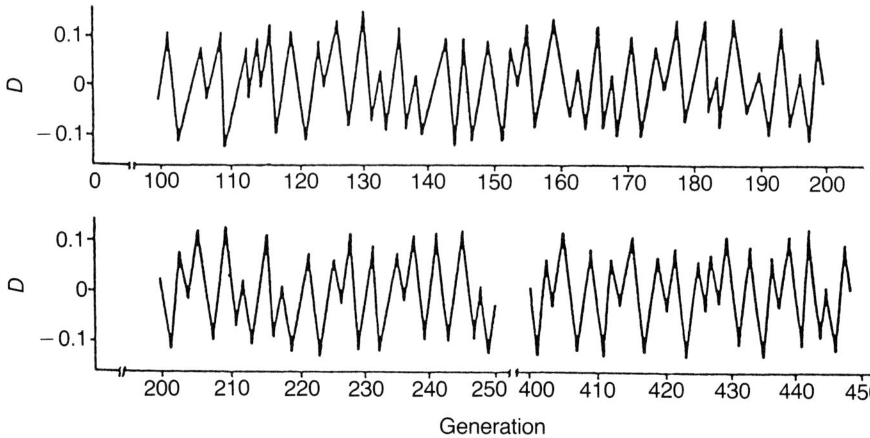

Figure 8.6 An example of chaotic behavior of D_{ab}: $\varphi = 6(1-p)$, $\psi = 1 \pm 0.9|2X-1|^{0.1}$, $k=10$, $rf=0.05$.

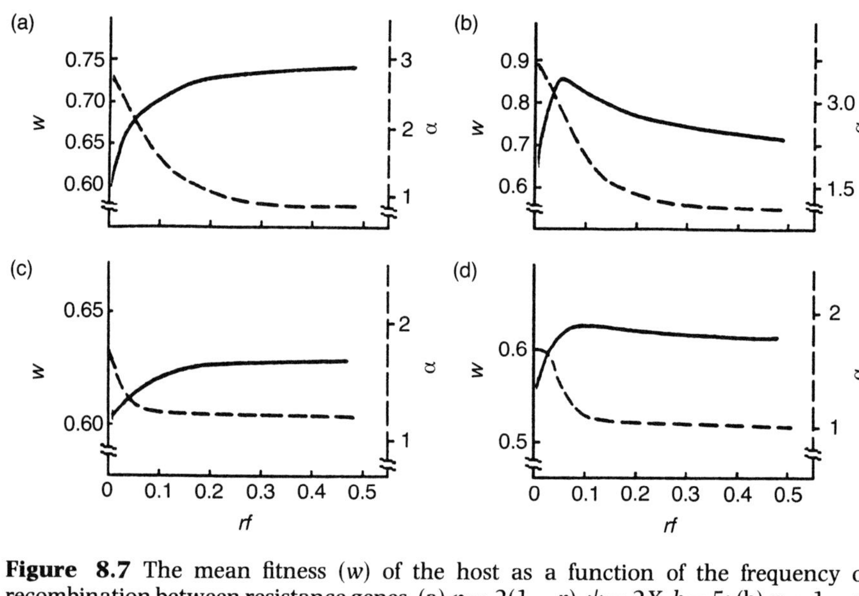

Figure 8.7 The mean fitness (w) of the host as a function of the frequency of recombination between resistance genes. (a) $\varphi = 2(1-p)$, $\psi = 2X$, $k=5$; (b) $\varphi = 1-p$, $\psi = 1 \pm 0.9|2X-1|^{0.5}$, $k=5$; (c) $\varphi = 2(1-p)$, $\psi = e^x - 0.5$, $k=1$; (d) $\varphi = 1-p$, $\psi = 1 + (4/\pi)\tan^{-1}(2X-1)$, $k=1$. Here $\alpha = w_{max}/w_{min}$.

may be selectively advantageous. In cases like these, fitness fluctuations (w_{max}/w_{min}) reflect the amount of loss sustained by the host owing to the continually increasing and decreasing correspondence between the genetic constitution of its population and the parasite population structure that has been formed by a given moment. The attainment of the correspondence is assured by selection, but this will take some time, during which the parasite succeeds in 'readjusting itself'. The rate of response by the host population is enhanced with increasing *rf*. Provided selection (specified by functions φ and ψ and parameter k) is not too intense and/or the initial values of *rf* are not too small (e.g. greater than 0.05), then values of w_{max} / w_{min} are usually also rather small (e.g. less than 1.5–2.0). Thus, for case (a) (see Figure 8.7), corresponding to $rf \geqslant 10\%$ is $w_{max} / w_{min} < 1.68$, i.e. a loss of less than 40%. This is sufficient for an increase in frequency of the rec^+ allele, resulting in $rf = r_3 = 0.3$ (see further Table 8.2). A still lower 'cost' is required for eliminating the allele for zero (or very low) *rf* when the selection regime corresponds to that shown in Figure 8.7(c). It will be recalled that in the model for the *rec* system evolution by selection against harmful mutations, the load required for eliminating the zero recombination allele could be rather high (Kondrashov, 1988a).

However, it would be wrong to assume that the loss resulting from adaptation of the host to the parasite is sustained only because of exceedingly tight linkage between the host resistance genes. Provided $w(rf)$ is an increasing function, one may expect selection to favor increased recombination. If that is the case, losses resulting from a low rate of response by the mechanism of the host gene pool repatterning should be reduced, and the value of w_{min} will rise. The value of w_{max} will concurrently decline, however, since the higher the recombination the more difficult it will be for the population to be, at least for a short period of time, almost entirely composed of optimum genotypes. This, naturally, reflects the known contradiction between the current fitness and genetic flexibility; increased recombination enhances 'reliability' by increasing w_{min} (a lower risk of the host population being eliminated during the periods of maximum non-correspondence between the interacting species structures), but this is achieved at the cost of producing non-optimum genotypes even in the best periods of the host population life. Incidentally, the mean fitness of a host population is generally increased.

A model describing interactions between multiple loci of host plant resistance and multiple loci of parasite virulence has been developed by Ostergard and Hovmoller (1991) for solving the problems of measuring resistance in real populations. The dynamics of virulence gene frequencies and linkage disequilibria depended on the selection regimes.

8.2.3 Co-selection on the *rec* locus

Numerical analysis of the *rec* modifier behavior indicates that the host–parasite interaction causing oscillations of D for the host resistance genes may indeed be a factor ensuring an evolutionary advantage of increased recombination. Consider the pattern of variation of D_{ab} when the *rec* locus is polymorphic. Let the values of $rf = r_3$, r_2 and r_1 correspond to the three states, *MM*, *Mm* and *mm*, for which the dynamics of D is oscillatory, with amplitudes

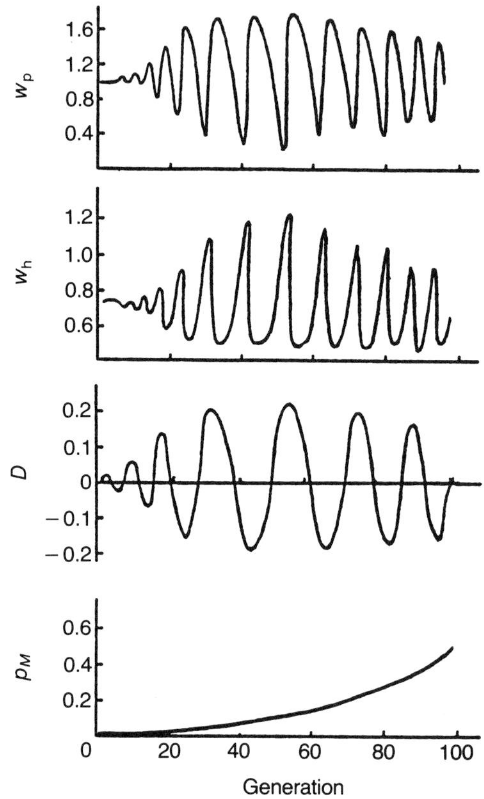

Figure 8.8 A typical dynamics of the key parameters of host and parasite populations: the selection regime is defined by $\varphi = 2(1-p)$, $\psi = 2X$ and $k = 5$, and the effect of the *rec* locus by $(r_1, r_2, r_3) = (0, 0.1, 0.3)$.

depending on *rf*. Then, the typical variation pattern for D_{ab} will be the one shown in Figure 8.8.

In a certain region of the trajectory with a fairly large amplitude of oscillation of D_{ab}, allele *M* becomes favored by selection; an increase in its frequency results in a higher rf_{ab}. Provided the differences between r_1 and r_3 are large, such that corresponding to r_1 is a large amplitude of oscillation of D_{ab} and to r_3 the absence of oscillations or a small amplitude of these, then an increase in the frequency of allele *M* will be accompanied by a reduction in the range of variation of *D*. When a certain level of p_M is attained, the mean rate of crossing over, rf_{ab}, will reach a point corresponding to a critical amplitude of the D_{ab} oscillation, below which changes in the sign of *D* with a given period length are no longer capable of producing changes in p_M. A polymorphism at the locus modifying recombination will be established in the population: a reduction in rf_{ab} would mean an increased amplitude of oscillation and 'switching on' of induced selection for a higher p_M, i.e. restoration of the attained level of rf_{ab}; with a random (owing to fluctuations) increase in the allele M frequency, selection in the opposite direction is impossible. Thus, the following rather unusual behavior may be characteristic of the system: when the differences between alleles M and m with respect to rf are very large, the rate of response to induced selection is high, but the locus M/m may remain polymorphic. Provided the differences between M and m (r_3 and r_1) are small, then, with the amplitude of the D_{ab} oscillation maintained at a sufficiently high level, allele M can attain fixation, in spite of the lower mean rate of the process.

Some examples are given in Table 8.2, from which it follows that selection in favor of allele M is possible even with small effects of the *rec* locus. As with ordinary loci subjected to selection, the behavior of the *rec* locus strongly depends on the relations of dominance–recessiveness, particularly when the frequency of one of the alleles is low: if the allele M for high recombination is recessive then, with a low initial frequency, $p_M(0)$, its rate of spread through the population is also low. It is clear from the foregoing that it is possible to obtain situations in which interallelic relations affect not only the rate but also the direction of change of the modifier. In particular, when r_1 is rather small and the recombination enhancer M is recessive, strong selection may result in quenching of the D_{ab} oscillations (or in larger amplitude and period), so that the rec^+ allele will tend to be eliminated by induced selection. Under high level of dominance of M, the oscillation of D_{ab} will be more 'moderate' and rec^+ will increase in frequency. It is such a case that we observed for $\varphi = 1 - p$, $\psi = 1 \pm 0.9|2X - 1|^{0.5}$ (see variants marked with a brace in Table 8.2). When rec^+ is dominant, oscillations with a period of $T \approx 30$ generations and $|D_{ab}| = 0.18$ are set up, the frequency of M increasing from 4% to 38% in 400 generations. When M is recessive, the period is enlarged ($T = 42$), the amplitude of oscillation of D_{ab} rises up to 0.22 and the frequency of allele M decreases by 1% in 400 generations.

An interesting result has been obtained in evaluating the role of random fluctuations of the parasite allele frequencies in situations in which the oscillations of D_{ab} are insufficiently large. In cases such as these, the introduction of a stochastic component into the system (an abiotic component of the parasite fitness specified by normally distributed pseudorandom numbers with variance σ^2) results in the rec^+ increasing in frequency. But if fluctuation variance σ^2 is too large, the effect may be the opposite one. The same situation is observed when spontaneous oscillations of D_{ab} are fairly large: introduction of noise into the system can result in quenching of oscillations. In general, the dynamics of the modifier, like that of the entire system, proves to be insensitive to small fluctuations.

Consider now the situation in which host and parasite are assumed to be three-locus systems (each consisting of two fitness loci and modifier). It has been shown that species interactions in this model can result in a simultaneous increase or decrease in frequency of the high-recombination allele in both

Table 8.2 Selection on the *rec* locus induced by host–parasite interaction

Selection in								
Host	*Parasite*	r_1	r_2	r_3	r	k	n	$p_M(n)$
$2(1-p)$	$2X$	0.01	0.02	0.10	0.5	1	400	0.31
		0.01	0.02	0.10	0.5	5	400	0.62
		0.10	0.20	0.30	0.5	5	400	0.69
		0.10	0.20	0.30	0.5	10	400	0.75
	$\sigma = 0.2^{\dagger}$	0.10	0.20	0.30	0.5	10	400	0.60
$1-p$	$1 \pm 0.9\lvert 2x-1\rvert^{0.5}$	0.00	0.00	0.05	0.5	5	400	0.03 }
		0.00	0.05	0.05	0.5	5	400	0.38 }
		0.01	0.02	0.10	0.5	1	400	0.06
		0.01	0.02	0.10	0.05	1	400	0.96
		0.00	0.20	0.30	0.5	5	400	0.41
	$\sigma = 0.2^{\dagger}$	0.00	0.20	0.30	0.5	5	400	0.50
		0.00	0.05	0.50	0.5	10	400	0.42
	$\sigma = 0.2^{\dagger}$	0.00	0.05	0.50	0.5	10	400	0.02
		0.05	0.10	0.20	0.5	1	400	0.19
		0.05	0.10	0.20	0.05	1	400	0.52
$1 \pm 0.9\lvert 2p-1\rvert^{0.5}$	$1 \pm 0.9\lvert 2x-1\rvert^{0.5}$	0.01	0.02	0.05	0.5	5	800	0.19
		0.01	0.02	0.05	0.05	5	800	0.92
$1-p$	$1 + 4\tan^{-1}(2x-1)/\pi$	0.05	0.10	0.20	0.5	1	400	0.10
		0.05	0.10	0.20	0.05	1	400	0.59
$1-p$	$e^x - 0.5$	0.05	0.10	0.20	0.5	1	400	0.02
		0.05	0.10	0.20	0.05	1	400	0.99
		0.05	0.10	0.20	0.5	5	400	0.75

The initial frequency of the rec^+ allele $p_M(0) = 0.04$.

†The introduction of a stochastic factor ('noise') with the standard deviation σ into the parasite coefficients of fitness.

populations as well as an increase in one of them and a decrease in the other. The existence of situations in which *rec* alleles are permanently advantageous in both populations seems to be the most interesting of these results. Bell and Maynard Smith (1987) considered the same problem using another type of formalization of species interactions. They found that a simultaneous increase in frequencies of the rec^+ alleles in host and parasite is only possible as an initial tendency.

8.2.4 Discussion

From the data presented, it can be concluded that the host–parasite interactions resulting in weak selection of various combinations of the host resistance genes may promote increased recombination for only tightly linked genes, while those inducing moderate selection enhance *rf* under both tight and relatively loose linkage. Finally, interactions resulting in strong selection promote blocking of recombination between rather tightly linked loci and further increase in *rf* between moderately linked genes. If selection in nature is strong, we are justified in expecting various genes for resistance to one parasite to be usually organized in blocks or distributed over different chromosomes (or chromosome arms).

This prediction is comparatively easy to test, but the data on the genomic organization of resistance systems are still very scarce and concern mainly cultivated plants. For example, genes for resistance to rust in maize and to powdery mildew in barley do form blocks. In contrast, 15 loci for resistance to rust in wheat are located on 10 chromosomes (Van der Plank, 1978). In tomato, resistance to *Cladosporium fulvum* is determined by tens of loci scattered over various chromosomes. One could believe the block organization of genes to be the result of successive tandem duplications. But this is only a part of the explanation, since the mechanisms responsible for dispersed organization of resistance genes remain unclear. The conclusions that follow from our analysis relate the differences between the two gene organization patterns in question to the selection regime resulting from 'host–parasite' interactions.

Our conclusion, that increased intensity of selection experienced by the host with increased *k* (the number of parasite cycles per one host generation) can induce selection for lower recombination, may be relevant to recent attempts at testing theoretical explanations of the maintenance of recombination. Thus, the absence of positive correlation between recombination level and generation time is considered as an indicator against the 'host–parasite' model (e.g. Koella, 1993). The above results show that this conclusion, based on an assumption of positive monotonic relationship between the intensity of selection (host lifetime, i.e. *k*) and advantage of recombination, may be not true. For example, in many cases an ∩-shaped dependence of recombination on generation time seems to be more suitable.

The host–parasite interactions can be modulated by selection for other ecologically significant traits, the genetic determinants of which are associated with the host resistance genes as a result of linkage or non-random mating (Parker, 1991). In this case, the system will show an even more complicated behavior (especially under varying environmental conditions). The direction of dynamics will be dependent on the relative significance of selective effects produced by abiotic and biotic environmental components and on the presence of a sufficient number of heterozygous loci for recombinants to be formed. In man, the components of the major histocompatibility complex *HLA* are organized in the form of two physically tightly linked gene clusters. Recent evidence suggests that these, like the MHC in mammals in general, play an important role in resistance to infectious diseases, (e.g. Hill *et al.*, 1991; Nevo and Beiles, 1992). The MHC region is known to be a hotspot of recombination, presumably because of the presence of minisatellites (see Kobori *et al.*, 1986; Johnston and Rich, 1989).

In the models described above, no mention is made of the sizes of interacting populations. Meanwhile, it is clear that introduction into the model of epidemiological parameters characterizing the dynamics of population numbers can have a dramatic effect on the system's behavior, including that of the *rec* modifier. Taking into account the population size of the parasite appears to be of utmost importance, since it is microorganisms that usually exhibit significant fluctuations in numbers and, consequently, drastic changes in the degree of pressure upon the host population, particularly in cases of reinfection after the elimination of a given race of the parasite. It is also necessary to assess possible effects of spatial heterogeneity of interacting populations, gene flow and drift (Burdon and Jarosz, 1991; Thompson and Burdon, 1992).

One aspect of the dependence of biotic selection intensity on the parasite numbers (to be more exact, on its mean fitness) has been shown in our model. It was assumed that if mean fitness of the parasite is lower than a certain critical level, w_{crit}, then the parasite pressure on the host population may be ignored. It turned out that, in a number of situations with non-oscillatory behavior of the system, allowing for this factor results in stable oscillations. Thus, we can expect that the qualitative conclusion about the strong effect of biotic interactions on the evolution of recombination will continue to hold, with further complexity of the model resulting from increasingly better allowance for epidemiological factors.

Most of our discussion on the role of species interactions as a factor behind the evolution of recombination has been, for the sake of unambiguousness, in terms characteristic of the 'host–parasite' systems (hence the interest in situations in which corresponding to each generation of one component are k generations of the other). Nevertheless, since the results hold for $k = 1$ as well, the models considered are also applicable to situations like 'predator–prey', 'plant–herbivore' and, conceivably, 'plant–pollinator' and 'two species competing for the common resource'. Interactions in such systems often play a decisive role in the life of the component species. Therefore, recombinational organization of groups of loci controlling relations with the 'partner' (parasite, predator, prey, competitor) is the reflection (and the product) of the previous history of these relations, on the one hand, and a factor of subsequent evolution of organization of not only controlling loci themselves but, presumably, also the genome as a whole, on the other hand.

Earlier (section 6.1), the factors responsible for the evolution of recombination were grouped under two classes – internal and external (with respect to the genetic system). The above analysis enables species interactions to be referred to a third class that occupies an intermediate position between the first two. Indeed, in our model the parasite acts as an external factor of the evolution of recombination in the host. But it can only play this part provided a mechanism of genetic feedback is operating, i.e. there is a continuous alternation of states of correspondence and non-correspondence between genetic structures of the host and parasite populations. The efficiency and the very possibility of functioning of such a mechanism depend, as we have seen, on the genetic organization of the host, i.e. on the properties of its resistance genes and, which is no less important, *rec* genes.

8.3 A MODEL OF CO-EVOLUTION WITH SELECTION ON THE CHARACTERS 'RESISTANCE' AND 'VIRULENCE'

In the model detailed above, the host resistance was assumed to be determined by two loci (the third locus controlling recombination between the two) and the parasite virulence by a single locus. This approach, however, is rather limited if one assumes the host resistance and the parasite virulence to be controlled by many segregating loci. Attempts, in a situation like this, to specify interactions between antagonistic species at the level of the genotype will inevitably encounter difficulties associated with a great number of various genotypes in both populations. At the same time, there is evidence for a complex genetic structure of a number of systems controlling resistance to pests and diseases (see Bremermann, 1987; Burdon, 1987; Simms and Rausher, 1987; Thompson and Burdon, 1992). Thus, already Flor (1956) has found that 27 genes are concerned with resistance to rust in flax. As mentioned above, in wheat, resistance to rust is controlled by 15 loci located on 10 chromosomes (Van der Plank, 1978). Highly polymorphic systems determining resistance to diseases have been demonstrated in animals (May and Anderson, 1983). Furthermore, the evolution of immune systems in vertebrates (multigene families of immunoglobulins) can only be explained by invoking multilocus models. The genetic basis of antagonistic interactions in predator–prey systems also seems to be very complex.

Therefore, adequate models of antagonistic species interactions should be multilocus ones. The resulting 'curse of dimensionality' can partially be overcome by only passing to models in which selection in the interacting populations is not on gene combinations but on characters, e.g. of resistance in the host and virulence in the parasite. In addition to having the merit of permitting a relatively easy transition to multilocus situations, formalizations of this type (with selection on the characters 'resistance' and 'virulence') are also more realistic. Described below are the models for the co-evolution of host and parasite based on selection on characters. In the first model, it is only the host resistance system that is assumed to be multilocus. The second model is a generalization of the first one; in it, virulence of the parasite is also controlled by many loci. Studies of such systems show that interactions between antagonistic species may result in selection for increased recombination in host and/or parasite.

8.3.1 Multilocus host–one-locus parasite

The proposed model (see also Preygel and Korol, 1990) deals with infinite random-mating diploid populations. The host resistance is controlled by n loci with alleles *0*/*1* at each locus. The *rec* modifier, M/m, is selectively neutral and determines the rate of crossing over between resistance genes, with allele M increasing and allele m reducing the frequency of exchanges. The behavior of a modifier, both linked and unlinked to resistance genes, was analyzed in simulation experiments. The host phenotype 'resistance to the parasite' is calculated either as the sum of all individual alleles at the selected loci (additive case) or as the number of loci at which at least one of the alleles is *1* (dominance case). The parasite virulence is controlled by a single diallelic locus, C/c; no phenotype concept was introduced for the parasite (which, of course, makes no difference in the one-locus case). As in the previous model (section 8.2), heterozygotes Cc are assumed to show no preference in attacking host phenotypes, i.e. fitness of the Cc individuals is constant for any phenotypic composition of the host population. To assign fitnesses to homozygotes CC and cc, let us introduce the concept of mean phenotype (z):

$$z = \sum_{i=1}^{n} z_i p_i,$$

where n is the number of various phenotypic classes in the host, z_i the value of the ith phenotype of the host (with the method of calculating phenotypes chosen, $z_i = 0, 1, \ldots, n-1$) and p_i the frequency of the ith phenotype in the population.

Suppose that fitnesses of homozygotes CC and cc (let these be denoted by W_C and W_c, respectively) are the functions of the mean phenotype value in the host population. For convenience, a normalized value, $z^* = z/(n-1)$, was used as the argument:

$$W_C = W_C(z^*), \quad W_c = W_c(z^*), \quad 0 \leqslant z^* \leqslant 1.$$

Two qualitatively different with respect to the type of functions W_C and W_c situations were studied in simulation experiments (Figure 8.9).

1. Let individuals *CC* be more successful in parasitizing host individuals with small phenotypic values and homozygotes *cc* be better at 'exterminating' individuals with large phenotypic values. Then, if the genetic constitution of the host population is such that prevailing in it are individuals with small phenotypic values, allele *C* will spread through the parasite population. Conversely, if the mean phenotype of the host is large, fitnesses of individuals *cc* will be higher than those of *CC*. That is to say, here fitness of the *CC* homozygotes is a monotonically decreasing and that of the *cc* individuals a monotonically increasing function of z^* $(0 \leqslant z^* \leqslant 1)$. By contrast, the fitness w_{z_i} of host individuals with small phenotypic values z_i is higher the lower the frequency of the *CC* homozygotes parasitizing them. In other words, if z_i is small, then the fitness w_{z_i} is a monotonically decreasing function of the *C* allele frequency, p_C. For hosts with large phenotypic values, it is advantageous to have the highest possible frequency of allele *C* in the parasite population, i.e. for large values of z_i, the function $w_{z_i}(p_C)$ is a monotonically increasing one. The above arguments are illustrated in Figure 8.9.

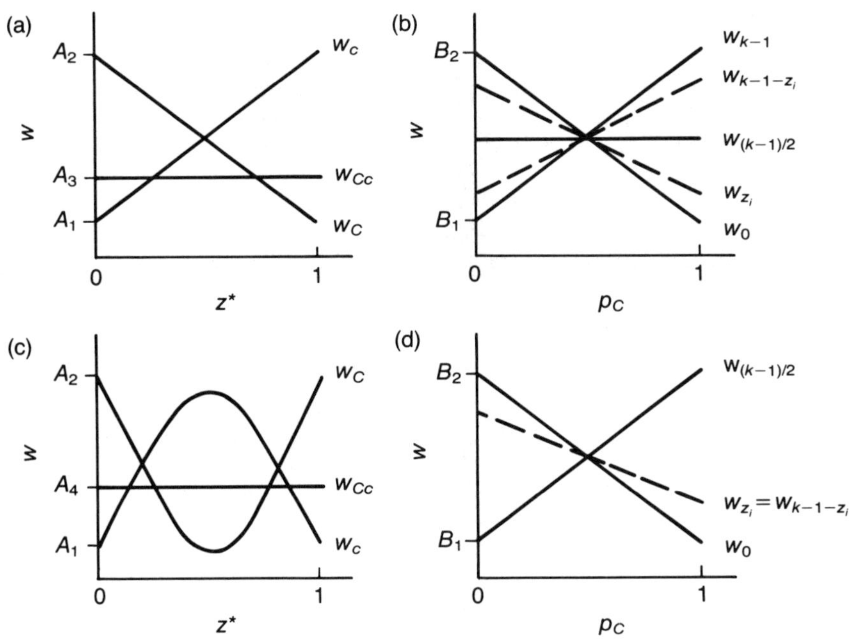

Figure 8.9 Fitness functions $W_C(z^*)$ and $W_c(z^*)$ for the parasite (a and c) and $w_{z_i}(p_c)$ for the host (b and d).

2. While in the first situation it was assumed that each of the parasite alleles 'exterminates' most effectively one of the 'tails' of the host phenotypic range, now assume that homozygotes *CC* are adapted to parasitizing extreme and *cc* intermediate host phenotypes. Clearly, in this case functions $W_C(z^*)$ and $W_c(z^*)$ are no longer monotonic . As in the previous model with selection on gene combinations, it is assumed that the parasite life cycle is k times that of the host ($k \geqslant 1$). The initial population structure of the host is such that linkage disequilibria of all orders both between the *rec* locus and resistance genes and within the resistance system are zero. Gamete frequency changes from generation to generation in the host population were calculated by the method of 'genetic operators' (section 7.2). The parasite population dynamics is defined by an ordinary equation for a one-locus random-mating population with selection.

One can expect that given a certain relationship among the parameters, oscillations will be set up in the system described, including those with respect to the mean phenotype (z) of the host. To put it another way, it is natural to assume that the parasite pressure will induce selection with a fluctuating optimum in the host population. Selection of this type may result in increased recombination (see the previous chapter).

8.3.2 Phenomenology of recombinational behavior of the system

The effect of selection on recombination may, in principle, depend on all parameters specifying the system dynamics, i.e. the form of fitness functions W_C, W_c and w_{zi}, the nature of effects of host genes on resistance, life cycle length ratio of host and parasite (k), recombination rates r_M and r_m, degree of linkage between the *rec* locus and resistance genes (rf), etc. Numerical experiments have shown that in the above model selection for increased recombination may occur under certain combinations of parameters (Tables 8.3 and 8.4).

An increase in the frequency of allele M was observed with both monotonic functions W_C and W_c (type 1 selection) and non-monotonic functions W_C and W_c (type 2 selection). It will be recalled that the former type of selection corresponds to differential 'extermination' by the parasite homozygotes of different 'tails' of the host phenotypic range. With the latter type of selection, genotypes *CC* are more successful in parasitizing extreme, and *cc* intermediate, phenotypes of the host. The advantage of increased recombination is also revealed both under additive effects of the host genes on resistance and with dominance of these loci. The greater the number of generations of the parasite (k) that succeed one another during a single life cycle of the host, the more intense is the selection in the host population and the greater the advantage of recombination resulting from this selection. Let the host resistance be controlled by two loci with additive effects: allele M of the unlinked modifier is recessive and causes free recombination between the selected loci and allele

Table 8.3 Frequency of allele M for increased recombination in the host after 100 cycles of the host–parasite co-evolution (selection of type 1 with parameters: $A_1 = 0$, $A_2 = 4$, $A_3 = 1$, $B_1 = 0$, $B_2 = 6$)

Number of loci	*k*	*Loci effect*	$R = (r_m, r_M)$	*rf*	$p_M(0)$	$p_M(100)$
2	1	A	$(0^+, 0.5)$	0.5	0.5	0.603
			$(0^+, 0.1)$	0.5	0.5	0.549
		D	$(0^+, 0.5)$	0.5	0.5	0.594
			$(0^+, 0.5)$	0.05	0.5	0.749
	2	A	$(0^+, 0.5)$	0.5	0.5	0.616
	6	A	$(0^+, 0.5)$	0.5	0.5	0.643
	7	A	$(0, 0.5^+)$	0.5	0.5	Q
			$(0.4, 0.5^+)$	0.5	0.5	0.511
			$(0.25, 0.5^+)$	0.5	0.5	Q
			$(0.25, 0.5^+)$	0.5	0.75	0.760
3	1	A	$(0^+, 0.5)$	0.5	0.5	0.576
			$(0^+, 0.5)$	0.05	0.5	0.927
		D	$(0^+, 0.5)$	0.5	0.5	0.593
			$(0^+, 0.5)$	0.05	0.5	0.773
	3	A	$(0^+, 0.5)$	0.5	0.5	Q

Here and in Table 8.4, k is the number of the parasite generations per host life cycle.
A (D), additivity (dominance) of host resistance loci; r_M and r_m, recombination rate between adjacent resistance genes as determined by alleles M and m;
$^+$denotes the dominant *rec* allele; *rf*, recombination between the *rec* locus and the resistance system; $p_M(0)$, $p_M(100)$, frequency of allele M at the start and at the 100th cycle respectively; Q, quenching of oscillations.
The initial frequency of allele C in the parasite was 0.2.

Table 8.4 Frequency of the allele M for increased recombination after 100 cycles of the 'host–parasite' system dynamics (selection of type 2 with parameters: $A_1 = 0.1$, $A_2 = 2$, $A_3 = 0.75$, $A_4 = 1$, $B_1 = 0$, $B_2 = 4$)

Number of loci	*k*	*Loci effect*	$R = (r_m, r_M)$	*rf*	$p_M(0)$	$p_M(100)$
2	1	A	$(0^+, 0.5)$	0.5	0.5	0.548
			$(0^+, 0.5)$	0.05	0.5	0.737
		D	$(0^+, 0.5)$	0.5	0.5	0.618
	2	A	$(0^+, 0.5)$	0.5	0.5	0.584
			$(0, 0.5^+)$	0.5	0.5	0.526
			$(0.05, 0.5^+)$	0.5	0.5	0.520
			$(0.05, 0.5^+)$	0.05	0.5	0.552
	3	A	$(0^+, 0.5)$	0.5	0.5	0.593
4	3	A	$(0^+, 0.5)$	0.5	0.5	0.553
		D	$(0^+, 0.5)$	0.5	0.5	0.621

m is dominant and completely suppresses recombination between these. Then for a chosen set of parameters (see Table 8.3), we found for $k = 1$ an increase in frequency of allele M in 100 cycles of the host, from 0.5 to 0.603, for $k = 2$ to 0.616 and for $k = 6$ to 0.643.

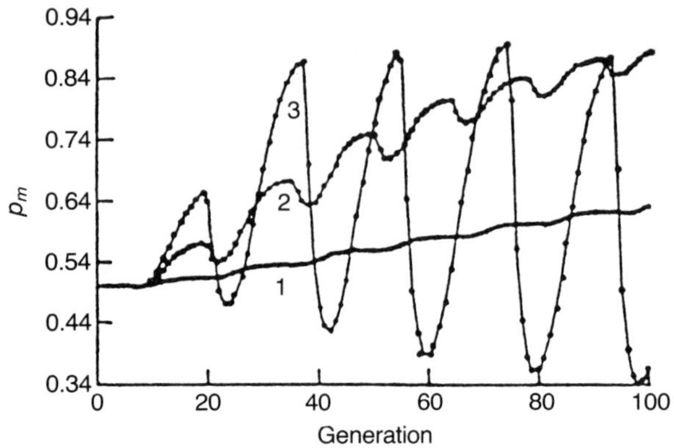

Figure 8.10 The effect of recombination (*rf*) between the *rec* modifier and the selected system on changes in the frequency of allele *M* (the case of a two-locus additive control of resistance). Selection of type 1 with parameters $A_1 = 0$, $A_2 = 4$, $A_3 = 1$, $B_1 = 0$, $B_2 = 6$, $k = 3$, $r_M = 0.5$, $r_m = 0$. (1) $rf = 0.5$; (2) $rf = 0.15$; (3) $rf = 0.01$.

With a further increase in the value of parameter *k*, the system behavior becomes qualitatively different: while previously oscillations of the genetic constitution occurred in both populations, now, after a short unsteady-state process, one allele (say *C*) becomes fixed in the parasite population and the host population is composed of a single type (say *11/11*). Note that if the pre-existing mean frequency of recombination in the population is high enough, there may be no quenching of oscillations, and allele *M* will increase in frequency (as in the situation considered in section 8.2).

Numerical experiments have shown that in each situation there is a certain degree of linkage between the modifier and the selected system optimal in terms of efficiency of selection on the *rec* locus. This is not surprising since selection with a fluctuating optimum is operating in the host population (see Chapter 7). In the majority of the cases considered, the most significant increase in the *M* allele frequency is observed under an intermediate degree of linkage between the modifier and resistance genes (a typical situation is shown in Figure 8.10).

8.3.3 The mechanism of selective advantage of increased recombination

In all the situations considered, the advantage of recombination is a consequence of selection with a fluctuating optimum occurring in the host population. To gain an insight into the mechanism causing allele *M* to increase in frequency, consider in more detail the simplest case – a two-locus control of the host resistance. A typical pattern of behavior of the system is shown in Figure 8.11 (for type 1 selection).

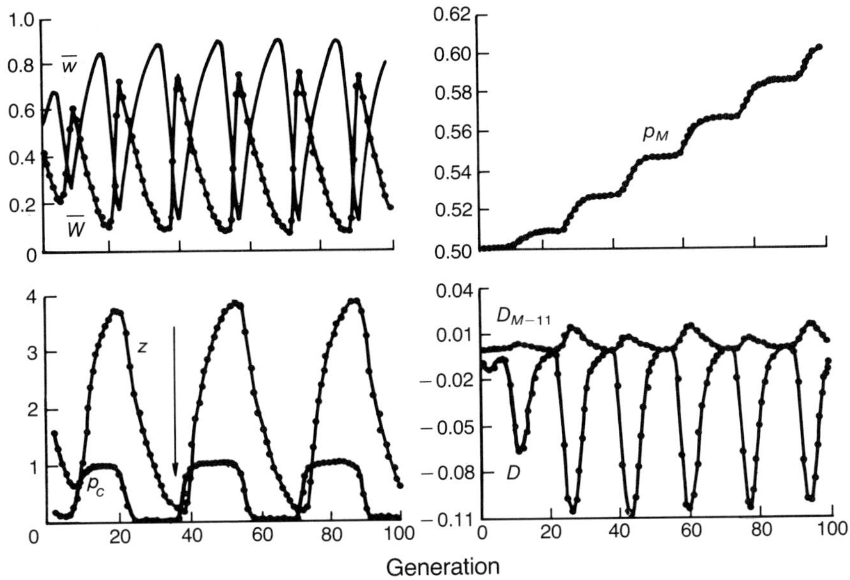

Figure 8.11 The behavior of key variables of the system (parameters as in Figure 8.10, except that $k = 1$, $rf = 0.5$). The arrow marks the beginning of the part of the trajectory discussed in the text.

At the beginning of the marked section of the trajectory, the host population consists almost entirely of the *00/00* individuals (the mean phenotype value, z^*, is close to zero); the parasite population is mainly represented by homozygotes *cc*. It will be remembered that under selection of this type homozygotes *cc* are successful in parasitizing host individuals with large phenotypic values. Therefore, at this stage, mean fitness of the host population (w_{host}) is high and that of the parasite (W_{par}) low. Consequently, it is the parasite that will 'lead the game' at the first stage. Indeed, frequency of allele *C* rapidly increases, and consequently so does the value of W_{par}. Changes in the genetic constitution of the parasite population induce selection in the host, which starts 'to run away' from the parasite by producing individuals with large phenotypic values (which are unsuitable for *CC*). The proportion of combinations *01* and *10* gradually increases and, as a result, negative linkage disequilibrium develops between them (as is clear from Figure 8.11, at this stage $D < 0$). Notice that the increase in frequency of *trans*-combinations mainly occurs not by recombination but as a result of selection since at the initial moment the frequency of *cis*-combinations *00* was close to unity. At this stage, the frequency of the allele *M* for high recombination is, therefore, not increased.

A rapid increase in the frequency of allele *C* has the result that at a certain moment the parasite population comes to consist almost entirely of homozygotes *CC*. Prevailing in the host population meanwhile are individuals with low phenotypic values. It is the host that is now 'at a disadvantage'. At this second stage it is the host that plays the leading role in the evolutionary game: it begins to actively accumulate combinations *11* resistant to the *CC* homozygotes of the parasite. Crossing over plays a significant role in the formation of these combinations. Breaking up combinations of the *01* and *10* type, the allele *M* for high *rf* spreads through the population by hitching a ride on a favorable combination *11*. Positive *D* develops at this stage between allele *M* and combination *11* ($D_{M-11} = p(M \cdot 11) - p_M p_{11} > 0$), which is particularly significant at the beginning of transition from combinations *01* and *10* to *11*.

Thus, at the end of the second stage the host population consists almost entirely of homozygotes *11/11* (z^* is large), while in the parasite population individuals *CC* prevail; fitness of the host population, w_{host}, is high, and that of the parasite, W_{par}, is low. This situation is very similar to the initial one; 'the evolutionary dance' goes on according to the same rules.

The proposed explanation for the mechanism of spread of the high-recombination allele is, in its main features, also applicable to models with selection of the second type when functions $W_C(z^*)$ and $W_c(z^*)$ are not monotonic. In this case, changes in the frequency of *rec* alleles are also the result of either co-selection (when the modifier is unlinked) or co-selection and hitchhiking (when the *rec* locus is linked to resistance genes). While with selection of the first type the population most of the time produces an excess of gametes in repulsion and the high-recombination allele periodically associates with either *11* or *00* combinations, in models of the second type the situation is somewhat different. In these, selection in the host population induced by the parasite pressure alternately favors *cis*- and *trans*-combinations and, consequently, in some portions of the trajectory allele *M* hitches a ride on combinations *00* and *11*, in others on *01* and *10*. Thus, the main distinction between models with selection of the first and second type is only in the combinations of the selected alleles the modifier obtains a hitch from. Clearly, the difference is not a vital one.

8.3.4 Multilocus host–multilocus parasite

An attempt to study the behavior of *rec* modifiers in both antagonistic populations has recently been made by Bell and Maynard Smith (1987). In their model, both interacting populations are haploid. The host resistance is controlled by two loci. The structure of the parasite virulence system is similar to that of the host resistance. Each population contains a *rec* locus: one of the alleles determines free recombination between all three of the loci; the other completely blocks recombination over the entire chromosome. Interactions between host and parasite are defined by the so-called response matrices, which determine fitnesses of host and parasite phenotypes, assuming that the alternative population is represented by a single phenotype. It is also assumed that the probability of the host being infected by a given type of parasite is proportional to the frequency of that type in the parasite population. The overall fitness of a given genotype of host is calculated as the sum of its responses to all types of parasite, weighted by their frequencies. Such a method of defining interactions is very sensitive to the number of genotypes in each of the interacting populations and can hardly be applied in case of multilocus diploids or even multilocus haploids.

Bell and Maynard Smith considered two types (models) of interaction. In the first, the parasite performance in infesting a given host type is determined by the absolute value of the difference between their phenotypes, individuals *ab* having a phenotype of *0*, *Ab* and *aB* of *1* and *AB* of *2*, i.e. the effects of loci were assumed to be additive. The second type reflects the situation when for every virulence gene in the parasite there is a corresponding resistance gene in the host (a 'gene-for-gene' mode of interaction). The main result of this work is that the spread of the rec^+ allele is only possible in one of the two interacting populations: an increase in crossing-over rate in the host is accompanied by its suppression in the parasite. We shall show that, with other types of selection in the 'host–parasite' system, there may be a simultaneous increase in the frequency of alleles enhancing recombination in both host and parasite. It is worth noting at this point that many pathogens have extremely effective systems for generating variability, including recombinational ones. An example is provided by the flu virus (Kaplan and Webster, 1977).

Let the host resistance and the parasite virulence be polygenic traits. Introduced for both populations are the concepts of mean phenotype of a population and normalized mean phenotype. Let kp be the number of phenotypes in the parasite population and kh in the host. Let, also, zp_i denote the value of the ith phenotype in the parasite and zh_j that of the jth in the host. Suppose that their frequencies are p_i and q_j, respectively. Then mean phenotype values are

$$zp = \sum_{i=1}^{kp} p_i zp_i \quad \text{for the parasite;}$$

$$zh = \sum_{j=1}^{kh} q_j zh_j \quad \text{for the host.}$$

Let also zp^* and zh^* denote the corresponding values after normalization (i.e. dividing by the value of the maximum phenotype in the population).

Selection in the parasite population is defined by a family of functions $\{W_{zp_i}(zh^*),\ zp_i = 0, \ldots, kp-1\}$. Host fitnesses depend on the value of mean phenotype in the parasite, zp^*: $\{w_{zh_j}(zp^*),\ zh_j = 0, \ldots, kh-1\}$. As in the model with a one-locus parasite, monotonic and non-monotonic functions W_{zp_i} are

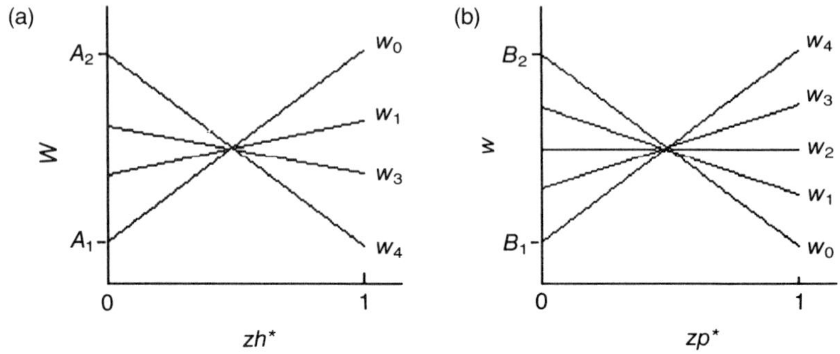

Figure 8.12 Fitness functions $W_{zp_i}(zh^*)$ for parasite (a) and $w_{zh_j}(zp^*)$ for host (b) (a case of two-locus additive control of resistance and virulence).

considered. For simplicity, we will discuss only the case of linear selection with two-locus additive control of resistance and virulence (Figure 8.12).

It turns out that there exist sets of parameters under which each of the interacting populations induces selection with a fluctuating optimum in the other member of the system. This confers an advantage of increased recombination simultaneously on host and parasite (Table 8.5). The spread of alleles enhancing recombination proceeds by the same rules as in the above situation. As before, host and parasite play alternately the role of the leading population (which most drastically alters its structure while chasing 'the runaway' antagonist) in the 'evolutionary game'. It is during this period of the greatest

Table 8.5 Frequencies, p_{MH} and p_{MP}, of high-recombination alleles after 200 cycles of dynamics under two-locus additive control of resistance–virulence (selection of type 1 with parameters: $A_1 = 0$, $A_2 = 6$, $B_1 = 0$, $B_2 = 6$)

k	$R_{par} = (r_{mp}, r_{MP})$	$R_{host} = (r_{mh}, r_{MH})$	$p_{MP}(200)$	$p_{MH}(200)$
1	(0, 0.1)	(0, 0.1)	0.642	0.560
	(0, 0.1)	(0, 0.5)	0.660	0.600
	(0, 0.5)	(0, 0.1)	0.725	0.561
	(0, 0.5)	(0, 0.5)	0.750	0.604
	(0, 0.01)	(0, 0.01)	0.536	0.512
3	(0, 0.1)	(0, 0.1)	0.597	0.609
	(0, 0.5)	(0, 0.5)	0.711	0.676
5	(0, 0.5)	(0, 0.5)	0.638	0.707
	(0.05, 0.25)	(0.05, 0.25)	0.571	0.611

k, number of parasite generations during a single generation of the host; r_{MP}, r_{mp}, levels of recombination corresponding to alleles *MP* and *mp* of the *rec* modifier in the parasite; r_{MH}, r_{mh}, the corresponding values for the host. Allele *M* is recessive in both populations; *rec* modifiers are assumed to be unlinked to the selected system; in the initial population, $p_{MP}(0) = p_{MH}(0) = 0.5$.

change in the population that selection for increased recombination occurs. Consequently, the periods during which the rec^+ alleles increase in frequency alternate in the host and parasite.

On the whole, the results presented in this chapter give further support to the hypothesis of a significant role of biotic interactions in the evolution of recombination. We have considered some models for the co-evolution of two antagonistic species (host and parasite) in which the genetic constitution of antagonists is defined explicitly and selection is on the polygenic traits 'resistance' and 'virulence'. The main result of these simulations is that, under interactions of this type, the alleles enhancing recombination spread through the host and/or parasite population. The possibility of the rec^+ alleles increasing in frequency in both populations has not been previously mentioned in the literature.

Increased recombination was observed in the population with both linked and unlinked modifier. This suggests that, in addition to 'hitch-hiking' in the traditional sense (due to linkage), there is another type of hitch-hiking reflecting the presence of feedback between the selected system and the rec^+ modifier. We suggest that this non-traditional 'hitch-hiking' be called 'co-selection of modifiers'.

As in the previous model (section 8.2), a reduction in the recombination fraction between the selected loci results in an increased intensity of selection in the system (notice that implied here are not only levels of recombination determined by alleles at the *rec* locus, but also frequencies of these alleles, i.e. mean recombination frequency). Selection for increased recombination was observed not only for $k > 1$ (when the host life cycle is longer than that of the parasite) but also for $k = 1$ (when corresponding to each generation of the host is one generation of the parasite). This suggests that the models proposed can be used to explain the evolution of recombination under antagonistic interactions in both the host–parasite system and systems whose two components have equally long life cycles (e.g. predator and prey).

In the above models, a selective advantage of increased crossing-over rate was generated by species interactions producing oscillatory changes in the population composition. This is equivalent to the scheme of selection in a cyclically varying environment discussed in Chapter 7. In that chapter, another possible version of such a model, with a monotonically changing optimum, was also mentioned (see Maynard Smith, 1980, 1988b; Bergman and Feldman, 1990; Charlesworth, 1993). A similar situation may also occur under species interactions producing, over long periods of time, the directional dynamics of the system components. Serebrovsky (1973) called evolution of this type 'orthogenesis on the basis of interspecific antagonistic interactions'. Although a process like this ensures directional changes in recombination, it must be 'fueled' by mutational variability, otherwise the dynamics will be terminated fairly soon after the fixation of selected systems (see Bergman and Feldman, 1990). Species interactions promoting orthoselection are therefore likely to cause, along with the evolution of the *rec* system, co-selection of modifiers of mutation rate.

8.4 THE *rec* SYSTEM EVOLUTION UNDER INTRASPECIFIC COMPETITION

Viewing intraspecific interactions and genotype–environment effects as a factor conferring evolutionary advantages to sex and recombination, Bell (1987) proceeded from the assumption that this factor is largely the result of genotype-selective rank-order spatial changes, as in sib competition models (G. Williams, 1975; Maynard Smith, 1978). With interspecific interactions of the host–parasite type, the advantages of sex and recombination result from temporal variation in selection coefficients. The last mechanism of recombination evolution can also be induced by intrapopulation competition in the form of density-dependent selection and a combination of density- and frequency-dependent selection (Korol and Mustyatsa, 1991; Kandel and Korol, 1991). In our ecological genetic models, competition generates fluctuations of the population genetic constitution, which may be accompanied by an increase in *rf*.

Consider a situation in which the population is subdivided into a number of competing ecological groups, but the intergroup competition is less intense than the intragroup competition. Whether an individual belongs to one ecological group or another depends on the allelic state at the selected system of loci (SSL). The rate of crossing over between the adjacent loci is controlled by a modifier, with alleles M and m determining respectively high and low *rf*. The case of a random-mating diploid population with a two-component SSL was analyzed under various dominance relations both within the selected loci and at the *rec* locus. Fitness of individuals of the ith phenotypic class was calculated as

$$w_i = 1 + r_i\left(1 - \sum_{j=1}^{n} \frac{N_j}{K_i + S_{ij}}\right),$$

where r_i, K_i and N_i are respectively the reproduction rate, the carrying capacity of the ecological subniche and the size of the ith group. Coefficients S_{ij} specify the reduction in intensity of intergroup competition relative to the intragroup competition intensity; n is the total number of groups.

Computer simulations show that the system just described can exhibit patterns of behavior preventing the high-recombination allele (M) from being eliminated or causing it to increase in frequency. Situations like these arise within a certain range of reproduction rate values at which numbers and, consequently, fitnesses of various phenotypic classes exhibit considerable oscillations. Changes in the frequency of alleles at the *rec*-modifying locus are influenced, to a certain extent, by the chosen values of S_{ij} as well as by dominance relations at the SSL and at the *rec* locus (Kandel and Korol, 1991).

As with some other systems of 'induced' selection for recombination, stable auto-oscillations of the SSL are possible in this model.

The proposed mechanism assumes no special pattern of environmental variation (temporal and/or spatial) or of interspecific interaction, and in this it is similar to the sib competition model (see Maynard Smith, 1978). The difference is that in the latter the factor promoting the maintenance of recombination is the relations among individuals which are dependent on the degree of relatedness. In our model, the factor ensuring selection for non-zero recombination is the auto-oscillations of phenotypic (ecological) group sizes within the same population. Ecological subdivision of populations seems to be a rather common phenomenon in nature. Therefore, the suggested mechanism may play a role in the *rec* system dynamics, at least in species with sufficiently high reproduction rates.

Roughgarden (1971) proposed an ecological–genetic model of a population polymorphic for a locus with alleles *A* and *a*, the fitnesses of genotypes *AA*, *Aa* and *aa* being dependent on population density. The behavior of this system has been studied and the possibility of gene frequency and population size oscillatory changes shown (Charlesworth, 1971; Roughgarden, 1971). We used this approach to study possible effects of density-dependent selection on the *rec* system dynamics (Korol and Mustyatsa, 1991). A three-locus panmictic population with non-overlapping generations was considered. Two loci, *A*/*a* and *B*/*b* are supposed to affect a quantitative trait related to fitness, say the *r* and/or *K* parameters (birth rate and carrying capacity). Then, the fitness of the *j*th genotype can be specified as a function of the population size:

$$w_j = \exp[r_j + (1 - N/K_j)],$$

where $N = \Sigma N_j$ and $j = \overline{1, 5}$ (for the case of two fitness loci with equal effects the number of different genotypes is 5). A third locus, *rec* modifier, with alleles *M* and *m*, determines the rate of recombination between *A*/*a* and *B*/*b*. As above, the fate of the modifier locus was studied numerically, based on the method of 'genetic operators' (see p. 201). In addition to density dependence, possible effects of frequency-dependent selection modulated by the population density were also taken into account, e.g. in the form of:

$$w_j = \exp[r_j + (1 - N/K_j)]\{s - N/K[s_1 g_j - s_2(g_{j-1} + g_{j+1})]\}.$$

Here s denotes the general density modulation of frequency-dependent effects, s_1 the selection dependence of the genotype on its own frequency g_i, and s_2 its dependence on the frequency of its closest 'phenotypic neighbors'.

The behavior of the system strongly depends on the fitness functions, i.e. on the effect of loci *A*/*a* and *B*/*b* on r_j and K_j. Among other regimes, oscillations, including chaotic ones, of linkage disequilibrium coefficient between fitness loci and of population size have been observed. As in the model considered above (p. 256), these oscillations can confer a short-term advantage of increased recombination. In the example presented in Figure 8.13, the fre-

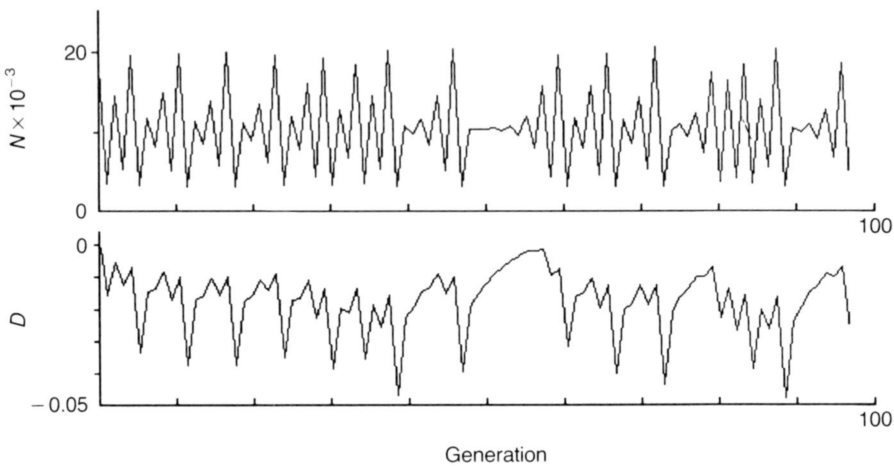

Figure 8.13 Chaotic-like dynamics of the population size and linkage disequilibrium due to density-dependent competition. In the model, the reproduction rate r is dependent on two loci, A/a and B/b, with equal effects; *rf* between A/a and B/b depends on a selectively neutral *rec* modifier M/m. In the example presented, the value of r was 3.5 for the genotypes *AABB*, *AABb* and *AaBB* and 2.0 for all others; the level of *rf* was 0, 0.1 and 0.5 for *mm*, *Mm* and *MM* respectively; the modifier was assumed to be unlinked to the selected loci. Other parameters were as follows: $K = 8000$ for all of the genotypes, $s = 1$, $s_1 = 0.05$ and $s_2 = 0.025$.

quency of allele M for increased *rf* has changed from 0.5 to 0.52 in 100 generations.

As we have seen, both inter- and intraspecific interactions could play an important role in the *rec* system evolution. Their relative contributions towards the evolution of recombination properties are hard to assess. The answer will probably depend on the solution of a more general problem, that of comparing the roles of competition and interspecific interactions in the formation of the organism's population and ecosystem adaptations, which is the central problem in theoretical and experimental ecology and in evolution theory (Stenseth and Maynard Smith, 1984; Sih *et al.*, 1985). In plants, intraspecific competition has been assumed to be generally more severe than interspecific (see Jones and Harvey, 1987), although this is unlikely to be the general rule. In any case, it is difficult not to agree with Bierzychudek (1987) that in experimental analysis of factors behind the evolution of sex and recombination progress can be made not by trying to find the single ultimate cause but by estimating the relative contributions and the frequency of occurrence of various causal factors corresponding to different theoretical models. Recent efforts to correlate recombination properties of animal and plants with their ecological, reproductive and life history characteristics (e.g. Vysotskaya, Bugrov and Stebaev, 1983; Burt and Bell, 1987; Bugrov and

Vysotskaya, 1988; Koella, 1993) could be considered, to our mind, as a new and a promising attempt to bridge experimental and field evidence and theoretical considerations of the recombination evolution, pioneered by Stebbins (1950) and Grant (1958).

In order to be more realistic, the ecological–genetic models of the evolution of the *rec* system, and the whole *R*-system, should also incorporate epidemiological parameters, such as population dynamics, density modulation of frequency-dependent selection, spatial structure of populations, stochastic effects, and so on (May and Anderson, 1983; Allen and Ebeling, 1983; Seger and Hamilton, 1988; Burdon and Jarosz, 1991; Thompson and Burdon, 1992; Singer and Parmesan, 1993). However, it is probably still more important to take into account the evolution of within-community species interactions themselves and community structure in general to study the role of multi-species interactions at the same and at different trophic levels (Buckley, 1987; Levin and Castillo-Chavez, 1990). Recently, possible tendencies for an ecosystem to move toward reduced interactions between species have been shown theoretically (Ginzburg, Akcakaya and Kim, 1988). At each stage of this process, a community has a small proportion of strongly interacting components, the effects of interactions between the other species being weak. The process of speciation is accompanied by the formation of new strongly interacting pairs and by the weakening of the remaining interactions. Our results from models of the host–parasite type and those from studies by other authors indicate that the pattern of pairwise interactions between species strongly depends on the properties of their *rec* systems and, in turn, promotes changes in these properties. In view of this, it is of interest to evaluate the robustness of the conclusions from theoretical ecological studies to possible changes in interspecific interactions resulting from genetic parametric variability, i.e. from the 'non-preset' nature of recombination values generating additional feedback systems in the evolving community. But this problem is the subject of future research.

9 Evolutionary interpretation of recombination phenomenology

9.1 THE ROLE OF RECOMBINATION DEPENDENCE ON ENVIRONMENT

The system of *rec* genes and factors controlling the mode of reproduction are the most important among the components determining the species' strategy for adaptation to a varying environment. Some mechanisms for adaptive interactions of these systems with environment must have been developed in the course of evolution. It has been assumed that low recombination is sufficient for maintaining genetic flexibility of a population and, therefore, low *rf* values are preferable in the general case, particularly so in adaptively important chromosome regions (Darlington and Mather, 1949).

Since temporal and spatial variation in environment is largely unpredictable, the *R*-system should provide an opportunity for release of genetic variability (albeit at a relatively low level) even under optimal ecological conditions. There are various possible ways of realizing this principle. Thus, self-pollinating plant species exhibit relatively high chiasma frequencies and release of a considerable amount of genetic variability as a result of sporadic outcrossing, whereas in cross-pollinating species chiasma frequency is lower and there is a slow but steady flow of variability from generation to generation (Grant, 1958, 1975).

On the other hand, continuous dissipation of stored variability depletes the population gene pool and reduces the evolutionary chances of survival (Rees and Jones, 1977). It is natural, therefore, to assume the existence of mechanisms restricting such a dissipation under optimal life conditions of the species and increasing variability when the conditions are not suitable. In other words, the dependence of the variability on environment may be a crucial

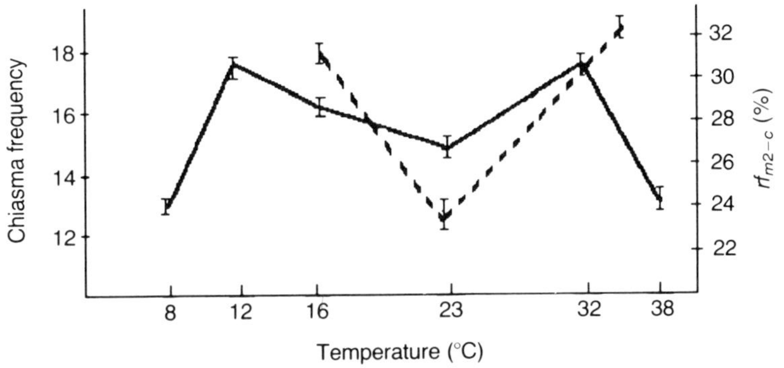

Figure 9.1 The dependence of recombination on temperature conditions during meiosis in tomato (from: Zhuchenko *et al.*, 1986; Korol, 1988). The decrease in chiasma frequency at extreme temperatures is explained by pairing disturbances.

factor reconciling the opposing requirements for maximized current fitness and for genetic flexibility of a population (Zhuchenko and Korol, 1982, 1983, 1985; Wills, 1983; Walbot and Cullis, 1985; Cullis, 1987, 1990; Korol, 1987). Another possibility is that of generating a wide range of genetic variability by the sexual process, clonal propagation (freezing) of the generated diversity and its evaluation due to natural selection (Vrijenhoek, 1990).

Environment can affect recombination in two ways: firstly, indirectly, when changes in the frequency of alleles at *rec* loci depend, through a feedback mechanism, on changes in the frequency of the selected loci (Chapters 6 to 8); secondly, directly, when in response to environmental factors the organism modifiers (within genetically determined limits) the value of *rf*, i.e. when not *rf* but the **norm of reaction** with respect to *rf* is genetically determined. The fact of direct influence of environment on recombination has long been known in genetics (Plough, 1917) and has been verified on a variety of organisms (see Zhuchenko and Korol, 1985; Korol, 1987; Parsons, 1988). Figure 9.1 provides an example of such an effect in tomato. However, in analyzing the problem of the evolutionary significance of recombination, the direct effects of environmental factors have not, until recently, been taken into consideration, with few exceptions (Huxley, 1942; Bodmer and Parsons, 1962; Borodin and Belyaev, 1980; Zhuchenko and Korol, 1982, 1983, 1985; Zhuchenko *et al.*, 1985; Borodin, 1987; Korol, 1987, 1988; Parsons, 1988; Hoffmann and Parsons, 1991). Here we suggest a simple model explaining how the dependence of recombination on environmental conditions could evolve.

9.1.1 Formulation in terms of the mean fitness

Consider a random-mating diploid population with two diallelic loci, A/a and B/b. Let frequencies of gametes AB, Ab, aB and ab be X_1, X_2, X_3 and X_4, respectively ($\Sigma X_i = 1, X_i \geqslant 0$ for $1 \leqslant i \leqslant 4$). Let $(w_{ij})^4_{i,j=1}$ be a matrix of genotype fitnesses and r the value of recombination frequency (rf) between the loci. For the case of non-overlapping generations, the dynamics equations relating gamete frequencies X_i in a given generation to frequencies X_i in the next one are (see, for example, Lewontin, 1974):

$$X'_1 = (w_1 X_1 - w_{14} rD)/w, \quad X'_2 = (w_2 X_2 + w_{23} rD)/w,$$
$$X'_3 = (w_3 X_3 + w_{23} rD)/w, \quad X'_4 = (w_4 X_4 - w_{14} rD)/w,$$

where

$$w_i = \sum_{j=1}^{4} w_{ij} X_j, \quad w = \sum_{i=1}^{4} w_i X_i, \quad D = X_1 X_4 - X_2 X_3.$$

As already mentioned, under constant environments (fitnesses w_{ij} are the same in all generations) the value of $rf = 0$ is the point of maximum of $w(rf)$ (Lewontin, 1974). If the environment varies in time in such a way that in some periods selection favors positive and in others negative linkage disequilibrium, i.e. a change in the environmental state is accompanied by a change in the sign of the equilibrium value of D, then non-zero rf may be the optimum level of recombination (Charlesworth, 1976).

In contrast to traditional formulations, we can assume that not only coefficients of fitness but also recombination values may vary in the same genotypes according to environment. The general scheme of study is as follows. First, an 'optimal' constant rf strategy is calculated, i.e. the value of rf^* for which the mean fitness averaged over a period of time (w^*) reaches a maximum. The next step is to ascertain whether in a set of variable (depending on environment) rf strategies there is a more 'advantageous' strategy for the population ensuring a higher averaged fitness than $w^* = w(rf^*)$ (Figure 9.2).

Consider the situation of periodic environment with two states. Let in each state selection be specified by the symmetrical matrix:

	AA	*Aa*	*aa*
BB	α	β	δ
Bb	γ	1	γ
bb	δ	β	α

Fitness parameters were chosen such that in the first state selection would favor positive and in the second negative linkage disequilibrium of the same magnitude, i.e. $\alpha_2 = \delta_1, \delta_2 = \alpha_1$. Let the population spend more time in one environmental state than in the other (hereafter, the first state will be referred to as 'typical' and the second one as 'atypical'). By a variable rf strategy we shall mean the pair $rf = (r_1, r_2)$, where r_1 is the rate of recombination in a typical and r_2 in an atypical environment.

In the numerical experiments, changes in the gamete frequencies, starting

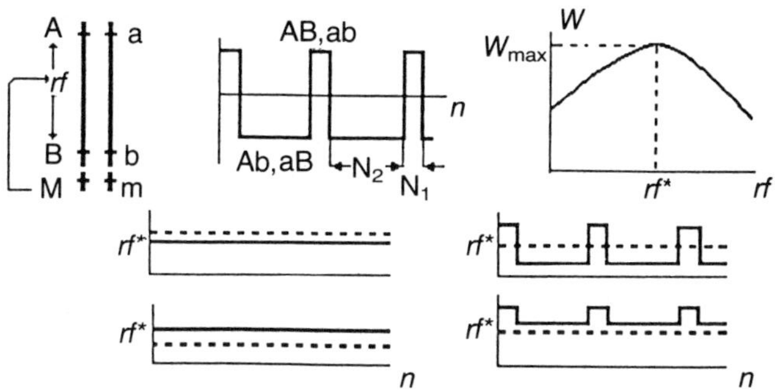

Figure 9.2 Comparing various types of *rf* strategies, selection regime as in Charlesworth (1976). The case $N_1 = 9$, $N_2 = 3$ is considered. The full line (——) represents an evolutionarily advantageous (established) and the broken line (---) the alternative (eliminated) *rf* strategy.

with a certain period, follow (with a given accuracy) those in the previous period. The mean fitness averaged (as the geometric mean) over the entire such quasistationary period was used as a population characteristic. As revealed by comparing various *rf* strategies for some patterns of environmental change, there are variable *rf* strategies ensuring a higher population fitness than the optimum stable level *rf** does (Figure 9.2 and Table 9.1). The presence of such an effect and the degree of its expression depend on the length of the period $N_1 + N_2$ and on the environmental asymmetry, N_1:N_2.

Table 9.1 Comparison of *rf* strategies with respect to mean fitness (a periodically changing environment with two states)

Period $N_1 + N_2$	*Variation pattern* N_1–N_2	*rf**	*w**	$rf = (r_1; r_2)$	$w = w(rf)$
15	12–3	0.01	0.7816	(0.00; 0.50)†	0.7765
	10–5	0.32	0.7667	(0.10; 0.50)†	0.7709
				(0.35; 0.50)	0.7675
	8–7	0.35	0.7666	(0.20; 0.50)	0.7663
				(0.10; 0.50)†	0.7671
25	20–5	0.04	0.7816	(0.00; 0.50)†	0.7931
				(0.05; 0.50)	0.7903
				(0.10; 0.50)	0.7865
				(0.00; 0.40)	0.7908
				(0.50; 0.00)	0.7670

* The best constant strategy.
† The best variable strategy.
N_i, number of generations; r_i, the level of *rf* in the *i*th environmental state.

Computer simulations have shown that not only 'fringing' ($r_1 < rf^* < r_2$) but also 'non-fringing' ($r_1, r_2 > rf^*$ or $r_1, r_2 < rf^*$) strategies increase mean fitness. Thus, for instance, for $N_1 = 20$, $N_2 = 5$ the optimal recombination is $rf^* = 0.04$, $w^* = 0.7816$. Under the 'non-fringing' strategy $rf = (0.05, 0.5)$, the mean fitness is $w = 0.7903$. Moreover, situations have been found in which 'non-fringing' strategies are the most advantageous among the variable *rf* strategies. Thus, for $N_1 = 9, N_2 = 3, rf^* = 0.01$ ($w^* = 0.7657$) turns out to be the best within the constant strategies. The 'fringing' strategy $rf = (0.0, 0.5)$ increases fitness up to the value of $w = 0.7667$. But the most advantageous in this case is the 'non-fringing' strategy $rf = (0.1, 0.5)$ ($w = 0.7679$). This finding is particularly interesting in view of the problem of 'why doesn't the genome congeal?', since the 'non-fringing' strategies just considered assume a higher level of recombination than rf^* even in typical environments.

9.1.2 Formulation with a modifier locus

In the model considered above, *rf* is the same in all members of the population. Incorporating in the model a modifier of recombination allows the arguments about whether or not recombination is beneficial to the population to be replaced by a more precise question: What is the pattern of changes in allele frequencies at the *rec* locus?

As already noted, the relationship between the population mean fitness and changes in the frequency of a modifier locus is determined by the mean fitness principle (Karlin and McGregor, 1974): under panmixia, the fate of modifier m with alleles m_1 and m_2 in a constant environment is determined by the relationship $w(m_1)$ and $w(m_2)$. Selection will favor the allele enhancing the population mean fitness. However, this is not always the case. It is known, for example, that restricted panmixia, migration or meiotic drive cause violations of this principle. It is also unclear to what extent it is applicable to situations with a varying environment. In this connection it is worth asking whether the conclusion about the advantage of a variable *rf* strategy still holds if one argues not in terms of mean fitness but at the level of modifier selection.

We have simulated the dynamics of a random-mating diploid population with three loci: selected A/a and B/b and a modifier, M/m, determining the recombination rate between the first two (see Figure 9.2). Equations for the dynamics of gamete frequencies were taken from Feldman (1972). In our case, in contrast to traditional models with a *rec* locus, alleles M and m determine different types of *rf* strategy rather than two stable levels of *rf*. Let allele m determine the evolutionarily stable constant strategy $rf = rf^*$, so that m eliminates from the population any other *rec* allele if the latter gives $rf \neq rf^*$. It is clear that optimal stable recombination in this formulation could, in general, differ from that which provides maximum mean fitness $w(rf)$. Let then allele M control a variable strategy, $rf = (r_1, r_2)$, where r_1 is the rate of recombination in the first (typical) and r_2 in the second (atypical) environmental state. The situations with a modifier linked ($r_{bm} = 0.05$) and unlinked ($r_{bm} = 0.5$) to the selected system, with dominant and recessive allele M, were examined.

In the majority of cases, the allele for a variable strategy increases in frequency, whereas the alternative allele for an optimal constant strategy is eliminated from the population (Table 9.2), but stable polymorphism at the modifier locus is also possible. The final frequency of the allele for the variable *rf* strategy depends, in some cases, on its initial frequency. In general, selected at the modifier locus are, as a rule, those alleles that increase mean fitness. This direction of selection on the *rec* locus is observed not only for the cases when one of the alleles determines the stable and the other the variable *rf* strategy but also when both alleles determine strategies of the same type.

Consider in more detail a situation in which $N_1 = 9$, $N_2 = 3$ (Figure 9.2). The previously formed conclusion about the optimality (in terms of the mean fitness) of the $rf^* = 0.01$ strategy within the class of stable strategies is also confirmed in the model with a modifier: allele m ($rf_m = rf^*$) substitutes the allele M for any alternative stable strategy $rf_M \neq rf^*$. Consider now an allele M which defines a variable strategy such that recombination in the 'rare' environment is increased. Then M can be at an advantage even when the rate of recombination in **both** environments exceeds rf^*. Thus, if $rf_M = (0.1, 0.5)$ and $rf_M = rf^* = 0.01$, then M increases in frequency. Furthermore, allele m is eliminated even when it determines a variable but 'fringing' strategy, for example $rf_m = (0.0, 0.5)$. The results suggest that ignoring the dependence of *rf* on environment may lead to erroneous conclusions about the direction of selection for recombination.

9.1.3 Discussion

Numerical analysis shows that the direct effect of environment of *rf* may be an important evolutionary factor. The recombination system ensuring the

Table 9.2 Frequency (p_M) of the allele for a variable *rf* strategy after 1000 cycles of dynamics (a periodically changing environment with two states)

		Modifier characteristics			
		Linked		*Unlinked*	
$rf = (r_1, r_2)$	$p_M(0)$	*Dom.*	*Recess.*	*Dom.*	*Recess.*
$N_1 - N_2 = 20 - 5$, $rf^* = 0.04$					
(0.00, 0.50)	0.1	0.994	1.000	0.990	1.000
(0.05, 0.50)	0.1	0.991	1.000	0.980	1.000
(0.10, 0.50)	0.5	1.000	1.000	0.887	1.000
(0.50, 0.00)	0.9	0.000	0.003	0.000	0.005
$N_1 - N_2 = 20 - 10$, $rf^* = 0.13$					
(0.00, 0.50)	0.1	0.990	1.000	0.987	1.000

dependence of *rf* on environment reconciles the opposing requirements for maximum current fitness and for the ability of the population to meet future environmental challenges (Zhuchenko and Korol, 1982, 1983, 1985). Variable strategies ensuring a more flexible regulation of genetic variability have an advantage over the stable ones at both inter- and intrapopulation levels.

The environment typical for a given population is the one that the population most frequently finds itself in in the course of evolution, and therefore the population comes to be better adapted to it than to the alternative environmental state. A typical environment, understood in this way, can therefore be regarded as an optimal one for a given population. The model-derived optimality of a lower *rf* in a typical environment compared with that in an atypical one shows, when interpreted in this way, good agreement with a number of experimental data. As early as 1917 Plough observed, in experiments with *Drosophila*, a U-shaped dependence of *rf* on temperature: recombination was minimum at optimum temperature and it increased with the departure of temperature from the optimum value. Similar data indicating an increase in recombination level with deteriorating environment (temperature increase or decrease, starvation, stress) have been reported in other works (Neel, 1941; McNelly-Ingle, Lamb and Frost, 1966; Lamb, 1969; Mock, 1973; King and Hayman, 1978; Borodin and Belyaev, 1980; Borodin, 1987; Henderson, 1988; Parsons, 1988; Lebel *et al.*, 1993), as well as in experiments carried out in our laboratory on tomato, maize, *Arabidopsis* and *Drosophila* (Zhuchenko and Korol, 1985; Korol, 1987, 1988) (see also Figure 9.1). Extreme environmental conditions will also tend to reduce interference (Grell, 1978; Kruleva *et al.*, 1992). Therefore, the genetic system of the species can be said to be assessing the 'typicalness' of environment, responding to deviations from the optimum by corresponding changes in recombination level and, possibly, spectrum (Zhuchenko and Korol, 1982, 1983, 1985).

Such a dependence of recombination on environment reminds one of another type of variable strategy of the *R*-system, that of cyclical parthenogenesis, characteristic of 15 000 animal species (Bell, 1982; Herbert, 1987; Vrijenhoek, 1990). In particular, under seasonal parthenogenesis, a switch to sexual reproduction occurs when there is a selective premium on genetic variability, i.e. upon ecological saturation of the habitat or prior to the advent of a dry summer or cold winter season. In general, cyclical parthenogenesis can be regarded as a very reasonable compromise between the conflicting requirements for maintaining a high reproductive potential and for genetic flexibility in the face of unpredictable environmental changes (Shnoll, 1979; Lynch and Gabriel, 1983; Roff, 1986).

The available evidence for high recombination activity of many mutagenic agents can also be considered from this standpoint. An increase in recombination seems to be, in this case, an adaptive response of the *R*-system resulting in a higher intrapopulation variance for the number of deleterious mutations per chromosome, thereby increasing the efficiency of purifying selection. There is evidence that stressful conditions result in increased genetic variability of not only recombinational but also mutational nature (Zujtin, 1938, 1941; Tsapygina, Dayev and Nikonov, 1981; Gorlov and Borodin, 1986), in induced transpositions of mobile genetic elements, DNA amplification, etc. (Echols, 1981; McDonald, 1983; Wills, 1984; Paquin and Williamson, 1984; Bachmann, Chambers and Price, 1985; Golubovsky, 1985; Cullis, 1987, 1990; Agur and Kerszberg, 1987; Parsons, 1987; Pollard, 1987; Al-Saheal and Larik, 1987; Grandbastien, 1992). In experiments by Vasiljeva *et al.* (1987), exposure to extreme temperatures resulted in significant heritable variation of phenotypic characters in *Drosophila* with a concurrent change in the pattern of localization of mobile elements (however, it should be mentioned that in other experiments with *Drosophila*, by Arnault and Biemont (1989), heat shock caused no transpositions of mobile elements).

Obviously, the above types of environmental influence on the genetic constitution of the next generation cannot be thought of as a **direct adaptation**. What is implied here is a certain expediency, non-randomness, of transition of potential genetic variability to free (accessible to selection) variability, resulting from the dependence of the frequency and genomic distribution of recombination events on environment. Variation in recombination frequency in response to stressful ecological conditions (whatever the interpretation of its mechanisms) is an established fact for many experimental organisms. The pattern of such variation is controlled by the genotype. A similar mechanism may exist for mutation-dependent variability, and recent experiments on microorganisms show that such a possibility could be real (Foster and Cairns, 1992; but see also a critical review of Lenski and Mittler, 1993 and a recent discussion in *Science*, **260**, 1058–60). Consequently, environment in our model acts not only in the form of selection, 'sorting out' variability according to Darwinian fitness, but also as a factor affecting the amount and spectrum of genetic variation. The latter effect is the product of selection in the past, on both the *F*- and *R*-systems. Thus, the role of environment in gene pool variability is much more active than usually follows from the conventional scheme of evolution.

Abundant evidence has been accumulated by now which is rather hard to interpret in terms of random mutations and recombination (i.e. with frequency and spectrum independent on environment), and selection. It has been known, for example, that insects and some other animal species evolve resistance to drugs very rapidly (Zelikman, 1967; Devonshire and Sawicki, 1979; Shwarz, 1980; Georgievsky, 1981; Flexon and Rodell, 1982; Ferrari and Georghiou, 1991). A similar situation has been reported in weeds, which become herbicide resistant in a few generations. Cases of appearance of genotrophs (in flax and tobacco) – saltatory changes induced by phosphorous and nitrogenous fertilizers – are well-known (see Cullis, 1990, for review).

These changes have been shown to be heritable and accompanied by DNA sequence rearrangements and protein electrophoretic spectrum changes.

Experiments with the *in vitro* culture of the mouse cell lineages have revealed that resistance to certain toxins (to methotrexate, in particular) may develop as a result of amplification of the dihydrochlorate reductase gene (Schimke, 1984). In the example in question, it is the structural gene coding for the enzyme inhibited by methotrexate that is amplified (up to 200 copies). A similar situation is observed with increasing resistance of cells to cadmium, colchicine, actinomycin D, adenosine and uridine (Beach and Palmiter, 1981; Kopnin and Gudkov, 1983; Fogel *et al.*, 1983). In these cases, amplification of the selected gene is frequently accompanied by co-amplification of flanking DNA sequences, i.e. what we have here seems to be DNA amplification rather than amplification of individual genes (see Fox, 1984). A particularly large body of evidence has been obtained in the last decade in plant genetics, which shows that stress conditions caused by *in vitro* cultivation can result in a broad spectrum of genomic alterations, including changes in methylation pattern, DNA amplification, selective loss of chromatin, chromosome rearrangements, etc. (e.g. Cullis, 1990; Cecchini *et al.*, 1992; Arnholdt-Schmidt, 1993). Therefore, there are mechanisms ensuring rapid genetic (occasionally stable) changes of an organism or cell, at least in response to environmental changes.

The foregoing discussion allows one to assume, by way of a working hypothesis, that some known cases of rapid heritable changes provoked by environmental conditions can be attributed to 'directional' effects of these conditions on recombination. In fact, it is now a general belief that the evolutionary formation of a new gene by mutation–selection interaction is preceded by the duplication of the genetic material. Local duplications may result from unequal crossing over, imprecise sister chromatid exchange, polymerase slippage or replicative transposition. These events are known to be dependent on environmental conditions. Earlier (pp. 127–9) we discussed the relationship between transcription and crossing over, in particular Whitehouse's (1966) hypothesis of transcription initiation sites coinciding with recombinators. Even if this model is incorrect in details, it can be proposed that the high template activity of a particular genomic region during premeiotic mitoses increases its accessibility to the recombination machinery (Zhuchenko and Korol, 1982, 1985; Voelkel-Meiman, Keil and Roeder, 1987; Vizir and Korol, 1990). Indirect evidence that such a 'memory' could exist has been obtained in *Drosophila* (Zujtin, 1938; Neel, 1941; Zhuchenko and Korol, 1985; Table 9.3) and fungi (McNelly-Ingle, Lamb and Frost, 1966; Landner, 1970).

One may assume that the effect of transcription on recombination in multicellular organisms may be due to an enhanced probability of the transcribed (or adjacent) regions remaining unfolded up to meiosis and being incorporated into the central region of the synaptonemal complex at meiosis. This will result in a higher crossing-over rate in the active region and, at the

Table 9.3 Effect of heat treatment of *D. melanogaster* females (line *b cn vg*) on recombination in their *b cn vg*/ + + + daughters (from Zhuchenko and Korol, 1985)

$t^{\dagger}$ (°C)	*Repeat*[‡]	*replicate*	*Progeny size*[§]	*rf* (%) *in the segment* *b–cn*	*cn–vg*
25	I	1–5	3260	7.95 ± 0.91	7.88 ± 0.96
36		1, 4, 5	2073	10.75 ± 0.88*	8.53 ± 0.72
		2	517	8.32 ± 1.21	0.00 ± 0.37***
		3	652	7.82 ± 1.05	4.57 ± 0.82**
25	II	1–5	3913	7.41 ± 0.52	6.65 ± 0.80
36		1, 5	1040	11.54 ± 1.32**	0.47 ± 0.28***
		2, 3, 4	2472	10.06 ± 0.62***	4.67 ± 0.46*

† Heat treatment was applied to virgin females for 6 hours; after 48 hours they were crossed with normal males and the resultant heterozygous females were tested for recombination.
‡ Two identical repeat experiments were carried out with five replicates in each. Owing to high heterogeneity of the resulting changes in the *cn–vg* segment the data were treated by groups, according to the values of rf_{cn-vg}. Control values were homogeneous.
§ First two 3-day broods were used in the progeny test; deviation from the control is significant at $^{*}P < 0.05$, $^{**}P < 0.01$ and $^{***}P < 0.001$.

same time, in a proportional increase in the rate of unequal exchanges. Therefore, a limitation with respect to a certain cellular reaction can increase, by virtue of enhancing transcription within a particular genomic region, the probability of crossing over and, hence, duplication of a gene or gene block located in this region. With respect to repair–mutation processes a similar scheme is discussed by Auerbach (1976), Tchaikovsky (1976), Krasilov (1979) and Yeom, Kuniya and Artzt (1992). This phenomenon can be regarded, on the one hand, as a genetic response reducing the limitation and, on the other, as a prerequisite for further improvement of a metabolic chain by mutation, recombination and selection. New findings show that DNA repair may be coupled with transcription, also because of the dependence on common enzymes (Buratowski, 1993; Bootsoma and Hoeijmakers, 1993). Several mechanisms for environmental induction of mutations and duplications dependent on the recombination–repair system have been suggested for procaryotes (*E. coli*) (Echols, 1981; Dimpfl and Echols, 1989; Foster and Cairns, 1992).

Of particular interest is the role in population evolution of genotypes exhibiting low adaptability to their environments (Bauer, 1935; Zhuchenko and Korol, 1985). Individuals maladapted to certain environmental conditions, the 'genetic load', can play no less important a role in increasing population adaptation than do the adaptive norm and genetic 'elite'. Their sensitivity to environmental conditions confers on them the capacity for generating, at increased frequencies, mutations, recombinations and duplications which could constitute the raw material of further genetic changes.

An increase in recombination in response to stressful conditions ensures, on the one hand, a higher degree of genetic variability in the progeny and, hence, greater opportunities for genotypic adaptation (prospective function) and, on the other hand, a larger variance in the per chromosome frequency of stress-induced harmful mutations and, consequently, a higher efficiency of purifying selection. The last mechanism is considered by many authors as a major factor maintaining recombination and sex in general. In this respect, one rather general feature of the *rec* system response to stresses – a higher rate of exchanges in the regions normally exhibiting low recombination – is worth mentioning.

As pointed out by McDonald (1983), the greater the deviation of environmental conditions from the standard ones, the less likely is the adaptation based on pre-existing genetic variability and the more important the variability generated *de novo*. Parsons (1987) considers the effects of stressors to be an important factor accelerating evolution. Kapitonov and Kolchanov (1988) have shown, in a mathematical model describing the dependence of the behavior of mobile elements on environmental states, that under sudden but rare environmental changes the probability of a population surviving may be greater if these changes induce transpositions of mobile elements. Inge-Vechtomov, Repnevskaya and Karpova (1987) have established the possibility of the UV induction of mating type-switching in yeast, which may affect the composition of progeny. Therefore, modificative variability of parameters of the *R*-system plays an important role in determining the genetic constitution of a population and can be viewed as one of the principal feedback mechanisms controlling genetic variability (Zhuchenko and Korol, 1982, 1983, 1985; Korol, 1987, 1988; Parsons, 1988; Hoffmann and Parsons, 1991). Moreover, variation of these parameters in response to stressful factors can be modulated by the organism's fitness: higher fitnesses promote a better preservation of genes and gene complexes that have been formed in previous generations (Zhuchenko and Korol, 1982, 1985).

Experimental evidence for this principle has been obtained on *Drosophila* (Tucic, Ayala and Marinkovic, 1981; Zhuchenko and Korol, 1985; Cvetkovic and Tucic, 1986) and tomato (Zhuchenko *et al.*, 1986; Korol, 1988). In *Drosophila*, the increase in male recombination induced by a heat treatment is negatively correlated with the resistance of flies to high temperatures (Table 9.4). In tomato the effect of stressful conditions on recombination depends on the degree of adaptation of genotypes to these conditions (Figure 9.3). Low temperature did not increase the exchange frequency in cold-resistant varieties and in their F_1 hybrids, while the reverse was true for cold-sensitive genotypes. By contrast, elevated temperatures caused an increase in recombination only in cold-resistant varieties.

Our simulation experiments help to understand how mechanisms controlling variability on the basis of negative feedback, restricting dissipation of genetic variability in an optimum environment and increasing the released

Table 9.4 Dependence of the recombinogenic effect of heat treatment in chromosome 2 of *D. melanogaster* males on the tolerance of the treated genotype

Genotype[†]	*Variant*[‡]	*Progeny size*	*Recombinants*	rf_{δ}(%)	*T/C*
R	C (25 °C)	4640	10	0.215	
	T (32 °C)	3805	16	0.419	1.9
S	C (25 °C)	3031	2	0.066	
	T (32 °C)	2799	24	0.850	12.9[§]

[†] Two lines, *T32* and *GK80*, resistant and susceptible to high temperature, were crossed with a multiple marked strain *al b pr c px sp*.
[‡] The resultant hybrid males, R and S, were exposed to high temperature for 24 hours. The progeny obtained from mating of the males with tester females for 5–10 days after the treatment was analyzed.
[§] The difference between C and T variants is significant at $P < 0.025$.

variability with deterioration of the conditions have become established in the course of evolution of the *rec* system. Another, no less interesting, conclusion emerging from our model is that allowing for the dependence of the crossing over rate on the environment makes it possible to account for the maintenance in populations of higher levels of recombination than predicted from conven-

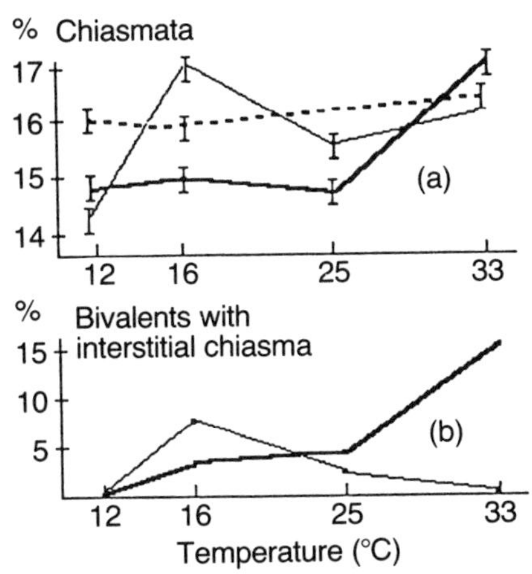

Figure 9.3 The dependence of recombinogenic effect of temperature conditions in tomato on genotype resistance to these conditions, as estimated by (a) chiasma frequency at diakinesis and (b) proportion of bivalents of 'eight-type' with an interstitial chiasma. The genotypes used were: cultivar *Gloria* (thin lines), which is resistant to high temperature and susceptible to low temperature; cultivar *Sever* (solid lines) susceptible to high temperature and resistant to low temperature; hybrid F_1 (*Sever* × *Gloria*), tolerant to both high and low temperatures (broken line) (based on data from Zhuchenko *et al.*, 1986; Korol, 1988).

tional models with constant *rf* values. Variable *rf* strategies, in which recombination even in optimal (and more so in extreme) environmental conditions is higher than the evolutionary stable (ES) constant level *rf**, may be evolutionarily more advantageous. Consequently, non-genetic variability of *rf* should be taken into consideration in discussing the evolution of recombination as one of the important factors promoting the maintenance of a relatively high rate of crossing over in natural populations.

9.2 THE EVOLUTION OF DOMINANCE OF *rec* GENES

R. Fisher was one of the first to attempt to solve the problem of the evolution of dominance. According to him, within-locus relations between alleles differing in selective value are determined by special modifiers of dominance. In the course of evolution, changes occur in the system of modifiers which reduce the expression of harmful alleles at the selected loci when these are in heterozygous form, i.e. 'transform' them into recessive ones (Fisher, 1930). This hypothesis was tested both experimentally (Kamshilov, 1939; Sheppard and Ford, 1966) and with mathematical models (Ewens, 1965, 1966; Mayo, 1966; Crosby, 1963; Wagner, 1981).

The most serious objection to Fisher's concept of the evolution of dominance goes back to disputes of the 1930s and was again discussed in the 1980s in terms of the regulation of activity of enzymes in complex metabolic systems (Kacser and Burns, 1981; Kacser, 1987). The authors have shown that if the control of the synthesis of some product is distributed among a number of enzymes, then even a reduction of 50% in the activity of one of these owing to mutation (in heterozygotes) will result in only a slight change in the net result. The reason for this is in the assumption that real enzymes in real organisms act at the level substantially lower than the saturation point, i.e. with a considerable reserve. Consequently, any newly arising mutation must be recessive. There are, however, works (experimental and theoretical) disproving the general significance of such conclusions (Cornish-Bowden, 1987). In other words, 'dominance is not inevitable' purely metabolically, and attempts at evolutionary explanation of the known genetic phenomenon are fairly justified (see also Timofeef-Ressovsky, Yablokov and Glotov, 1973; Mitrofanov, 1977).

Dominance relations are not peculiar to the selected loci. The existence of dominance–recessiveness relations between genes controlling recombination has been demonstrated on a variety of organisms. The suppressor alleles generally dominate over the enhancer alleles (Kidwell, 1972; Dewees, 1975), although there are also cases of partial dominance of suppressors and of co-dominance (Chinnici, 1971; Turner, 1979; Ebinuma and Yoshitake, 1981), and even of heterosis, both positive and negative (Rees and Thompson, 1956; Hinton, 1966; Srivastava, 1980; Catcheside, 1981).

A whole range of effects associated with dominance relations at the *rec* loci was observed by Charlesworth and Charlesworth (1985) in an experiment with artificial selection for altered crossing over between markers *Gl* and *Sb* on chromosome 3 of *D. melanogaster*. A comparison between a high-recombination strain (H) and the unselected control population (C) showed that differences between them with respect to *rf* within the *Gl–Sb* segment are determined by all three large chromosomes. Measuring the crossing-over rate between *Gl* and *Sb* in genotypes resulting from combining chromosomes from strains H and C enabled the dominance relations to be determined for the *rec* genes. It turned out that there are strong epistatic interactions between modifiers of recombination from different chromosomes. Enhancers of recombination on chromosome 2 are dominant if the individuals are homozygous for the other chromosomes of the complement and they come from strain H, and recessive if the X chromosome and chromosome 3 come from strain C. Among the modifier genes on the X chromosome, there is complete or partial dominance of enhancers of recombination, depending on the genic content of chromosomes 2 and 3. Finally, dominance of recombination enhancers over the suppressors has been observed for the *rec* genes on chromosome 3. The fact that the dominance–recessiveness of the *rec* genes may depend on the genic content of other chromosomes is an important argument in favor of the hypothesis that dominance relations at the *rec* loci are not 'intrinsic' properties of the *rec* alleles themselves, and, hence, one is justified in raising the question of their evolution.

Another point that is worth mentioning here is a possible dependence of the rate and even direction of allele frequency changes at the *rec* locus on dominance. Thus, Felsenstein and Yokoyama (1976) have shown, with the aid of computer simulations, that in the model of selection against deleterious mutations the high-recombination allele spreads through a population only when it is recessive (see also Bergman and Feldman, 1990; Charlesworth, 1993).

The possible significance of dominance–recessiveness relations of *rec* alleles in controlling variability can readily be demonstrated by the following example. Consider a population in which alleles enhancing recombination are recessive. Suppose that there is overdominance for genes concerned with resistance to stressful conditions. Then, with deterioration of environment, one can expect the proportion of heterozygous genotypes to increase. On the other hand, because of environmental deterioration, population numbers may be sharply decreased, resulting in a higher level of inbreeding. If the rate of inbreeding is rather high and selective advantage of heterozygotes relatively small, then the proportion of homozygotes rec^+/rec^+ may increase in the population, resulting in a higher level of recombination. Thus, if the rec^+ allele is recessive, one can speak of variation in recombination frequencies within a population being regulated by the mechanism of negative feedback (Zhuchenko and Korol, 1982, 1985; Korol, 1987). This may, in part, explain the observed increase in variability of populations experiencing the bottleneck effect (Bryant, McCommas and Combs, 1986; Bryant and Meffert, 1988; Carson, 1990).

In all the models for the evolution of dominance considered to date, the question is taken up of changes in the relations between alleles at the loci directly affecting fitness, i.e. in components of the F-system. Modifiers at such loci affect the fitness of heterozygous genotypes and, therefore, are not selectively neutral, i.e. they can also be referred to the F-system. The situation arising in connection with the problem of dominance–recessiveness of the *rec* modifiers – components of the R-system – is qualitatively different from the one just considered. The genes affecting the rate of crossing over between the selected loci are usually considered as neutral for their immediate carriers and, therefore, the more so should be the modifiers of their dominance (let us call them '*dom* modifiers').

We have demonstrated that the modifier principle suggested by Fisher for explaining the evolution of dominance of the selected loci (components of the F-system) is applicable to the dominance of selectively neutral *rec* loci (Zhuchenko, Preygel and Korol, 1986). The model examined was that of an infinite, four-locus random-mating population (Figure 9.4). A neutral *rec* modifier (M/m) controls the amount of crossing over between the selected loci A/a and B/b; allele M is responsible for high rf_{ab} and allele m for low. The locus modifying dominance determines the value of rf_{ab} in the Mm heterozygotes. So, rf_{ab} is specified by the matrix

$$\begin{array}{c|ccc|}
 & DD & Dd & dd \\
MM & r_1 & r_1 & r_1 \\
Mm & r_2 & r_3 & r_4 \\
mm & r_5 & r_5 & r_5
\end{array} .$$

In running the model, it was assumed that allele D determines complete dominance of allele M (i.e. $r_2 = r_1$) and allele d causes the dominance of

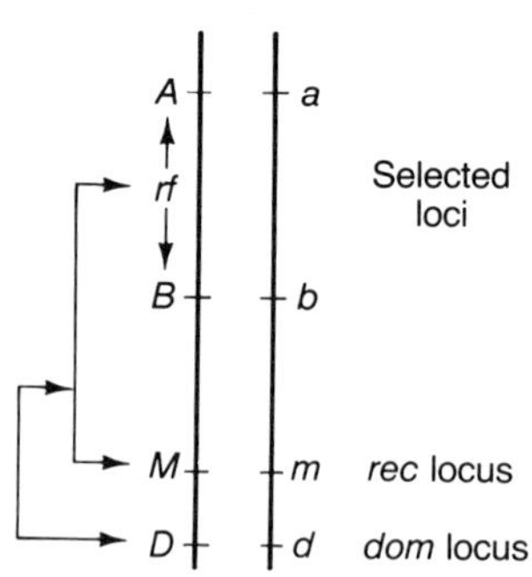

Figure 9.4 The hierarchic model for analyzing the evolution of dominance of the *rec* locus.

m ($r_4 = r_5$). With the aid of r_3, one can specify the degree of dominance at the *dom* locus itself: $r_3 = r_2$ implies complete dominance of *D*, and the equality $r_3 = r_4$ complete recessivity of *D*. Clearly, with increasing frequency of allele *D* in the population, the rate of crossing over between the selected loci will increase and, conversely, an increase in frequency of *d* will result in a lower rf_{ab}.

It should be noted that for Fisher's mechanism of the evolution of dominance to be effective, a high rate of mutation is required at the selected modified locus (otherwise the rate of changes in the frequency of alleles at the modifier locus will be insufficient, even on the evolutionary time scale). Analogous to this requirement, in our model for modification of interallelic relations at the *rec* locus, is the condition of polymorphism at the *rec* locus. The abundant evidence obtained in estimating variation in recombination frequency in natural populations as well as experiments on selection (direct or co-selection) for altered *rf* indicate that, very likely, the condition is satisfied.

Assume that the population is evolving in a constant environment. If that is the case, then the closest possible linkage is selectively advantageous (Feldman, 1972; Altenberg and Feldman, 1987). Therefore, selection on loci *A*/*a* and *B*/*b* will produce changes at the *rec* locus in favor of the allele *m* for low recombination. One may expect (although it does not follow a priori from anything) that these changes, in their turn, will result in an increase in the frequency of allele *d* concerned with dominance of *m*. This assumption has been confirmed by numerical experiments (Table 9.5) with a symmetrical

Table 9.5 Changes in allele frequencies, p_M and p_D, after 1500 generations of dynamics in a constant environment

			$p_D(0)$	0.2		0.8	
r_3	rf_{bm}	rf_{md}	$p_M(0)$	0.2	0.8	0.2	0.8
0.5	0.5	0.5	$p_M(n)$	0.0000	0.0177	0.0000	0.0000
			$p_D(n)$	0.1225	0.0222	0.7903	0.7045
0	0.5	0.5	$p_M(n)$	0.0200	0.0278	0.0000	0.0019
			$p_D(n)$	0.1560	0.1293	0.7356	0.4578
0.5	0.5	0.05	$p_M(n)$	0.0027	0.0308	0.0000	0.0000
			$p_D(n)$	0.0718	0.0043	0.7339	0.6088
0	0.5	0.05	$p_M(n)$	0.0228	0.0301	0.0000	0.0125
			$p_D(n)$	0.1308	0.1075	0.6905	0.2745
0.5	0.05	0.5	$p_M(n)$	0.0000	0.0000	0.0000	0.0000
			$p_D(n)$	0.1576	0.0756	0.7925	0.7314
0	0.05	0.5	$p_M(n)$	0.0002	0.0011	0.0000	0.0000
			$p_D(n)$	0.1767	0.1569	0.7586	0.5735
0.5	0.05	0.05	$p_M(n)$	0.0000	0.0030	0.0000	0.0000
			$p_D(n)$	0.0759	0.0055	0.7888	0.5527
0	0.05	0.05	$p_M(n)$	0.0024	0.0036	0.0000	0.0006
			$p_D(n)$	0.1308	0.1065	0.6590	0.2259

The coefficients of the fitness matrix were as follows: $\alpha_0 = 0.7$, $\alpha_1 = 0$, $\beta = 0.75$, $\gamma = 0.9$. Here and in Table 9.6, $rf_{ab} = 0.5$, corresponds to allele *M*, $rf_{ab} = 0$ to allele *m*; $r_3 = 0$ is equivalent to *m* being dominant over *M* in the *DdMm* heterozygote, $r_3 = 0.5$ to *M* being dominant over *m*.

fitness matrix:

$$\begin{array}{c|ccc|} & AA & Aa & aa \\ BB & \alpha_0 & \beta & \alpha_1 \\ Bb & \gamma & 1 & \gamma \\ bb & \alpha_1 & \beta & \alpha_0 \end{array}.$$

In a constant environment, allele M is eliminated from the population and the frequency of D declines as long as the transient polymorphism at the *rec* locus persists, i.e. the tendency for the low (in our case, zero) recombination allele to be dominant becomes stronger.

Simulations have shown that the rate of changes in the frequency of alleles of the *dom* modifier is much lower than that of the *rec* alleles (and even more so compared with the selected system). The greatest reduction in the frequency of allele D is observed with a low initial frequency of favorable allele m since in this case transient polymorphism at M/m persists longer and, hence, the longer is the time interval 'allotted' for the *dom* locus evolution.

Suppose now that the environment varies in time. Consider the cases of a cyclically changing environment with two states, such that in the first state selection within a two-locus system (A/a–B/b) generates positive linkage disequilibrium (favoring combinations AB and ab) and in the second, negative disequilibrium (Ab and aB are favored by selection). With a certain period of environmental fluctuation of this type, the allele M for high *rf* turns out to be selectively more advantageous than the allele for low *rf*. In the simulation experiments we used the same selection regime as Charlesworth (1976): in the first environmental state it is specified by the parameters $\alpha_0 = 0.7, \alpha_1 = 0$, $\beta = 0.75, \gamma = 0.9$, in the second by parameters $\alpha_0 = 0, \alpha_1 = 0.7, \beta = 0.75$ and $\gamma = 0.9$; the period of fluctuations is 10 generations. Under the circumstances, one may expect the allele concerned with dominance of M over m to be increasing in frequency. The results obtained support this assumption (Table 9.6).

Table 9.6 Changes in allele frequencies, p_M and p_D, after 1000 generations of dynamics in a periodically changing environment

			$p_D(0)$	0.2		0.8	
r_3	rf_{bm}	rf_{md}	$p_M(0)$	0.2	0.8	0.2	0.8
0.5	0.5	0.5	$p_M(n)$	0.8129	0.9621	0.6037	0.8095
			$p_D(n)$	0.3148	0.2028	0.8107	0.7903
0	0.5	0.5	$p_M(n)$	0.9677	0.9959	0.6895	0.8995
			$p_D(n)$	0.2568	0.2061	0.8961	0.8059
0.5	0.5	0.05	$p_M(n)$	0.7402	0.9603	0.6017	0.8090
			$p_D(n)$	0.4832	0.2407	0.8365	0.8023
0	0.5	0.05	$p_M(n)$	0.9981	0.9958	0.6338	0.8929
			$p_D(n)$	0.3628	0.2130	0.9776	0.8721

Consequently, allele frequencies at the *dom* locus change in such a way that dominance of the *rec* locus allele, advantageous under the circumstances (*m* in a constant and *M* in a varying environment), is enhanced, although at a low rate of change. This is not unexpected: the *rec* locus, interallelic relations at which are modified by the *dom* locus, is selectively neutral and its modifier of dominance is of a still higher 'order of neutrality'. Thus, selection on an adaptively significant system can produce changes not only in the neutral *rec* locus but also in a 'still more neutral' modifier of dominance of the *rec* locus.

It is worth noting that despite a relatively low rate of frequency changes the allelic frequencies at the *dom* locus can have a significant effect on the rate of change of the *rec* locus and of the entire selected system. Thus, a high initial frequency of allele *D* (concerned with dominance of the allele *M* for non-zero recombination, which is selectively advantageous in a varying environment) perceptibly slows down the rate of allele frequency changes at the *rec* locus, enabling it to stay polymorphic for a longer period of time (see Table 9.6). The ratio of the *dom* allele frequencies may determine not only the rate but also the direction of allele frequency changes at the *rec* locus. All this leads to the conclusion that the modifier hypothesis of Fisher put forward by him to explain the formation of dominance relations at the selected loci is also applicable to *rec* loci – the components of the *R*-system.

Earlier, we mentioned the reduction principle established by Feldman and co-workers for modifiers of recombination in a constant environment (Feldman, 1972; Liberman and Feldman, 1986; Altenberg and Feldman, 1987): a new *rec* allele introduced into a panmictic population at equilibrium will only spread if it causes a reduction in recombination rate relative to the existing level. The obtained results on the *dom* locus dynamics in a constant environment suggest that the reduction principle can be extended to second-order modifiers as well. It seems plausible that this may be the case with the evolution of dominance of not only *rec* loci but also other *R*-system components, e.g. *dom* modifiers of mutation and migration controlling loci may obey the reduction principle.

9.3 LINKAGE BETWEEN THE *rec* MODIFIER AND ITS TARGET REGION: A POSSIBLE EVOLUTIONARY MODEL

The dependence of the dynamics of selectively neutral modifiers on linkage with the SSL has been the subject of a long-standing discussion in the literature. The critical role of linkage has been established in a number of models of evolution concerned with the control of inbreeding (Karlin and McGregor, 1974; Charlesworth, Morgan and Charlesworth, 1992), recombination level (Charlesworth, 1976; Feldman, Christiansen and Brooks, 1980; Holsinger and Feldman, 1983b; Uyenoyama and Bengtsson, 1989), mutation rate (Gillespie, 1981b; Twomey and Feldman, 1990), normal segregation (Eshel, 1985) and sex ratio (Lessard, 1987; Feldman and Otto, 1989). Usually, the linkage of a modifier with its target region is assumed to be a preset parameter that is independent of the genotype. One of the few exceptions is the study of Thomson and Feldman (1974) of the co-evolution of a meiotic drive modifier and a recombination modifier. The latter determines the rate of crossing over between the abnormally segregating locus and its modifier.

With respect to recombination, one point of view is that the most reasonable is to study the behavior of linked modifiers of recombination. Besides, it is assumed that the hitch-hiking due to the linkage of the modifier with the SSL is the main mechanism for the spread of the rec^+ allele through the population (Bell and Maynard Smith, 1987; Maynard Smith, 1988a, b; Nee, 1988). Other authors prefer to include unlinked modifiers in their models, explaining this choice by the possibility of assessing the 'pure' effect of feedback selection resulting from the potential of recombination to form favorable combinations of the selected genes (Felsenstein and Yokoyama, 1976; Hutson and Law, 1981; Kondrashov, 1984). Again, it is generally implied that linkage can merely enhance the advantage of the rec^+ allele due to the hitch-hiking mechanism.

As has been shown earlier (Korol, Preygel and Preygel, 1990; Chapter 7), the effect of linkage of the modifier with its target region on the efficiency of selection for recombination is far more complex than has been suspected. While studying the dynamics of a multilocus system in a fluctuating environment, we have found that in a frequently changing (serrated) environment selection for increased *rf* can only occur when the modifier is unlinked (or loosely linked) to the selected system. It is only in a relatively stable environment (e.g. rectangular oscillations of low frequency) that the largest equilibrium value of recombination and the highest rate of allele frequency changes at the *rec* locus are observed under tight linkage of the modifier with its target region, in accord with traditional understanding.

Such an ambiguous and strong influence of the modifier linkage on its fate as this leads to the question of how relative positions of the modifier and its 'target' are formed. Are they due to chance alone or to a newly arising mutation affecting the rate of crossing over in a certain region, or is the initial position 'corrected' by selection in the course of evolution? The construction presented below by way of an explanation, is, of course, an artificial one. However, it may be helpful in ascertaining whether the evolution of the intensity of linkage between the *rec* locus and its target region is in principle possible and, if so, what the features of this evolution are.

Consider an infinite panmictic diploid population. Assume that the selection regime is as in the model of Maynard Smith (Chapter 7). The recombination fraction between the loci of the selected system (SSL) is determined by the *rec* modifier with alleles M and m, allele M increasing and m decreasing the rate of exchanges. Let the linkage of the *rec* locus with the SSL be controlled by

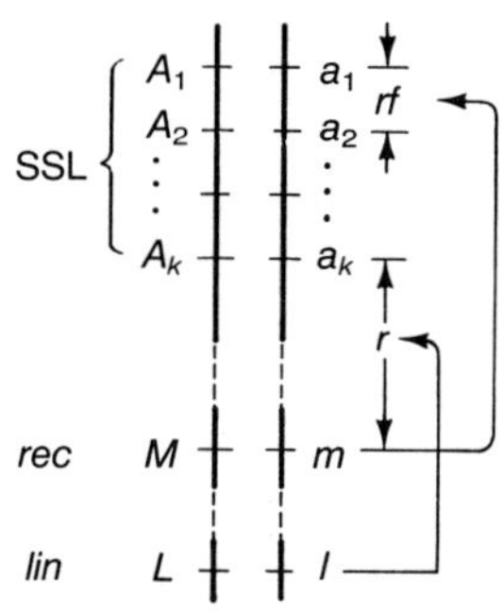

Figure 9.5 The scheme of a model involving a modifier of linkage (*lin*) between the *rec* locus and the selected system of loci (SSL).

a special locus – a modifier of linkage (*lin* locus) with alleles *L* and *l* (*L* increases and *l* decreases the exchange rate between the *rec* locus and the nearest locus of the selected system). The general scheme of the model is shown in Figure 9.5. It is clear a priori that the rate of allele frequency changes at the *lin* locus will be still lower than that at the *rec* locus. Besides, as with the hierarchical model of a *rec* locus dominance evolution (section 9.2), the time of possible evolution at the *lin* locus is limited by the existence of polymorphism for the *rec* modifier.

Computer simulations have shown that in a periodic serrated environment with a short cycle, there is selection at the *lin* locus in favor of the allele for high crossing over between the *rec* locus and the SSL. Consider, for example, changes in the frequency of the *rec* and *lin* loci in a population evolving in the cyclical environment with parameters $T = 4$, $h = 1.75$, $s = 0.75$ (see Chapter 7 for the meaning of the parameters). Let the *rec* allele *M* be recessive and determine free recombination between the adjacent loci of the SSL and allele *m* be dominant and completely suppress exchanges between them; in the initial population, $p_M = p_m = 0.5$. The rate of recombination, *r*, between the *rec* modifier and the SSL depends on the genotype at the *lin* locus:

	LL	*Ll*	*ll*
r	0.5	0.5	0.05

Clearly, the higher the frequency of allele *L* (p_L) in the population, the higher is the rate of exchanges between the *rec* locus and the SSL and, conversely, the greater the proportion of allele *l* in the population, the tighter is the linkage between the *rec* modifier and the SSL. Let $p_L = 0.1$ at the initial moment. In this case, individuals with a tightly linked modifier of recombination predominate in the population and, as would be expected, allele *M* is eliminated from the population: after 200 cycles of dynamics its frequency has changed from 0.5 to 0.072. At the same time, there is a tendency (albeit a weak one) toward a greater independence of the *rec* locus and the SSL: the frequency of allele *L* is increased up to 0.119. If the initial frequency of allele L is $p_L = 0.5$, then allele M increases in frequency, $p_M(200) = 0.997$. Moreover, the frequency of allele L is also increased: $p_L(200) = 0.544$. Consequently, depending on the initial frequency of L and dominance relations between alleles at the *lin* locus, **opposite directions of selection** at the *rec* locus are possible, which is not surprising since these factors determine the intensity of linkage of the *rec* locus with the SSL. However, with either direction of selection at the *rec* locus, allele L increases in frequency at the *lin* locus (note that in our model, the *rec* and *lin* modifiers are unlinked to each other). Thus, an examination of the model with a *rec* locus modifying linkage within the SSL and a modifier of linkage between the *rec* locus and the SSL shows that selection may favor those alleles at the *lin* locus that ensure an optimum (in terms of selection for *rf*) linkage of the *rec* locus with the selected system.

9.4 THE EVOLUTION OF SEX DIFFERENCES IN RECOMBINATION

9.4.1 The evidence

By the Haldane–Huxley rule, if crossing over is strongly suppressed or completely lacking in meiosis of one of the sexes, then that sex is, invariably, the heterogametic one. The effect of sex on recombination frequency has been established in a variety of animal species, the Haldane–Huxley rule holding in all cases of alternative (all-or-none type) sex differences. Provided crossing over occurs at a substantial frequency during the formation of both male and female gametes, i.e. sex differences in *rf* are not of alternative but of quantitative nature, then exchange frequency in the heterogametic sex may be either lower or higher than that in the homogametic one (for reviews, see Callan and Perry, 1977; Bell, 1982; Zhuchenko and Korol, 1985; Trivers, 1988; Burt, Bell and Harvey, 1991), the sign and magnitude of the difference $rf_♀ - rf_♂$ being segment specific. Thus, in the mouse, $rf_♀ < rf_♂$ in some regions, but in the majority of segments where $rf_♀ \neq rf_♂$, a higher frequency of exchanges has been reported during oogenesis (Dunn and Bennett, 1967; Blank *et al.*, 1988). Sex differences in exchange distribution pattern have also been revealed by cytological analysis. Thus, invariably proximal chiasma localization, excluding up to 90% of the chromosome length from exchange process, has been found in large autosomes of the grasshopper *Stethophyma grossum* males; in females, chiasma localization is not as strong as that (White, 1954; Perry and Jones, 1974).

Recently there has been rapid accumulation of data on human recombination owing to the use of molecular markers. Sex differences in *rf* have frequently been revealed (Donis-Keller *et al.*, 1987; White and Lalouel, 1988; Tanzi *et al.*, 1992). Thus, the total recombinational length of chromosome 1 estimated from female meiosis data is twice as long as that from male; in some regions (mainly subtelomeric ones), the opposite tendency is possible (O'Connell *et al.*, 1989). A higher degree of exchange localization and a lower crossing-over rate in a proximal region at male meiosis has previously been shown for the same chromosome (Keats, Morton and Rao, 1981). On the whole, sex differences are assumed to be a major source of biological variation in *rf* in humans (Rao *et al.*, 1978; Ott, 1985).

This problem, despite its high theoretical and practical value, has been poorly studied in higher plants (with a few exceptions, e.g. maize, *Arabidopsis* and tomato). Most of the maize genomic regions examined exhibit $rf_♀ < rf_♂$ or $rf_♀ \approx rf_♂$, although the extent of difference may be strongly dependent on the genotype (Robertson, 1984). A substantially higher crossing-over rate in microsporocytes than in macrosporocytes has been found in *Arabidopsis* (Vizir and Korol, 1990) (see also Figure 2.1). In tomato hybrids, both situations, $rf_♀ > rf_♂$ and $rf_♀ < rf_♂$, are possible depending on the segment in question and genotype, while cases with $rf_♀ > rf_♂$ prevail (Korol, 1987; DeVincente and Tanksley, 1991). *Fritillaria japonica* exhibits a high degree of sexual dimorphism in exchange frequency. Here, no chiasmata have been found in microsporocytes, despite the formation of a normal synaptonemal complex (Ito, Takegami and Noda, 1983).

9.4.2 Possible mechanisms

The mechanisms responsible for differences in *rf* between male and female meiosis are currently unknown. In most animal species, sex is determined by a certain chromosome. So, it can be suggested that the differences between male and female meiosis are controlled by genes located precisely on this chromosome, i.e. they are due not only to the features of the reacting system (e.g. pattern of chromatin coiling in oocytes and spermatocytes) but also to those of the controlling system. This is borne out by the experimental evidence. In particular, it has been found that genes affecting recombination in *D. ananassae* males are located on the Y chromosome (Moriwaki, Tobari and Matsuda, 1979). Another possibility is that the sex chromosome canalizes the development of rather general physiological and biochemical sex differences (including those in the hormonal status) in a toggle-like manner, with specific recombination features of male and female meiosis being determined as a secondary effect of the main regulator. The latter suggestion is supported by the data from the fish *Oryzias latipes*, in which the heterogametic sex is represented by the XY males (Yamamoto, 1961). Males XY were transformed into functional XY females by a hormonal treatment and then estimated for recombination in the sex chromosome. The crossing-over rate was found to be increased by a factor of 5(!) compared with the level in the normal males.

That the latter suggestion is justifiable is also borne out by extensive evidence for specific features of crossing over in plant micro- and macrosporogenesis since, in this case, the differences in exchange frequency and distribution are confined to the same organism or, frequently, the same flower. The epigenetically determined cell specialization seems to be responsible for the patterns of chromatin condensation, of temporal organization of meiosis and, possibly, of recombinase activity. This may lead to sex differences in the duration of individual stages of meiosis; in the temporal sequence of synaptic initiation and in the duration of synapsis in different genomic regions; in the localization of single- and double-strand DNA breaks and the degree of their involvement in the operation of molecular machinery of exchange and repair.

9.4.3 Evolutionary explanations

Evolutionary reasons for sex differences in recombination remain unclear. One possible 'physiological' explanation of the cause of sex differences is based on the assumption of different levels of DNA metabolism in male and female meiocytes (Fogwill, 1958; Bernstein, Hopf and Michod, 1988). According to this model, one can expect $rf_{♀} > rf_{♂}$. But as we have seen earlier, this prediction is far from always substantiated experimentally. For example, one cannot explain the evidence for predominantly higher exchange frequencies in microsporocytes of maize, *Arabidopsis*, *Allium* and other plant species on this basis (Gohil and Kaul, 1980; Robertson, 1984; Zhuchenko and Korol, 1985; Vizir and Korol, 1990). It seems more reasonable to look for functional (combinative) explanations for reasons underlying sex differences in crossing over rate.

The appearance of sexual dimorphism seems to result in strong selective pressure in favor of tighter linkage of genetic factors determining major sex differences (Haldane, 1922; Nei, 1969; Rice, 1987). According to Nei, given a sufficient diversity in the ancestral population of modifier genes with specific supressing effects on crossing over within the gonosomes (X chromosomes) in the heterogametic sex, the corresponding alleles become fixed, recombination in the rest of the genome is maintained and the gonosome behaves as a supergene in this sex. If modifiers with specific effects are insufficiently effective, then selection for reduced *rf* within the X chromosome will be on modifiers affecting the entire genome and may result in a total inhibition of exchanges in the heterogametic sex. It cannot be excluded that this is the reason for the lack of crossing over in *Drosophila* males and butterfly females (Turner, 1979). According to another point of view (Bull, 1983; Trivers, 1988), sexual differences in autosomal crossing over pre-existing in the ancestral population predetermine which of the sexes will become heterogametic. In the course of evolution one of the autosomal pairs evolves into a gonosomal supergene. **Sexual selection** is considered to be the driving force of such evolution (see below).

Suppose that Nei's hypothesis is true. How, then, can sex differences in *rf* for those species in which autosomal crossing over has been preserved in the heterogametic sex be explained? There seems to be more than one answer. On the one hand, even if selection against crossing over within sex chromosomes was largely on modifiers with specific effects, it is hard to believe that response to selection has not involved, even partially, *rec* genes with general effects. On the other hand, the 'specificity' of modifiers is rather relative in nature: while fixation of some of them, along with a reduction in crossing over in the sex chromosome, resulted in a correlated decline in *rf* within individual autosomes, the opposing selection of modifiers 'specific' to these autosomes could compensate (or overcompensate) for the previous reduction. Possible interchromosomal effects should also be taken into consideration. The combined actions of the indicated factors make the supposition about the segment specificity of sex differences in recombination frequency highly plausible, which may be manifested in the dependence of **exchange distribution** on sex (Fletcher and Hewitt, 1980; Santos, 1993).

From the adaptationist point of view, the existence of sexual differences in recombination can considerably enhance the regulatory potential of the *R*-system. The differences in responses to environmental changes as expressed in the amount and spectrum of genotypic variability generated during male and female meiosis make it possible, while keeping the control system complexity at almost the same level (i.e. involving the same controlling genes), to ensure a more flexible population regulation, to reconcile the opposing requirements for maximum current fitness and genetic flexibility.

The above model for the evolution of sex differences in recombination based on sexual selection (Trivers, 1988) is, on the whole, consistent with these general notions and describes the situations when $rf_{♀} > rf_{♂}$. In essence, this explanation is as follows. Since variation in the degree of male reproductive success is much wider than that of the female, the realized variants of male gametes can be expected to have more successful than average (i.e. than in realized variants of female gametes) gene combinations. Therefore, reduced recombination in male meiosis, increasing the correlation between offspring and fathers, results in a higher mean fitness. Or, in other words, successful males with lower exchange frequencies are more likely to transmit their beneficial gene combinations to progeny than are similar males with high *rf* (at least in the genomic regions affecting male reproductive success).

9.4.4 Taking account of gamete selection

The hypothesis of Trivers (1988) deals with the problem of **preserving adaptations**. We have proposed a complementary explanation which proceeds from the problem of the **formation of adaptations** and possibilities of testing these not only at the diploid level, but also on gametes. The latter may be particularly significant in the case of plants (see Zhuchenko and Korol, 1985; Mulcahy and Mulcahy, 1987; Snow and Spira, 1991). Of paramount importance here are the differences in the functions of male and female gametes in reproduction, adaptive radiation and evolution of the population (species). The ratio of male to female gametes ($n_{♂}/n_{♀}$) produced at both individual level and the population level varies widely between species, reaching values of the order 10^4 to $\approx 10^{10}$. It is known that while recombination generates new valuable gene combinations and gene systems, it leads also to breaking up of adaptive complexes, resulting in continuous erosion of the attained degree of fitness. Therefore, the experimentation of a species in creating new variants will be less wasteful if their frequencies in female gametogenesis are lower than those in male. This is because, generally, the value of a single female gamete to the population is much higher than that of a male gamete (considering the ratio $n_♂/n_♀$).

The same arguments being applied to mutability predict higher mutation rates in male germ cells than in females, which corresponds to the reality, at least in humans (Shimmin, Chang and Li, 1993). In these respects, our model is similar to Geodakyan's (1987) hypothesis of the prospective role of the male sex as such. If that is the case, female meiosis seems to ensure the production of a higher percentage of viable gametes (because of a reduced total amount of genetic variability and, consequently, a better preservation of beneficial gene combinations). A possible negative correlation between *rf* and male gamete fertility in mammals was pointed out by Cohen (1973, 1975; but see Wallace, 1974).

A potential advantage of increased $rf_♂$ is the possiblity of improving adaptation to environmental conditions by selection on male gametes (which seem to be less influenced by sampling effects because of their higher number). For cross-pollinated plants, an increased male recombination is also justified because of a lower expected correlation between ecological conditions of the progeny and pollen parent than between those of the progeny and pistillate parent. Furthermore, selection of male gametes for correspondence to the egg cell genotype based on recognition systems associated with numerous membrane properties including immune responses can play an important role in determining the progeny composition, including the maintenance of polymorphism for systems of co-adapted genes (Cohen, 1975; Bodmer, 1977; Hartung, 1980; Zhuchenko and Korol, 1985; Bremermann, 1987).

An important advantage of our explanation is the possibility of a functionally significant interpretation for a number of cases of sex differences in recombination in hermaphroditic organisms (primarily plants) to which Nei's model is not applicable at all (nor is the model of Trivers (1988) applicable when $rf_♀ < rf_♂$) (see, for example, our data on *Arabidopsis*, in which these differences may be as large as 200–300%; Vizir and Korol, 1990; Figure 2.1). This is also true for some dioecious species which exhibit higher degrees of linkage and/or more localized pattern of exchange distribution in the homogametic sex than in the heterogametic one, for example in some mammals (Bennett, Hayman and Hope, 1986; Hayman, Moore and Evans, 1988) and in a number of orthopteran species (Cano, Jones and Santos, 1987). From this viewpoint it is also easy to understand why in grasshoppers some genomic changes (supernumerary segments) affect chiasma distribution and increase chiasma frequency in male meiosis and are non-effective in female meiosis (Santos, 1993).

The genomic regions with high concentrations of genes controlling the competitive ability of male gametes are likely to exhibit the opposite tendency – a reduction in $rf_♂$ relative to $rf_♀$. An example is provided by the *t*-locus in mouse: a strong distortion of segregation ratios in the progeny of heterozygous males resulting from reduced fertilizing capacity of sperm with the normal allele is accompanied by crossing-over suppression in the adjacent region in males heterozygous for the *t*-locus compared with that in both females and normal males (Dunn and Bennett, 1967; Demin and Safronova, 1980; Silver, 1993). Interestingly, the *t*-locus is linked to the major histocompatibility complex which, in mammals in general, is a recombination hotspot and is characterized by linkage disequilibrium between various sites, a large number of alleles and by persistence of high heterozygosity even with prolonged close inbreeding. It has been suggested that one of the main mechanisms of such a behavior is gametic selection occurring at syngamic stages (Bremermann, 1987). Therefore, sex-dependent regulation of recombination may be directed not only by sexual selection, but also by gamete selection, non-random fertilization, zygote selection, as well as by abortive regulation of progeny by maternal body (see Glaser, 1985; Zhuchenko and Korol, 1985; Marshall and Ellstrand, 1986; Charlesworth, Schemske and Sork 1987; Mulcahy and Mulcahy, 1987; Hormaza and Herrero, 1992).

It is possible to assume that sex differences in the degree of canalization of meiosis with respect to the frequency and distribution of recombination events may depend on the degree of fitness of genotypes to the environment. A population in the central part of the species area, where the density is high, can more easily 'afford' experimenting with not only male but also female gametes as well as zygotes, since strong stabilizing selection preserves the status quo by maintaining the population size at a stable level and perfecting the adaptive norm of the population. Therefore, sex differences in *rf* are not likely to be very large here. A contrasting situation may occur on the periphery of the species area. Owing to a high frequency and unusual localization of exchanges in male meiosis, basically new genotypic variants can be generated, making adaptive radiation possible. Moreover, canalization of crossover exchange in female meiosis and protection of macrogametes against environmental stresses promote the maintenance of genetic continuity, the establishment of the developed adaptations. Clearly, it is the difference in the norm of reaction of male and female meiocytes to environmental changes as well as differential viability of male and female gametes that play the key role in our model. An essential difference from the above-mentioned hypothesis of Geodakyan (1987) is that the regulatory role of sex differences in crossover frequency and distribution is preserved even in the absence of dioecy in the population, in particular under hermaphroditism, a breeding system of wide occurrence in plants. These group selection arguments could also be formulated and tested in terms of modifier evolution based on individual selection.

PART THREE

Conclusions

Conclusions

The material presented in this book as well as abundant evidence in the literature suggests large variation in the parameters characterizing recombination: exchange frequency and distribution. This variation can depend on such factors as genotype, sex, age and environment. Although it is generally accepted that recombination is one of the main sources of heritable variation indispensable for the efficiency of selection (both natural and artificial), the significance of variability of recombination itself and of the majority of its other features remains unknown. Nor are the possible reasons for evolving and maintaining these features clear. The above concerns, for example, such properties as high non-uniformity of exchanges per unit physical length (between-segment differences of 1–2 orders of magnitude), a narrow range of per chromosome chiasma frequency variation among higher eukaryotes (usually 1–4, while DNA content per chromosome varies over 1–3 orders of magnitude), the ubiquity of genetic interference, high intraspecific polymorphism and non-heritable variability of recombination parameters, etc. In view of this, two groups of interrelated questions of fundamental importance for understanding the problem of heredity in general can be recognized:

- Are the recombination features and properties observed in extant species of any functional significance for population adaptation and evolution as well as for breeding programs?
- Can these properties be assumed to have spread and been established because they improve the efficiency of transformations of genetic variability available for selection to act upon? Or are they a by-product of evolutionary processes not associated directly with combinative functions of recombination (processes which have determined the formation of sexual reproduction, DNA replication and repair, as well as the genome structure)?

Analysis of the data in literature and our own results seems to warrant positive answers to the questions about the adaptive significance and adaptive evolution of recombination properties. Darlington (1939) considered the process of reconciling the opposing requirements for maximum fitness to existing conditions and for retaining by a species the capacity for adapting to future environmental variations to be the 'leitmotif' of such an evolution.

If the environment is relatively stable or predictable, then the attainment of maximum fitness in the present is possible provided the genetic system of population adaptation (or the *R*-system, in our terminology) generates a narrow spectrum of more or less optimum genotypes. In the extreme case, this means a nearly complete restriction of chance in transmission of genetic information from generation to generation, ensuring a minimum loss of fitness. Such determinism, the 'freezing' of optimum genotypes, can be achieved through various forms of apomixis, parthenogenesis or vegetative reproduction. This increases the 'current' fitness at the expense of future genetic flexibility. In a less extreme form, this 'conservative' mode is realized with various forms of inbreeding (including partial or complete selfing in plants, various restrictions on panmixia in higher animals), increasing the probability of evolutionarily tested valuable gene complexes being preserved (Shields, 1982, 1988). In the long run, the very existense of discrete species as reproductively isolated gene pools is a critical factor restricting recombinational variability (Dobzhansky, 1959; Mayr, 1963; Darlington, 1971). If there were no species differentiation, sexual reproduction would render any essential adaptive specialization of gene complexes impossible. The existence of reproductive barriers between species ensures a reduction in losses resulting from the formation of imbalanced combinations (Grant, 1981). This may be of vital importance for local adaptation in a 'coarse-grained' environment (in particular, for plants that exhibit low individual mobility).

However, a high incidence of these modes of reproduction in the margin of the species area points to a basic intrinsic contradiction. Breeding systems realize two fundamental properties of living organisms, namely heritable variability and propagation. Sex and reproduction (propagation) may be considered as initially independent processes which have only become coupled in the course of evolution of multicellular organisms (Ghiselin, 1988; Ruvinsky, 1991). Unfavorable conditions and their unpredictability increase the value of genetic variability and, hence, of sex (an idea going back to Weismann). On the other hand, in such an environment, the probability of successful sexual reproduction via outcrossing is reduced (owing to vulnerability of meiosis and gametogenesis in general to stressful factors and poor chance of finding a partner under low population density). Such an opportunistic strategy of the genetic system of many species, which increases 'quantity' at the expense of 'quality', seems to be an all but inevitable cost of the ancient 'decision' to combine, in multicellular organisms, the heritable variability functions with those of propagation.

The mode of adaptation which is an opposite of the conservative one is based on an open R-system producing a broad spectrum of variability exceeding significantly the range of genotypes adapted to existing environmental conditions and compatible with a given *Bauplan* (basic structure) of the organism. Here, the main factor of non-randomness (preservation of organization) is not the R-system but stabilizing selection. Examples of such organisms are provided by many woody plants exhibiting enormous seed overproduction. Darlington (1939) believed that it is in the capacity of generating random variability that the power of the genetic system to 'foresee' lies, because randomness is an optimum strategy in unpredictable environments. Such a preponderance of chance ensuring a population's survival under sudden environmental changes in the future results in a continuous break-up of co-adapted gene complexes, thereby reducing the immediate fitness (Mather, 1943). As an adaptation strategy, it is only acceptable under enormous overproduction of zygotes. This way of resolving the contradiction between the requirements for maximum fitness to existing conditions and for genetic flexibility can hardly be regarded as a universal one, and still less as an optimum one.

It appears that a more successful relationship between the F- and R-systems is the one ensuring a balance between random and deterministic components, both in generating variability by the R-system and in realizing it at the F-system level. Moreover, the adaptivity of the processes of genetic variability storage, release and realization may be ensured by a number of feedforwards and feedbacks within the triad 'R-system $\leftrightarrow$ F-system $\leftrightarrow$ environment': the amount and spectrum of free genetic variability in the F-system components are regulated by the R-system, the functioning of 'the regulator' itself (the R-system) being to a certain extent dependent on the regulated object (the F-system) and on environment (Zhuchenko and Korol, 1982, 1983, 1985; Korol, 1987).

This provides a new insight into the role of environment in evolutionary reorganization of the population gene pool. In the traditional scheme of evolution, environment plays the role of a key non-random factor restricting the flow of variability from generation to generation by 'sorting out' genotypes according to their fitnesses (natural selection). Environmental variation has, however, been known to affect the R-system parameters by altering frequency and distribution of crossovers, rate and spectrum of mutations, frequency of mobile element transpositions, rate of outcrossing, etc. Such a relationship may ensure a certain expediency (non-randomness) of population gene pool responses to environmental influences, by increasing variability precisely at the time of maximum requirement for it. Therefore, environment not only selects the fittest but also determines, to an extent, just when the highest and lowest diversity will be subjected to selection (Figure C.1).

The moderately deviating (from the 'normal structure') variants that have not yet been eliminated by selection can, according to the proposed scheme,

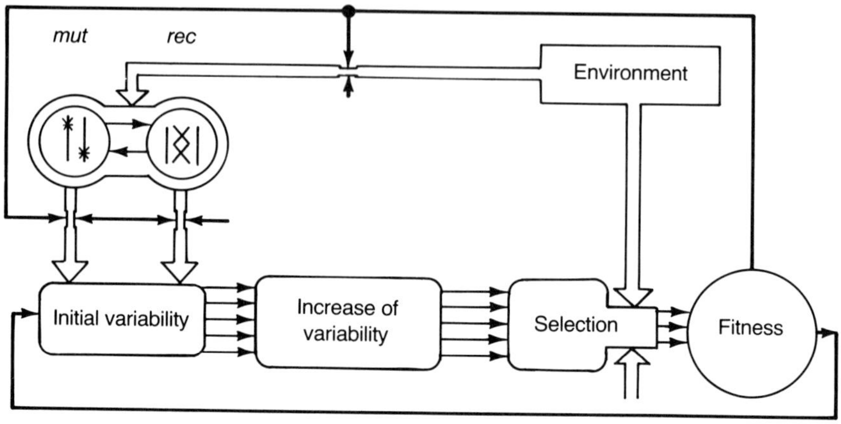

Figure C.1 Possible effects of environments on the microevolutionary processes.

generate novel genetic variability with a higher evolutionary potential. Bauer (1935) put forward a similar and somewhat paradoxical idea according to which it is 'losers rather than winners in the struggle for existence which provide the raw material of evolutionary change'. The general reason for such an increased production of genetic variability can be seen in the 'genomic stress' caused by external and internal environment (McClintock, 1984; Zhuchenko and Korol, 1985; Cullis, 1990; Hoffmann and Parsons, 1991). How common such a system of variability control based on negative feedback (with a corresponding reduction in the costs of mutation and recombination) is in nature, has yet to be ascertained.

The evolutionary advantage of the dependence of recombination on environment ('variable *rf* strategy') and the possibility of its spread through and fixation in a population have been shown in our simple computer simulations. Similarly, the indirect effects of selection for fitness genes on interallelic relations at *rec* loci and on the relative positions of the *rec* loci and their target regions (by selection on the corresponding modifiers) has been demonstrated. The analysis has revealed that selection on adaptive gene systems induced by environmental variation may be an effective factor changing allele frequencies at selectively neutral loci affecting various recombination properties even if these genes are unlinked to the selected loci. Computer simulations have shown this factor to exhibit a rather high rate of response, which makes attempts at **experimental modeling** of the *rec* system microevolution in laboratory populations feasible. In reproducible experiments with large cage populations of *Drosophila*, selection for resistance to adverse environmental factors has been shown to result, rather soon (a matter of a few dozens of generations), in drastic changes in the *rec* system, including segment-specific increases in crossing-over frequency, changes in dominance–recessiveness relations of *rec* genes, a lower degree of genetic interference. By contrast, exchange frequency became reduced in stable conditions. Thus, Fisher's prediction of linkage tightening formulated more than half a century ago has been for the first time substantiated experimentally. Furthermore, the stage has been set for a new promising direction in recombination research, namely experimental modeling of the *rec* system microevolution.

References

Abdullah, U.F. and Charlesworth, B. (1974) Selection for reduced crossing over in *Drosophila melanogaster. Genetics*, **76**, 447–51.

Acharya, S.N. and Jana, S. (1982) Genetic variability in a heterogeneous population of barley (*Hordeum vulgare*). *Can. J. Genet. Cytol.*, **24**, 436–40.

Acton, A.P. (1961) An unsuccessful attempt to reduce recombination by selection. *Am. Nat.*, **95**, 119–20.

Aeschbacher, H.V., Vuataz, L., Sotck, J. and Stadler, R. (1977) Use of the beta-binomial distribution in dominant-lethal testing for 'weak mutagenic activity'. *Mutat. Res.*, **44**, 369–90.

Aguilera, A. and Klein, H.L. (1988) Genetic control of intrachromosomal recombination in *Saccharomyces cerevisiae*. I. Isolation and genetic characterization of hyper-recombination mutations. *Genetics*, **119**, 779–90.

Agur, Z. and Kerszberg, M. (1987) The emergence of phenotypic novelties through progressive genetic change. *Am. Nat.*, **129**, 862–75.

Ahokas, H. (1989) Transfection of germinating barley seed electrophoretically with exogenous DNA. *Theor. Appl. Genet.*, **77**, 469–72.

Akam, M. (1989) *Hox* and HOM: homologous gene clusters in insects and vertebrates. *Cell*, **57**, 347–49.

Akin, E. (1979) *The Geometry of Population Genetics*, Springer, Berlin.

Akin, E. (1990) The differential geometry of population genetics and evolutionary games, in *Mathematical and Statistical Developments of Evolutionary Theory* (ed. S. Lessard), Kluwer Academic, Dordrecht, pp. 1–93.

Albini, S.M. and Jones, G.H. (1988) Synaptonemal complex spreading in *Allium cepa* and *Allium fistulosum*. II. Pachytene observation: the SC karyotype and the correspondence of late recombination nodules and chiasmata. *Genome*, **30**, 399–410.

Allard, R.W. (1956) Formulae and tables to facilitate the calculation of recombination values in heredity. *Hilgardia*, **24**, 235–78.

Allard, R.W. (1963) Evidence for genetic restriction of recombination in the lima bean. *Genetics*, **48**, 1389–95.

Allard, R.W. (1975) The mating system and microevolution. *Genetics*, **79** (Suppl.), 115–26.

Allard, R.W. (1988) Genetic changes associated with the evolution of adaptedness in cultivated plants and their wild progenitors. *J. Heredity*, **79**, 225–38.

Allard, R.W. (1990) The genetics of host-pathogen coevolution: implications for genetic resource conservation. *J. Heredity*, **81**, 1–6.

Allard, R.W. and Kahler, A.L. (1974) Multilocus genetic organization and morphogenesis. *Brookhaven Symp. Biol.*, **25**, 239–43.

Allard, R.W., Babbel, G.R., Clegg, M.T. and Kahler, S.L. (1972) Evidence for coadaptation in *Avena barbata*. *Proc. Natl Acad. Sci. USA*, **69**, 3043–48.

Allen, P.M. and Ebeling, W. (1983) Evolution and the stochastic description of simple ecosystems. *Biosystems*, **16**, 113–26.

Al-Saheal, Y.A. and Larik, A.S. (1987) Genetic control of environmentally induced DNA variation in flax genotrophs. *Genome*, **29**, 643–46.

Altenberg, L. and Feldman, M.W. (1987) Selection, generalized transmission and the evolution of modifier genes. I. The reduction principle. *Genetics*, **117**, 559–72.

Altukhov, Y.P. (1989) *Genetic Processes in Populations*, 2nd edn, Nauka, Moscow.

Amos, B., Barrett, J. and Dover, G.A. (1991) Breeding behaviour of pilot whales revealed by DNA fingerprinting. *Heredity*, **67**, 49–55.

Anderson, C.L., Carew, E.A. and Powell, J.R. (1993) Evolution of the *Adh* locus in *Drosophila willistoni* group: the loss of an intron, and shift in codon usage. *Mol. Biol. Evol.*, **10**, 605–18.

Anderson, E.G. (1949) *Introgressive Hybridization*, Wiley, NY.

Anderson, W.F. (1992) Human gene therapy. *Science*, **256**, 808–13.

Anderson, W.W. (1983) Achieving synthesis in population biology, in *Population Biology: Retrospect and Prospect* (eds C.E. King and P.S. Dawson), Columbia University Press, New York, pp. 189–205.

Andre, D., Colau, D., Schell, J. *et al.* (1986) Gene tagging in plants by a *T*-DNA insertion mutagen that generates *APH(3')II*-plant gene fusions. *Mol. Gen. Genet.*, **204**, 512–18.

Antolin, M.F. and Strobeck, C. (1985) The population genetics of somatic mutation in plants, *Am. Nat.*, **126**, 52–62.

Antonovics, J. and Ellstrand, N.C. (1984) On the measurement of natural and sexual selection: theory. *Evolution*, **38**, 709–19.

Aquadro, C.F. (1992) Why is the genome variable? Insights from *Drosophila*. *Trends Genet.*, **8**, 355–61.

Aquadro, C.F., Lado, K.M. and Noon, W.A. (1988) The *rosy* region of *Drosophila melanogaster* and *Drosophila simulans*. I. Contrasting levels of naturally occurring DNA restriction map variation and divergence. *Genetics*, **119**, 875–88.

Arnault, C. and Biemont, C. (1989) Heat shock does not mobilize mobile elements in genomes of *Drosophila melanogaster* inbred lines. *J. Mol. Evol.*, **28**, 388–90.

Arnheim, N., White, T. and Rainey, W. (1990) Application of the polymerase chain reaction to organismal and population biology. *BioScience*, **40**, 174–82.

Arnholdt-Schmidt, B. (1993) Rapid changes in amplification and methylation pattern of genomic DNA in cultured carrot root explants (*Daucus carota* L.) *Theor. Appl. Genet.*, **85**, 793–800.

Arnold, J., Asmussen, M.A. and Avise, J.C. (1988) An epistatic mating system model can produce permanent cytonuclear disequilibria in a hybrid zone. *Proc. Natl Acad. Sci. USA*, **85**, 1893–96.

Arnold, M.L., Contreras, N., Shaw, D.D. (1988) Biased gene conversion and asymmetrical introgression between subspecies. *Chromosoma*, **96**, 368–71.

Arques, D.G. and Michel, C.J. (1987) Periodicities in introns. *Nucleic Acids Res.*, **15**, 7581–93.

Arrand, J.E., Squires, S., Bone, N.M. and Johnson, R.T. (1987) Restoration of UV-induced excision repair in *Xeroderma D* cells transfected with the *den*V gene of bacteriophage *T4*. *EMBO J.*, **6**, 3125–31.

Arroyo, M.T.K. (1973) Chiasma frequency evidence on the evolution of autogamy in *Limnanthes floccosa (Limnanthacrae). Evolution*, **27**, 679–88.

Asmussen, M.A. and Arnold, J. (1991) The effects of admixture and population subdivision on cytonuclear disequilibria. *Theor. Pop. Biol.*, **39**, 273–300.

Asmussen, M.A., Arnold, J. and Avise, J.C. (1989) The effects of assortative mating and migration on cytonuclear associations in hybrid zones. *Genetics*, **122**, 923–34.

Attia, T. and Lelley, T. (1987) Effects of constitutive heterochromatin and genotype on distribution of chiasmata in the seven individual rye bivalents. *Theor. Appl. Genet.*, **74**, 527–30.

Auerbach, Ch. (1976) *Mutation Research. Problems, Results and Perspectives*, Chapman & Hall, London.

Avner, P., Amar, L., Dandolo, L. and Cuenet, J.L. (1988) Genetic analysis of the mouse using interspecific crosses. *Trends Genet.*, **4**, 18–23.

Ayala, F.J. (1981) Mechanisms of evolution, in *Evolution* (ed. M.V. Mina), Myr, Moscow, pp. 33–65.

Ayala, F.J. (1983) Genetic polymorphism: from electrophoresis to DNA sequences. *Experientia*, **39**, 813–23.

Ayala, F.J. and Kiger, J.A. Jr (1984) *Modern Genetics*, Vol. 3, Benjamin/Cummings, Menlo Park, CA.

Bachmann, K., Chambers, K.L. and Price, H.J. (1985) Genome size and natural selection: observations and experiments in plants, in *The Evolution of Genome Size* (ed. T. Cavalier-Smith), John Wiley, New York, pp. 267–276.

Bailey, N.T.J. (1961) *An Introduction to the Mathematical Theory of Genetic Linkage*, Clarendon Press, Oxford.

Bailey, N.T.J. (1967) *The Mathematical Approach to Biology and Medicine*, Wiley, London.

Baker, B.S. and Hall, J.C. (1976) Meiotic mutants: genetic control of meiotic recombination and chromosome segregation, in *The Genetics and Biology of Drosophila*, Vol. 1A (eds M.A. Ashburner and E. Novitski), Academic Press, New York, pp. 351–434.

Baker, B.S., Boyd, J.B., Carpenter, A.T.C. *et al.* (1976) Genetic control of meiotic recombination and somatic DNA metabolism in *Drosophila melanogaster. Proc. Natl Acad. Sci. USA*, **73**, 4140–44.

Baker, W.K. and Kaeding, E.A. (1981) Linkage disequilibrium at the alphaesterase loci in a population of *Drosophila montana* from Utah. *Am. Nat.*, **117**, 804–9.

Banga, S.S., Boyd, J.B., Valerie, K. *et al.* (1989) *den*V gene of bacteriophage *T4* restores DNA excision repair to *mei-9* and *mus201* mutants of *Drosophila melanogaster. Proc. Natl Acad. Sci. USA*, **86**, 3227–31.

Barker, J.S.F., East, P.D. and Weir, B.S. (1986) Temporal and microgeographic variation in allozyme frequencies in a natural population of *Drosophila buzzatii. Genetics*, **112**, 577–611.

Barker, J.S.F., Starmer, W.T., Vacek, D.C. (1987) Analysis of spatial and temporal variation in the community structure of yeasts associated with decaying *Opuntia cactus. Microbial Ecol.*, 267–76.

Barlow, D.P. (1993) Methylation and imprinting: from host defense to gene regulation. *Science*, **260**, 309–10.

Bartlett, M.S. and Haldane, J.B.S. (1935) The theory of inbreeding with forced heterozygosis. *J. Genet.* **31**, 327–40.

Barton, N.H. (1983) Multilocus clines. *Evolution*, **37**, 454–71.

Barton, N.H. (1986) The maintenance of polygenic variation through a balance between mutations and stabilizing selection. *Genet. Res.*, **47**, 209–16.

Barton, N.H. (1990) Pleiotropic models of quantitative variation. *Genetics*, **124**, 773–82.

Barton, N.H. and Clark, A. (1990) Population structure and process in evolution, in *Population Biology. Ecological and Evolutionary Viewpoints* (eds K. Wohrmann and S.K. Jain), Springer, Berlin, pp. 115–73.

Barton, N.H. and Hewitt, G.M. (1985) Analysis of hybrid zones. *Annu. Rev. Ecol. Syst.*, **16**, 113–48.

Barton, N.H. and Hewitt, G.M. (1989) Adaptation, speciation and hybrid zones. *Nature*, **341**, 497–503.

Basten, C.J. and Ohta, T. (1992) Simulation study of a multigene family, with special reference to the evolution of compensatory advantageous mutations. *Genetics*, **132**, 247–52.

Bauer, E.S. (1935) *Theoretical Biology*, VIZM Press, Moscow.

Baur, E. (1924) Untersuchungen uber des Wesen, die Enstehung und die Vererbung von Rassenunterschieden bei *Antirrhinum majus*. *Bibliogr. Genet.*, **4**, 1–170.

Beach, L.R. and Palmiter, R.D. (1981) Amplification of the metallothionein-1 gene in cadmium resistant mouse cells. *Proc. Natl Acad. Sci. USA*, **78**, 2110–14.

Beadle, G.W. (1932) A possible influence of the spindle fibre on crossingover in *Drosophila*. *Proc. Natl Acad. Sci. USA*, **18**, 160–65.

Beavis, W.D. and Grant, D. (1991) A linkage map based on information from four F_2 populations of maize (*Zea mays*). *Theor. Appl. Genet.*, **82**, 636–44.

Beavis, W.D., Grant, D., Albertsen, M. and Fincher, R. (1991) Quantitative trait loci for plant height in four maize populations and their associations with qualitative genetic loci. *Theor. Appl. Genet.*, **83**, 141–44.

Beckmann, J.S. and Soller, M. (1983) Restriction fragment length polymorphisms in genetic improvement: methodologies, mapping and costs. *Theor. Appl. Genet.*, **67**, 35–43.

Beckmann, J.S. and Soller, M. (1986) Restriction fragment length polymorphisms in plant genetic improvement. *Oxford Surveys Plant Mol. Cell. Biol.*, **3**, 199–250.

Beckmann, J.S. and Soller, M. (1988) Detection of linkage between marker loci and loci affecting quantitative traits in crosses between segregating populations. *Theor. Appl. Genet.*, **76**, 228–36.

Beckmann, J.S. and Soller, M. (1990) Toward a unified approach to genetic mapping of eukaryotes based on sequenced tagged microsatellite sites. *Biotechnology*, **8**, 930–32.

Begun, D.J. and Aquadro, C.F. (1992) Levels of naturally occurring DNA polymorphism correlate with recombination rates in *D.melanogaster*. *Nature*, **356**, 519–20.

Beier, D.R., Dushkin, H., Sussman, D.J. (1992) Mapping genes in the mouse using single-strand conformation polymorphism analysis of recombinant inbred strains and interspecific crosses. *Proc. Natl Acad. Sci. USA*, **89**, 9102–06.

Bell, G. (1982) *The Masterpiece of Nature: The Evolution and Genetics of Sexuality*, University of California Press, Berkeley.

Bell, G. (1987) Two theories of sex and variation, in *Evolution of Sex and its Consequences* (ed. S.C. Stearns), Birkhauser, Basle, pp. 93–134.

Bell, G. (1988) Uniformity and diversity in the evolution of sex, in *The Evolution of Sex: An Examination of Current Ideas* (eds R.E. Michod and B.R. Levin), Sinauer, Sunderland, MA, pp. 87–105.

Bell, G. (1989) *Sex and Death in Protozoa*, Cambridge University Press, Cambridge.

Bell, G. and Maynard Smith, J. (1987) Short-term selection for recombination among mutually antagonistic species. *Nature*, **328**, 66–68.

Bengtsson, B.O. and Uyenoyama, M.K. (1990) Evolution of the segregation ratio: modification of gene conversion and meiotic drive. *Theor. Pop. Biol.*, **38**, 192–218.

Bennett, D.M. (1982) Nucleotypic basis of the spatial ordering of chromosomes in eukaryotes and the implications of the order for genome evolution and phenotypic variation, in *Genome Evolution* (eds G.A. Dover and R.B.L. Flavell), Academic Press, San Diego, CA, pp. 239–261.

Bennett, J.H., Hayman, D.L., Hope, R.M. (1986) Novel sex differences in linkage values and meiotic chromosome behaviour in a marsupial. *Nature*, **323**, 59–60.

Bennetzen, J.L. and Freeling, M. (1993) Grasses as a single genetic system: genome composition, collinearity and compatibility. *Trends Genet.*, **9**, 259–61.

Berg, R.L. (1948) On interrelations between mutability and selection in natural populations of *Drosophila melanogaster*. *J. Gen. Biol.* (USSR), **9**, 299–313.

Bergman, A. and Feldman, M.W. (1990) More on selection for and against recombination. *Theor. Pop. Biol.*, **38**, 68–92.

Bernatzky, R. and Tanksley, S.D. (1986) Majority of random cDNA clones correspond to single loci in the tomato genome. *Mol. Gen. Genet.*, **203**, 8–14.

Bernstein, C. (1979) Why are babies young? Meiosis may prevent aging of the germ line. *Persp. Biol. Med.*, **22**, 539–44.

Bernstein, C. (1987) Damage in DNA of an infecting phage *T4* shifts reproduction from asexual to sexual allowing rescue of its genes. *Genet. Res.*, **49**, 183–89.

Bernstein, H. (1983) Recombinational repair may be an important function of sexual reproduction. *BioScience*, **33**, 326–31.

Bernstein, H., Hopf, F.A. and Michod, R.E. (1988) Is meiotic recombination an adaptation for repairing DNA, producing genetic variation, or both?, in *The Evolution of Sex: An Examination of Current Ideas* (eds R.E. Michod and B.R. Levin), Sinauer, Sunderland, MA, pp. 139–60.

Bernstein, H., Byerly, H., Hopf, F.A. and Michod, R.E. (1985) Genetic damage, mutation, and the evolution of sex. *Science*, **229**, 1277–80.

Bierzychudek, P. (1987) Resolving the paradox of sexual reproduction: a review of experimental tests, in *The Evolution of Sex and its Consequences* (ed. S.C. Stearns), Birkhauser, Basle, pp. 163–73.

Bierzychudek, P. (1989) Environmental sensitivity of sexual and apomictic *Antennaria*: do apomicts have general-purpose genotypes? *Evolution*, **43**, 1456–66.

Binelli, G., Gianfranceschi, L., Pe, M.E. *et al.* (1992) Similarity of maize and sorghum genomes as revealed by maize RFLP probes. *Theor. Appl. Genet.*, **84**, 10–16.

Birky, C.W. and Walsh, J.B. (1988) Effects of linkage on rates of molecular evolution. *Proc. Natl Acad. Sci. USA*, **85**, 6414–18.

Bishop, D.K., Park, D., Xu, L. and Kleckner, N. (1992) *DMC1*: a meiosis-specific yeast homolog of *E. coli recA* required for recombination, synaptonemal complex formation, and cell cycle progression. *Cell*, **69**, 439–56.

Blank, R.D., Campbell, G.R., Calabro, P. and D'Eustachio, P. (1988) A linkage map of mouse chromosome 12: localization of *Igh* and effects of sex and interference on recombination. *Genetics*, **120**, 1073–83.

Bodmer, W.F. (1972) Evolutionary significance of the HLA system. *Nature*, **237**, 138–45.

Bodmer, W.F. (ed.) (1977) Genetics of the Cell Surface. *Proc. R. Soc. London B*, **202**.

Bodmer, W.F. and Felsenstein, J. (1967) Linkage and selection: theoretical analysis of the deterministic two-locus random mating model. *Genetics*, **57**, 237–65.

Bodmer, W.F. and Parsons, P.A. (1962) Linkage and recombination in evolution. *Adv. Genet.*, **11**, 1–100.

Boerwinkle, E., Chakraborty, R. and Sing, C.F. (1986) The use of measured genotype information in the analysis of quantitative phenotypes in man. *Ann. Hum. Genet.*, **50**, 181–94.

Bonierbale, M.W., Plaisted, R.L. and Tanksley, S.D. (1988) RFLP maps based on a common set of clones reveal modes of chromosomal evolution in potato and tomato. *Genetics*, **120**, 1095–103.

Bootsoma, D. and Hoeijmakers, J.H.J. (1993) Engagement with transcription. *Nature*, **362**, 114–15.

Borodin, P.M. (1987) Stress and genetic variability. *Genetika* (USSR), **23**, 1003–10.

Borodin, P.M. and Belyaev, D.K. (1980) The influence of stress on crossing over frequency in the second chromosome of house mouse. *Dokl. Akad. Nauk SSSR*, **253**, 727–29.

Borst, P. and Greaves, D.R. (1987) Programmed gene rearrangements altering gene expression. *Science*, **235**, 658–67.

Borstnik, B., Pumpernik, D. and Hofacker, G.L. (1987) Point mutations as an optimal search process in biological evolution. *J. Theor. Biol*, **125**, 249–68.

Bourd, S.B. and Klimenko, V.V. (1990) Recombination nodules in spread synaptonemal complexes of silkworm (*Bombyx mori L.*). *Tsytol. Genet.* (USSR), **24**, 17–24.

Bradshaw, A.D. (1965) Evolutionary significance of phenotypic plasticity in plants. *Adv. Genet.*, **13**, 115–56.

Bregliano, J.C., Picard, G., Bucheton, A. *et al.* (1980) Hybrid dysgenesis in *Drosophila melanogaster*. *Science*, **207**, 606–11.

Bremermann, H.J. (1987) The adaptive significance of sexuality, in *The Evolution of Sex and its Consequences* (ed. S.C. Stearns), Birkhauser, Basle, pp. 135–61.

Bridges, C.B. (1914) The chromosome hypothesis of linkage applied to cases in peas and *Primula*. *Am. Nat.*, **48**, 524–34.

Brockdorff, N., Amar, L.C. and Brown, S.D.M. (1989) Pulse-field linkage of the *P3*, *G6pd* and *Cf-8* genes on the mouse *X* chromosome: demonstration of synteny at the physical level. *Nucleic Acids Res.*, **17**, 1315–26.

Bronstein, S.I. and Korol, A.B. (1984) Computer modelling of the evolution of recombination in the 'host-parasite-environment' system, in *Ecological Genetics of Plants and Animals* (Conference Proceedings) (ed. A.A. Zhuchenko), Shtiintsa, Kishinev, pp. 65–67.

Brooks, L.D. (1988) The evolution of recombination rates, in *The Evolution of Sex: An Examination of Current Ideas* (eds R.E. Michod and B.R. Levin), Sinauer, Sunderland, MA pp. 87–105.

Brooks, L.D. and Marks, W.R. (1986) The organization of genetic variation for recombination in *Drosophila melanogaster*. *Genetics*, **114**, 525–47.

Brooks, R.J. (1984) Approximate likelihood ratio tests in the analysis of beta binomial data. *Appl. Stat.*, **33**, 285–89.

Brown, A.H.D. (1979) Enzyme polymorphism in plant populations. *Theor. Pop. Biol.*, **15**, 1–42.

Brown, A.H.D., Feldman, M.W. and Nevo, E. (1980) Multilocus structure of natural populations of *Hordeum spontaneum*. *Genetics*, **96**, 523–36.

Brown, S.D.M. (1992) The mouse genome project and human genetics. *Genomics*, **13**, 490–92.

Brummer, E.C., Kochert, G. and Bouton, J.H. (1991) RFLP variation in diploid and tetraploid alfalfa. *Theor. Appl. Genet.*, **83**, 89–96.

Bryant, E.H., McCommas, S.A. and Combs, L.M. (1986) The effects of an experimental bottleneck upon quantitative genetics variation in the housefly. *Genetics*, **114**, 1191–211.

Bryant, E.H. and Meffert, L.M. (1988) Effect of an experimental bottleneck on morphological integration in the housefly. *Evolution*, **42**, 698–707.

Buckley, R.C. (1987) Interactions involving plants, homoptera and ants. *Annu. Rev. Ecol. Syst.*, **18**, 111–35.

Buetow, K.H. and Chakravarty, A. (1987) Multipoint gene mapping using seriation. I. General methods. *Am. J. Hum. Genet.*, **41**, 180–88.

Bugrov, A.G. and Vysotskaya, L.V. (1988) The frequency and localization of chiasmata as an index of species ecological specialization in locusts (*Orthoptera, Acrididae*), in *Landscape Ecology of Insects* (ed. G.S. Zolotarenko), Nauka, Novosibirsk, pp. 56–63.

Bull, J.J. (1983) *Evolution of Sex Determining Mechanisms*, Benjamin/Cummings, Menlo Park, CA.

Bulmer, M.G. (1985) *The Mathematical Theory of Quantitative Genetics*, 2nd edn, Oxford University Press, Oxford.

Buratowski, S. (1993) DNA repair and transcription: the helicase connection. *Science*, **260**, 37–38.

Burdon, J.J. (1987) *Diseases and Plant Population Biology*, Cambridge University Press, Cambridge.

Burdon, J.J. and Jarosz, A.M. (1991) Host–pathogen interactions in natural populations of *Linum marginale and Melampsora lini:* I. Patterns of resistance and racial variation in a large host population. *Evolution*, **45**, 205–17.

Burke, T., Dolf, G., Jeffreys, A.J. and Wolf, R. (eds) (1991) *DNA Fingerprinting: Approaches and Applications.* Birkhauser, Basle.

Burnham, C.R., Stout, J.T. and Weinheimer, W.H. *et al.* (1972) Chromosome pairing in maize. *Genetics*, **71**, 111–29.

Burr, B. and Burr, F.A. (1991) Recombinant inbreds for molecular mapping in maize: theoretical and practical considerations. *Trends Genet.*, **7**, 55–60.

Burr, B., Burr, F.A., Thompson, K. *et al.* (1988) Gene mapping with recombinant inbreds in maize. *Genetics*, **118**, 519–26.

Burt, A. and Bell, G. (1987) Mammalian chiasma frequencies as a test of two theories of recombination. *Nature*, **326**, 803–805.

Burt, A., Bell, G., Harvey, P.H. (1991) Sex differences in recombination. *J. Evol. Biol.*, **4**, 259–77.

Butler, L. (1952) The linkage map of the tomato. *J. Heredity*, **43**, 25–35.

Butler, L. (1965) The effects of coupling and repulsion on linkage values when woolly is segregating. *Tomato Genet. Coop. Rep.*, **15**, 27–28.

Butler, L. (1973) Modified monogenic ratios. *Tomato Genet. Coop. Rep.*, **23**, 15.
Butler, L. (1977) Recombination values and their errors. *Can. J. Genet. Cytol.*, **19**, 521–29.
Caligari, P.D.S. and Mather, K. (1975) Genotype–environment interaction. III. Interactions in *Drosophila melanogaster*. *Proc. R. Soc. London B*, **191**, 387–411.
Callan, H.G. and Perry, P.E. (1977) Recombination in male and female meiocytes contrasted. *Phil. Trans. R. Soc. London B*, **277**, 227–33.
Campbell, A. (1984) Types of recombination: common problems and common strategies. *Cold Spring Harbor Symp. Quant. Biol.*, **49**, 839–44.
Campbell, D.R. and Waser, N.M. (1987) The evolution of plant mating systems: multilocus simulations of pollen dispersal. *Am. Nat.*, **129**, 593–609.
Campbell, J.H. (1982) Autonomy in evolution, in *Perspectives in Evolution* (ed. R. Milkman), Sinauer, Sunderland, MA, pp. 190–201.
Cano, M.I., Jones, G.H. and Santos, J.L. (1987) Sex differences in chiasma frequency and distribution in natural populations of *Eyprepocnemis plorans* containing *B*-chromosomes. *Heredity*, **59**, 237–43.
Cano, R.J., Poinar, H.N., Pieniazek, N.J. *et al.* (1993) Amplification and sequencing of DNA from a 120–135-million-year-old weevil. *Nature*, **363**, 536–38.
Cantet, R.J.C. and Smith, C. (1991) Reduced animal model for marker assisted selection used best linear unbiased prediction. *Genet. Sel. Evol.*, **23**, 221–33.
Carbonell, E.A., Gerig, T.M., Balansard, E. and Asins, M.J. (1992) Interval mapping in the analysis of non-additive quantitative trait loci. *Biometrics*, **48**, 305–15.
Carlson, J.E., Tulsieram, L.K., Glaubitz, J.C. *et al.* (1991) Segregation of random amplified DNA markers in F_1 progeny of conifers. *Theor. Appl. Genet.*, **83**, 194–200.
Carpenter, A.T.C. (1975) Electron microscopy of meiosis in *Drosophila melanogaster* females. II. The recombination nodule – a recombination associated structure at pachytene? *Proc. Natl Acad. Sci. USA*, **72**, 3186–89.
Carpenter, A.T.C. (1988) Thoughts on recombination nodules, meiotic recombination, and chiasmata, in *Genetic Recombination* (eds R. Kucherlapati and G.R. Smith), American Society for Microbiology, Washington, DC, pp. 529–48.
Carpenter, A.T.C. and Baker, B.S. (1982) On the control of the distribution of meiotic exchange in *Drosophila melanogaster*. *Genetics*, **101**, 81–89.
Carpenter, A.T.C. and Sandler, L. (1974) On recombination-defective meiotic mutants in *Drosophila melanogaster*. *Genetics*, **76**, 453–75.
Carson, H.L. (1957) The species as a field for gene recombination, in *The Species Problem* (ed. E. Mayr), American Association for the Advancement of Science, Washington, DC, pp. 23–27.
Carson, H.L. (1975) The genetics of speciation at the diploid level. *Am. Nat.*, **109**, 83–92.
Carson, H.L. (1987) The genetic system, the deme, and the origin of species. *Annu. Rev. Genet.*, **21**, 405–23.
Carson, H.L. (1990) Increased genetic variance after a population bottleneck. *Trends Ecol. Evol.*, **5**, 228–30.
Carter, T.C. and Falconer, D.S. (1951) Stocks for detecting linkage in the mouse and the theory of their design, *J. Genet.*, **50**, 307–23.
Case, T.J. and Bender, E.A. (1981) Is recombination advantageous in fluctuating and spatially heterogeneous environments? *J. Theor. Biol.*, **90**, 181–90.
Catcheside, D. (1977) *The Genetics of Recombination*, Edward Arnold, London.

Catcheside, D. (1981) Genes in *Neurospora* that suppress recombination when they are heterozygous. *Genetics*, **98**, 55–76.

Cattanach, B.M. (1986) Parental origin effects in mice. *J. Embryol. Exp. Morph.*, **97** (Suppl.), 137–50.

Cavalier-Smith, T. (1985) Eucaryote gene numbers, non-coding DNA and genome size, in *The Evolution of Genome Size* (ed. T. Cavalier-Smith), John Wiley, New York, pp. 69–103.

Cecchini, E., Natali, L., Cavallini, A. and Durante, M. (1992) DNA variations in regenerated plants of pea (*Pisum sativum L.*). *Theor. Appl. Genet.*, **84**, 874–79.

Cederberg, H. (1985) Variation of high and low recombination frequencies in *Neurospora crassa* owing to genetic and environmental conditions. *Hereditas*, **103**, 135–41.

Chadov, B.F., Chadova, E.V. and Hotskina, E.A. (1985) The general scheme of meiotic pairing of chromosomes in *Drosophila melanogaster*. *Dokl. Akad. Nauk SSSR*, **281**, 164–68.

Chakraborty, R. (1987) Biochemical heterozygosity and phenotypic variability of polygenic traits. *Heredity*, **59**, 19–28.

Chakraborty, R. and Weiss, K.M. (1988) Admixture as a tool for finding linked genes and detecting that difference from allelic association between loci. *Proc. Natl Acad. Sci. USA*, **85**, 9119–23.

Chakravarti, A. (1983) Utility and efficiency of linked marker genes for genetic counseling. III. Proportion of informative families under linkage disequilibrium. *Am. J. Hum. Genet.*, **35**, 592–610.

Chandley, A.C. (1981) Does 'affinity' hold the key to fertility in the female mule? *Genet. Res.*, **37**, 105–9.

Chandley, A.C. and Mitchell, A.R. (1988) Hypervariable minisatellite regions are sites of crossing-over at meiosis in man. *Cytogenet. Cell Genet.*, **48**, 152–55.

Chang, C.C. and Kikudome, C.Y. (1974) The interaction of knobs and *B* chromosomes of maize in determining the level of recombination. *Genetics*, **77**, 45–54.

Chao, L. and Cox, E.C. (1983) Competition between high and low mutating strains of *Escherichia coli*. *Evolution*, **37**, 125–34.

Charles, D.R. (1938) The spatial distribution of crossovers in X-chromosome tetrads of *Drosophila melanogaster*. *J. Genet.*, **36**, 103–26.

Charlesworth, B. (1971) Selection in density-regulated populations. *Ecology*, **52**, 469–74.

Charlesworth, B. (1976) Recombination in a fluctuating environment. *Genetics*, **83**, 181–95.

Charlesworth, B. (1978) Model for evolution of *Y* chromosomes and dosage compensation. *Proc. Natl Acad. Sci. USA*, **75**, 5618–22.

Charlesworth, B. (1985) Recombination, genome size and chromosome number, in *The Evolution of Genome Size* (ed. T. Cavalier-Smith), John Wiley, New York, pp. 489–513.

Charlesworth, B. (1988) The evolution of mate choice in a fluctuating environment. *J. Theor. Biol.*, **130**, 191–204.

Charlesworth, B. (1990) Mutation–selection balance and the evolutionary advantage of sex and recombination. *Genet. Res.*, **55**, 199–221.

Charlesworth, B. (1993) Directional selection and the evolution of sex and recombination. *Genet. Res.*, **61**, 205–224.

Charlesworth, B. and Charlesworth, D. (1973) A study of linkage disequilibrium in populations of *Drosophila melanogaster*. *Genetics*, **73**, 351–59.

Charlesworth, B. and Charlesworth, D. (1985a) Genetic variation in recombination in *Drosophila. I.* Responses to selection and preliminary genetic analysis. *Heredity*, **54**, 71–83.

Charlesworth, B. and Charlesworth, D. (1985b) Genetic variation in recombination in *Drosophila*, II. Genetic analysis of a high recombination stock. *Heredity*, **54**, 85–98.

Charlesworth, B., Langley, C.H. (1986) The evolution of self-regulated transposition of transposable elements. *Genetics*, **112**, 359–83.

Charlesworth, B., Langley, C.H. and Stephan, W. (1986) The evolution of restricted recombination and the accumulation of repeated DNA sequences. *Genetics*, **112**, 947–62.

Charlesworth, B., Morgan, M.T. and Charlesworth, D. (1991) Multilocus models of inbreeding depression with synergistic selection and partial self-fertilization. *Genet. Res.*, **57**, 177–94.

Charlesworth, D. and Charlesworth, B. (1979) Selection on recombination in clines. *Genetics*, **91**, 581–89.

Charlesworth, D., Charlesworth, B. and Strobeck, C. (1979) Selection for recombination in partially self-fertilizing populations. *Genetics*, **93**, 237–44.

Charlesworth, D., Morgan, M.T. and Charlesworth, B. (1990) Inbreeding depression, genetic load, and the evolution of outcrossing rates in a multilocus system with no linkage. *Evolution*, **44**, 1469–89.

Charlesworth, D., Schemske, D.W. and Sork, V.L. (1987) The evolution of plant reproductive characters: sexual versus natural selection, in *Evolution of Sex and its Consequences* (ed. S.C. Stearns), Birkhauser, Basle, pp. 317–35.

Chesnokov, Yu.V., Vikonskaya, N.A., Korol, A.B. (1990) Genetic transformation of maize (abstract), in *Molecular Mechanisms of Genetic Processes*, 7th All-Union Symposium, Nauka, Moscow, p. 183.

Chetverikov, S.S. (1926) On certain aspects of the evolutionary process from the standpoint of modern genetics. *J. Exp. Biol.* (Russian) **A2**, 3–54. English translation (1961) *Proc. Am. Phil. Soc.*, **165**, 167–95.

Chinnici, J.P. (1971) Modification of recombination frequency in *Drosophila*. II. The polygene control of crossing over. *Genetics*, **69**, 85–96.

Christiansen, F.B. (1990) Selection in natural populations, in *Mathematical and Statistical Developments of Evolutionary Theory* (ed. S. Lessard), Kluwer, Dordrecht, pp. 121–76.

Clark, A.G. (1981) Estimation of linkage in the presence of multiplicative viability effects. *Genetics*, **99**, 157–68.

Clark, A.G. (1987) A test of multilocus interaction in *Drosophila melanogaster*. *Am. Nat.*, **130**, 283–99.

Clarke, B.C. (1979) The evolution of genetic diversity. *Proc. R. Soc. London B*, **205**, 453–74.

Clegg, M.T., Horch, C.R. and Kidwell, J.F. (1979) Dynamics of correlated genetic systems. VI. Variation in recombination rates in experimental populations of *Drosophila melanogaster*. *J. Heredity*, **70**, 297–300.

Clegg, M.T., Kahler, A.L. and Allard, R.W. (1978) Estimation of life cycle components of selection in an experimental plant population. *Genetics*, **89**, 765–92.

Clegg, M.T., Kidwell, J.F. and Horch, C.R. (1980) Dynamics of correlated genetic systems. V. Rates of decay of linkage disequilibrium in experimental populations of *Drosophila melanogaster. Genetics*, **94**, 217–34.

Clegg, M.T., Kidwell, J.F., Kidwell, M.G. and Daniel, N.J. (1976) Dynamics of correlated genetic systems. I. Selection in the region of the *glued* locus of *Drosophila melanogaster. Genetics*, **83**, 793–810.

Cobbs, G. (1978) Renewal process approach to the theory of genetic linkage: case of no chromatid interference. *Genetics*, **89**, 563–81.

Cockerham, C.C. and Weir, B.S. (1983) Linkage between a marker locus and a quantitative trait of sibs. *Am. J. Hum. Genet.*, **35**, 263–73.

Cohen, D. and Zohary, D. (1986) The selection operating on the evolution equilibrium of the frequency of sexual reproduction in predominantly asexual population, in *Evolutionary Processes and Theory* (eds S. Karlin and E. Nevo), Academic Press, Orlando, FL, pp. 765–82.

Cohen, J. (1973) Crossovers, sperm redundancy and their close association. *Heredity*, **31**, 408–13.

Cohen, J. (1975) Gamete redundancy – wastage or selection? in *Gamete Competition in Plants and Animals*, North-Holland, Amsterdam, pp. 99–112.

Conneally, R.M. and Rivas, M.L. (1980) Linkage analysis in man. *Adv. Hum. Genet.*, **10**, 209–66.

Conneally, R.M., Haines, J.L., Tanzi, R.E. *et al.* (1989) Huntington disease: no evidence for locus heterogeneity. *Genomics*, **5**, 304–8.

Cooper, W.S. (1984) Expected time to extinction and the concept of fundamental fitness. *J. Theor. Biol.*, **107**, 603–29.

Corcoran, L.M., Thompson, J.K., Walliker, D. and Kemp, D.J. (1988) Homologous recombination within subtelomeric repeat sequences generates chromosome size polymorphisms in *P. falciparum. Cell*, **53**, 807–13.

Cornish-Bowden, A. (1987) Dominance is not inevitable. *J. Theor. Biol.*, **125**, 333–38.

Coyne, J.A. and Charlesworth, B. (1986) Location of an X-linked factor causing sterility in male hybrids of *Drosophila simulans* and *D. mauritiana. Heredity*, **62**, 97–106.

Cramer, H. (1974) *Mathematical Methods of Statistics*, Princeton University Press, Princeton.

Croft, J.A. and Jones, G.H. (1989) Meiosis in *Mesostoma ehrenbergii*. IV. Recombination nodules in spermatocytes and a test of the correspondence of late recombination nodules and chiasmata. *Genetics*, **121**, 255–62.

Crosby, J.L. (1963) The evolution and nature of dominance. *J. Theor. Biol.*, **5**, 35–51.

Crow, J.F. (1988) The importance of recombination, in *The Evolution of Sex: An Examination of Current Ideas* (eds R.E. Michod and B.R. Levin), Sinauer, Sunderland, MA, pp. 56–73.

Crow, J.F. and Kimura, M. (1965) Evolution in sexual and asexual populations. *Am. Nat.*, **99**, 439–50.

Crowley, J.G. (1969) Chiasma frequency and longevity in *Lolium multiflorum. Heredity*, **24**, 181–88.

Cullis, C.A. (1987) The generation of somatic and heritable variation in response to stress. *Am. Nat.*, **130**, 62–73.

Cullis, C.A. (1990) DNA rearrangements in response to environmental stress. *Adv. Genet.*, **28**, 73–97.

Cvetkovic, D. and Tucic, N. (1986) Female recombination rates and fitness in *Drosophila melanogaster. Z. Zool. Syst. Evol.—Forsch.*, **24**, 198–207.

Damuth, J. and Lorraine, H.I. (1988) Alternative formulations of multilevel selection. *Biol. Phil.*, **3**, 407–30.

Darlington, C.D. (1939) *The Evolution of Genetic Systems*, Cambridge University Press, Cambridge.

Darlington, C.D. (1958) *The Evolution of Genetic Systems*, 2nd edn, Basic Books, New York.

Darlington, C.D. (1971) The evolution of polymorphic systems, in *Ecological Genetics and Evolution*, Blackwell Scientific Publications, Oxford, pp. 1–19.

Darlington, C.D. (1979) The chromosomes as feedback systems in evolution. *Kybernetes*, **8**, 275–84.

Darlington, C.D. (1981) Chromosomes and organisms: the evolutionary paradoxes. *Chrom. Today*, **7**, 1–6.

Darlington, C.D. and Mather, K. (1949) *The Elements of Genetics*, Allen & Unwin, London.

Darvasi, A. and Soller, M. (1992) Selective genotyping for determination of linkage between a marker locus and a quantitative trait locus. *Theor. Appl. Genet.*, **85**, 353–59.

Darvasi, A. and Weller, J.I. (1992) On the use of the moments method of estimation to obtain approximate maximum likelihood estimates of linkage between a genetic marker and a quantitative locus. *Heredity*, **68**, 43–46.

Darvasi, A., Weinreb, A., Minke, V., Weller, J.I. and Soller, M. (1993) Detecting marker-QTL linkage and estimating QTL gene effect and map location using a saturated genetic map. *Genetics*, **134**, 943–51.

Darnell, J.E. (1978) Implications of RNA: RNA splicing in evolution of eukaryotic cells. *Science*, **202**, 1257–60.

Das Gupta, R., Morrow, B., Marondel, I. *et al.* (1993) An integrated approach for identifying and mapping human genes. *Proc. Natl Acad. Sci. USA*, **90**, 4364–68.

Day, P.R. (1958) The genetics of *Coprinus lagopus*. *Rep. John Innes Hort. Inst.*, **49**, 16–18.

Day, T.H., Dobson, T., Hiller, P.C. *et al.* (1982) Associations of enzymic and chromosomal polymorphisms in the seaweed fly, *Coelora frigida*. *Heredity*, **48**, 35–44.

Dear, P.H. and Cook, P.R. (1989) Happy mapping: a proposal for linkage mapping the human genome. *Nucleic Acids Res.*, **17**, 6795–807.

Demenais, E., Lathrop, G.M. and Lalouel, J.M. (1988) Detection of linkage between a quantitative trait and marker locus by the lod scores method: sample size and sampling considerations. *Ann. Hum. Genet.*, **52**, 237–46.

Demin, Yu.S. and Safronova, L.D. (1980) Genetics of house mouse (*Mus musculus L.*) *T*-locus. *Adv. Modern Genet.* (USSR), **9**, 97–142.

Dennell, R.E. and Keppy, D.O. (1979) The nature of genetic recombination near the third chromosome centromere of *Drosophila melanogaster*. *Genetics*, **93**, 117–30.

Dentine, M.R. and Cowan, C.M. (1990) An analytical model for the estimation of chromosome substitution effects in the offspring of individuals heterozygous at a segregating marker locus. *Theor. Appl. Genet.* **79**, 775–80.

DeSalle, R., Gatesy, J., Wheeler, W. and Grimaldi, D. (1992) DNA sequences from a fossil termite in Oligo-Miocene amber and their philogenetic implications. *Science*, **257**, 1933–36.

Detlefsen, J.A. and Clemente, L.S. (1923) Genetic variation in linkage values. *Proc. Natl Acad. Sci. USA*, **9**, 149–56.

Detlefsen, J.A. and Roberts, E. (1921) Studies on crossing-over. I. The effect of selection on crossover values. *J. Exp. Zool.*, **32**, 333–54.

DeVincente, M.C. and Tanksley, S.D. (1991) Genome-wide reduction in recombination of backcross progeny derived from male versus female gametes in an interspecific cross of tomato. *Theor. Appl. Genet.*, **83**, 173–78.

DeVincente, M.C. and Tanksley, S.D. (1993) QTL analysis of transgressive segregation in an interspecific tomato cross. *Genetics*, **134**, 585–96.

Devonshire, A.L. and Sawicki, R.M. (1979) Insecticide-resistant *Myzus persicae* as an example of evolution by gene duplication. *Nature*, **280**, 140–41.

Devoret, R. (1988) Molecular aspects of genetic recombination, in *The Evolution of Sex: An Examination of Current Ideas* (eds R.E. Michod and B.R. Levin), Sinauer, Sunderland, MA, pp. 24–44.

Devos, K.M., Atkinson, M.D., Chinoy, C.N. *et al.* (1991) Chromosomal location and variability in wheat, barley and rye of a wheat gene encoding a *bZIP* protein (*EmBP-1*). *Theor. Appl. Genet.*, **82**, 665–67.

Devos, K.M., Atkinson, M.D., Chinoy, C.N. *et al.* (1992) RELP-based genetic map of the homologous group 3 chromosomes of wheat and rye. *Theor. Appl. Genet.*, **83**, 931–39.

Dewees, A.A. (1975) Genetic modification of recombination rate in *Tribolium castaneum*. *Genetics*, **81**, 537–52.

DiCaprio, L. and Hastings, P.J. (1976) Gene conversion and intragenic recombination at the *SUP6* locus and surrounding region in *Saccharomyces cerevisiae*. *Genetics*, **84**, 697–721.

Dickstein, R., Huh, N.D., Sandlie, I., Grossman, L. (1988) The expression of the *Escherichia coli uvrA* gene in human cells. *Mutat. Res.*, **193**, 75–86.

Diers, B.W., Keim, P., Fehr, W.R. and Shoemaker, R.C. (1992) RFLP analysis of soybean seed protein and oil content. *Theor. Appl. Genet.*, **83**, 608–12.

Dimpfl, J., Echols, H. (1989) Duplication mutation as an SOS response in *Escherichia coli*: enhanced duplication formation by a constitutively activated *RecA*. *Genetics*, **123**, 255–60.

Dingle, H. (1990) The evolution of life histories, in *Population Biology* (eds K. Worhmann and S.K. Jain), Springer, Berlin, pp. 267–89.

Dobzhansky, Th. (1937) *Genetics and the Origin of Species*, Columbia University Press, New York.

Dobzhansky, Th. (1946) Genetics of natural populations. XIII. Recombination and variability in populations of *Drosophila pseudoobscura*. *Genetics*, **31**, 269–90.

Dobzhansky, Th. (1959) Evolution of genes and genes in evolution. *Cold Spring Harbor Symp. Quant. Biol.*, **24**, 15–27.

Dobzhansky, Th. (1970) *Genetics of the Evolutionary Process*, Columbia University Press, New York.

Dobzhansky, Th., Spassky, B. and Anderson, W. (1965) Bichromosomal synthetic semilethals in *Drosophila pseudoobscura*. *Proc. Natl Acad. Sci. USA*, **53**, 345–48.

Doebley, J. and Stec, A. (1991) Genetic analysis of the morphological differences between maize and teosinte. *Genetics*, **129**, 285–95.

Doebley, J. and Stec, A. (1993) Inheritance of the morphological differences between maize and teosinte: comparison of results for two F_2 populations. *Genetics*, **134**, 559–70.

Doebley, J.D., Stec, A., Wendel, J. and Edwards, M. (1990) Genetic and morphological analysis of a maize-teosinte F_2 population: Implications for the origin of maize. *Proc. Natl Acad. Sci. USA*, **87**, 9888–92.

Donis, R.O. (1991) Muller's ratchet and flu virus. *Nature*, **353**, 308–9.

Donis-Keller, H., Green, P., Helms, C. *et al.* (1987) A genetic linkage map of the human genome. *Cell*, **51**, 319–37.

Doolittle, W.F. and Stoltzfus, A. (1993) Genes-in-pieces revisited. *Nature*, **361**, 403.

Dooner, H.K. and Kermicle, J.L. (1986) The transposable element *Ds* effects the pattern of intragenic recombination at the *bz* and *R* loci in maize. *Genetics*, **113**, 135–43.

Dover, A.G. (1986) Molecular drive in multigene families: how biological novelties arise, spread and are assimilated. *Trends Genet.*, **2**, 159–65.

Draetta, G., Beach, D. and Moran, E. (1988) Synthesis of *p34*, the mammalian homolog of the yeast *cdc2+/CDC 28* protein kinase, is stimulated during adenovirus-induced proliferation of primary baby rat kidney cells. *Oncogene*, **2**, 553–57.

Drake, J.W. (1989) The evolution of mutation rates: prospects for anti-mutagenesis. *Environ. Mol. Mutagenes*, **14** (Suppl.), 52.

Drake, J.W. (1991) Spontaneous mutation. *Annu. Rev. Genet.*, **25**, 125–46.

Drake, J.W. (1993) Rates of spontaneous mutation among RNA viruses. *Proc. Natl Acad. Sci. USA*, **90**, 4171–75.

Driscoll, D.J., Palmer, C.G. and Melman, A. (1979) Nonhomologous associations of *C*-heterochromatin at human male meiotic prophase. *Cyt. Cell Genet.*, **23**, 23–32.

Dubinin, N. P. (1932) About some main problems of genetics. *Biol. J.* (USSR), **2**, 112–146.

Dudley, J.W. (1992) Theory for identification of marker locus-QTL associations in population by line crosses. *Theor. Appl. Genet.*, **85**, 101–4.

Dudley, J.W., Saghai Maroof, M.A. and Rufener, G.K. (1992) Molecular marker information and selection of parents in corn breeding programs. *Crop Sci.*, **32**, 301–04.

Dul, J.L. and Drexler, H. (1988a) Transcription stimulates recombination. I. Specialized transduction of *Escherichia coli* by *trp* phages. *Virology*, **162**, 466–470.

Dul, J.L. and Drexler, H. (1988b) Transcription stimulates recombination. II. Generalized transduction of *Escherichia coli* by phages *T1* and *T4*. *Virology*, **162**, 471–77.

Dunn, L.C. and Bennett, D. (1967) Sex differences in recombination of linked genes in animals. *Genet. Res.*, **9**, 211–20.

Dunn, L.C. and Caspari, E. (1945) A case of neighboring loci with similar effects. *Genetics*, **30**, 543–68.

Ebinuma, H. (1987) Selective recombination system in *Bombyx mori*. I. Chromosome specificity of the modification effect. *Genetics*, **117**, 521–31.

Ebinuma, H. and Yoshitake, N. (1981) The genetic system controlling recombination in the silkworm. *Genetics*, **99**, 231–45.

Echols, H. (1981) *SOS* functions, cancer and inducible evolution. *Cell*, **25**, 1–2.

Edmonds, D.E. and Rose, M.R. (1988) Reproductive wastage and the evolution of genetic systems. *J. Theor. Biol.*, **132**, 247–50.

Edwards, J.H. (1991) The reliability of locus orderings. *Ann. Hum. Genet.*, **55**, 315–20.

Edwards, M.D., Helentjaris, T., Wright, S. and Stuber, C.W. (1992) Molecular marker-facilitated investigations of quantitative trait loci in maize. *Theor. Appl. Genet.*, **83**, 765–74.

Egel-Mitani, M., Olson, L.W. and Egel, R. (1982) Meiosis in *Aspergillus nidulans*: another example for lacking synaptonemal complexes in the absence of crossover interference. *Hereditas*, **97**, 179–87.

Eldredge, N. (1986) Information, economics, and evolution. *Annu. Rev. Ecol. Syst.*, **17**, 351–69.

Ellegren, H., Anderson, L., Johansson, M. and Sandberg, K. (1992) DNA fingerprinting in horses using a simple $(TG)_n$ probe and its application to population comparisons. *Animal Genet.*, **23**, 1–9.

Ellis, T.H.N. (1986) Restriction fragment length polymorphism markers in relation to quantitative characters. *Theor. Appl. Genet.*, **72**, 1–2.

Ellis, T.H.N., Turner, L., Hellens, R.P. *et al.* (1992) Linkage maps in pea. *Genetics*, **130**, 649–63.

Ellison, B.E. (1966) Limit theorems for random mating in infinite population. *J. Appl. Prob.*, **3**, 94–144.

Ellstrand, N.C. and Antonovics, J. (1985) Experimental studies of the evolutionary significance of sexual reproduction. II. A test of the density-dependent selection hypothesis. *Evolution*, **39**, 657–66.

Elston, R.C. (1966) On testing whether one locus can account for the genetic difference in susceptibility between two homozygous lines. *Genetics*, **54**, 89–94.

Elston, R.C. (1984) The genetic analysis of quantitative trait differences between two homozygous lines. *Genetics*, **108**, 733–44.

Elston, R.C. and Glassmann, E. (1967) An approach to the problem of whether clustering of functionally related genes occurs in organisms. *Genet. Res.*, **9**, 141–47.

Elston, R.C. and Stewart, J. (1973) The analysis of quantitative traits for simple genetic models from parental, F_1 and backcross data. *Genetics*, **73**, 695–711.

Endler, J.A. (1986) *Natural Selection in the Wild*, Princeton University Press, Princeton.

Endler, J.A. and McLellan, T. (1988) The processes of evolution: Toward a newer synthesis. *Annu. Rev. Ecol. Syst.*, **19**, 395–421.

Engh van den, G., Sachs, R. and Trask, B.J. (1992) Estimating genomic distance from DNA sequence location in cell nuclei by a random walk model. *Science*, **257**, 1410–12.

Epperson, B.K. and Allard, R.W. (1987) Linkage disequilibrium between allozymes in natural populations of lodgepole pine. *Genetics*, **115**, 341–52.

Erlich, H.A. and Arnheim, N. (1992) Genetic analysis using the polymerase chain reaction. *Annu. Rev. Genet.*, **26**, 479–506.

Erlich, H.A., Gelfand, D. and Sninsky, J. (1991) Recent advances in the polymerase chain reaction. *Science*, **252**, 1643–51.

Eshed, Y., Abu-Abied, M., Saranga, Y. and Zamir, D. (1992) *Lycopersicon esculentum* lines containing small overlapping introgressions from *L. pennellii*. *Theor. Appl. Genet.*, **83**, 1027–34.

Eshel, I. (1978) On the mutual maintenance of a polymorphism in two nonepistatic loci with partial selfing. *Theor. Pop. Biol.*, **13**, 99–111.

Eshel, I. (1985) Evolutionary genetic stability of Mendelian segregation and the role of free recombination in the chromosomal system. *Am. Nat.*, **125**, 412–20.

Eshel, I. and Feldman, M.W. (1970) On the evolutionary effect of recombination. *Theor. Pop. Biol.*, **1**, 88–100.

Eshel, I. and Weinshall, D. (1987) Sexual reproduction and viability of future offspring. *Am. Nat.*, **130**, 775–87.

Estivill, X., Farral, M., Williamson, R. *et al.* (1988) Linkage disequilibrium between cystic fibrosis and linked DNA polymorphisms in Italian families: a collaborative study. *Am. J. Hum. Genet.*, **43**, 23–28.

Evgenjev, M.B., Enikolopyan, G.N. and Peunova, N.I. (1982) Transposition of mobile dispersed genetic elements in *Drosophila*. *Dokl. Akad. Nauk SSSR*, **264**, 1495–96.

Ewens, W.J. (1965) Further notes on the evolution of dominance. *Heredity*, **20**, 443–50.

Ewens, W.J. (1966) Linkage and the evolution of dominance. *Heredity*, **21**, 363–70.

Ewens, W.J. (1968) A genetic model having complex linkage behaviour. *Theor. Appl. Genet.*, **38**, 140–43.

Ewens, W.F. (1979) *Mathematical Population Genetics*, Springer, New York.

Fayed, A.H., Mandour, A.S. and Ismail, M.A. (1984) Variation in chiasma frequency in various types of *Vicia faba* L. *Indian J. Genet.*, **44**, 460–64.

Fedoroff, N., Furtek, D. and Nelson, O. (1984) Cloning of the *bronze* locus in maize by a simple generalized procedure using the transposable element *Ac*. *Proc. Natl Acad. Sci. USA*, **81**, 3825–29.

Feldman, M.W. (1972) Selection for linkage modification. I. Random mating populations. *Theor. Pop. Biol.*, **3**, 324–46.

Feldman, M.W. and Balkau, B. (1973) Selection for linkage modification. II. A recombination balance for neutral modifiers. *Genetics*, **74**, 713–26.

Feldman, M.W. and Christiansen, F.B. (1975) The effect of population subdivision on two loci without selection. *Genet. Res.*, **24**, 151–62.

Feldman, M.W., Christiansen, F.B. and Brooks, L.D. (1980) Evolution of recombination in a constant environment. *Proc. Natl Acad. Sci. USA*, **77**, 4838–41.

Feldman, M.W. and Eshel, I. (1982) On the theory of parent-offspring conflict: a two-locus genetic model. *Am. Nat.*, **119**, 285–92.

Feldman, M.W. and Krakauer, J. (1976) Genetic modification, and modifier polymorphisms, in *Population Genetics and Ecology* (eds S. Karlin and E. Nevo), Academic Press, New York, pp. 547–83.

Feldman, M.W. and Liberman, U. (1979) On the number of stable equilibria and the simultaneous stability of fixation and polymorphism in two locus models. *Genetics*, **92**, 1355–60.

Feldman, M.W. and Otto, S.P. (1989) More on recombination and selection in the modifier theory of sex-ratio distortion. *Theor. Pop. Biol.*, **35**, 207–25.

Feldman, M.W. and Otto, S.P. (1991) A comparative approach to the population-genetics theory of segregation distortion. *Am. Nat.*, **137**, 443–56.

Feldmann, K.A., Marks, M.D., Christiansen, M.L. and Quatrano, R.S. (1989) A dwarf mutant of *Arabidopsis* generated by T-DNA insertion mutagenesis. *Science*, **243**, 1351–54.

Felsenstein, J. (1974) The evolutionary advantage of recombination. *Genetics*, **78**, 737–56.

Felsenstein, J. (1976) The theoretical population genetics of variable selection and migration. *Annu. Rev. Genet.*, **10**, 253–80.

Felsenstein, J. (1979) A mathematically tractable family of genetic mapping functions with different amounts of interference. *Genetics*, **91**, 769–75.

Felsenstein, J. (1985) Recombination and sex: is Maynard Smith necessary?, in *Evolution: Essays in Honour of John Maynard Smith* (eds P.J. Greenwood, P.H. Harvey and M. Slatkin), Cambridge University Press, Cambridge, pp. 209–19.

Felsenstein, J. (1988) Sex and the evolution of recombination, in *The Evolution of Sex: An Examination of Current Ideas* (eds R.E. Michod and B.R. Levin), Sinauer, Sunderland, MA, pp. 74–86.

Felsenstein, J. and Yokoyama, S. (1976) The evolutionary advantage of recombination. Individual selection for recombination. *Genetics*, **83**, 845–59.

Ferrari, J.A. and Georghiou, G.P. (1991) Quantitative genetic variation of esterase activity associated with a gene amplification in *Culex quinquefasciatus. Heredity*, **66**, 265–72.

Finnegan, D.J. (1989) Eukaryotic transposable elements and genome evolution. *Trends Genet.*, **5**, 103–07.

Fisher, R.A. (1918) The correlation between relatives on the supposition of Mendelian inheritance. *Trans. R. Soc. Edin.*, **53**, 399–433.

Fisher, R.A. (1930) *The Genetical Theory of Natural Selection*, Clarendon Press, Oxford.

Fisher, R.A. (1946) A system of scoring linkage data with special reference to the *pied* factors in mice. *Am. Nat.*, **80**, 568–78.

Fisher, R.A. (1948a) A quantitative theory of genetic recombination and chiasma formation. *Biometrics*, **4**, 1–13.

Fisher, R.A. (1948b) *Statistical Methods for Research Workers*, 10th edn, Oliver & Boyd, London.

Fisher, R.A. (1949) *The Theory of Inbreeding*, Oliver & Boyd, London.

Flamm, W.G. (1972) Highly repetitive sequences of DNA in chromosomes. *Int. Rev. Cytol.*, **32**, 1–50.

Flavell, R.B. (1986) Repetitive DNA and chromosome evolution in plants. *Phil. Trans. R. Soc. London B*, **312**, 227–42.

Fletcher, H.L. (1988) Sex and recombination. *Nature*, **331**, 492.

Fletcher, H.L. and Hewitt, G.M. (1980) A comparison of chiasma frequency and distribution between sexes in three species of grasshoppers. *Chromosoma*, **77**, 129–44.

Flexon, P.B. and Rodell, C.F. (1982) Genetic recombination and directional selection for DDT resistance in *Drosophila melanogaster. Nature*, **298**, 672–75.

Flor, H. (1956) The complementary genic systems in flax and flax rust. *Adv. Genet.*, **8**, 29–54.

Fogel, S., Welch, J.W., Cathala, G. and Karin, M. (1983) Gene amplification in yeast: *CUP1* copy number regulates copper resistance. *Curr. Genet.*, **7**, 347–55.

Fogle, H.W. and Currence, T.M. (1950) Inheritance of fruit weight and earliness in a tomato cross. *Genetics*, **35**, 363–75.

Fogwill, M. (1958) Differences in crossing over and chromosome size in the sex cells of *Lilium* and *Fritillaria. Chromosoma*, **9**, 493–504.

Fontdevila, A., Zapata, C., Alvarez, G. *et al.* (1983) Genetic coadaptation in the chromosomal polymorphism of *Drosophila subobscura*. I. Seasonal changes of gametic disequilibrium in a natural population. *Genetics*, **105**, 935–55.

Ford, E.B. (1971) *Ecological Genetics*, 3rd edn, Chapman & Hall, London.

Foss, E., Lande, R., Stahl, F.W. and Steinberg, C.M. (1993) Chiasma interference as a function of genetic distance. *Genetics*, **133**, 681–91.

Foster, G.G. and Whitten, M.J. (1991) Meiotic drive in *Lucilia cuprina* and chromosomal evolution. *Am. Nat.*, **137**, 403–15.

Foster, G.G., Whitten, M.J., Konovalov, C. *et al.* (1981) Autosomal genetic maps of the Australian sheep blowfly, *Lucilia cuprina dorsalis* R.-D. (*Diptera: Calliphoridae*), and possible correlations with the linkage maps of *Musca domestica* L. and *Drosophila melanogaster* (Mg). *Genet. Res.*, **37**, 55–69.

Foster, P.L., Cairns, J. (1992) Mechanisms of directed mutation. *Genetics*, **133**, 783–89.

Fox, M. (1984) Gene amplification and drug resistance. *Nature*, **307**, 212–13.

Franklin, J.R. and Feldman, M.W. (1977) Two loci with two alleles: linkage equilibrium and linkage disequilibrium can be simultaneously stable. *Theor. Pop. Biol.*, **12**, 95–113.

Franklin, J.R. and Lewontin, R.C. (1970) Is the gene the unit of selection? *Genetics*, **65**, 707–34.

Friedman, B.E., Bouchard, R.A. and Stern, H. (1982) DNA sequences repaired at pachytene exhibit strong homology among distantly related higher plants. *Chromosoma*, **87**, 409–24.

Gale, M.D. and Rees, H. (1970) Genes controlling chiasma frequency in *Hordeum*. *Heredity*, **25**, 393–410.

Ganal, M.W., Young, N.D. and Tanksley, S.D. (1989) Pulsed field gel electrophoresis and physical mapping of large DNA fragments in the *Tm-2a* region of chromosome 9 in tomato. *Mol. Gen. Genet.*, **215**, 395–400.

Gebhardt, C., Ritter, E., Debener, T. *et al.* (1989) RFLP analysis and linkage mapping in *Solanum tuberosum*. *Theor. Appl. Genet.*, **78**, 65–75.

Gebhardt, C., Ritter, E., Barone, A. *et al.* (1991) RFLP maps of potato and their alignment with homologous tomato genome. *Theor. Appl. Genet.*, **83**, 49–57.

Geiringer, H. (1944) On the probability theory of linkage in Mendelian heredity. *Ann. Math. Stat.*, **15**, 25–37.

Geldermann, H., Pieper, U. and Roth, B. (1985) Effects of marked chromosome sections on milk performance in cattle. *Theor. Appl. Genet.*, **70**, 138–46.

Geodakyan, V.A. (1987) *Evolutionary logic of sex differentiation during phylogenesis and ontogenesis*, Institute of Developmental Biology, Moscow. Dr. Sci. Thesis.

Georges, M., Lathrop, M., Hilbert, P., *et al.* (1990) On the use of DNA fingerprints for linkage studies in cattle. *Genomics*, **6**, 641–74.

Georgiev, G.P., Ilyin, Y.V., Ryskov, A.P. *et al.* (1977) Isolation of eukaryotic DNA fragments containing structural genes and the adjacent sequences. *Science*, **195**, 394–97.

Georgievsky, A.B. (1981) About elementary evolutionary phenomenon, in *Problems of the Modern History of Evolutionary Theory* (ed. Ya. M. Gall), Nauka, Leningrad, pp. 46–55.

Gershon, E.S., Martinez, M., Goldin, L.N., Gejman, P.V. (1990) Genetic mapping of common diseases: the challenges of manic–depressive illness and schizophrenia. *Trends Genet.*, **6**, 282–87.

Geyer, P.K., Richardson, K.L., Corces, V.G., Green, M.M. (1988) Genetic instability in *Drosophila melanogaster*: *P*-element mutagenesis by gene conversion. *Proc. Natl Acad. Sci. USA*, **85**, 6455–59.

Ghiselin, M. (1988) The evolution of sex: a history of competing points of view, in *The Evolution of Sex: An Examination of Current Ideas* (eds R.E. Michod and B.R. Levin), Sinauer, Sunderland, MA, pp. 7–23.

Gibson, J.B. and Thoday, I.M. (1962) An apparent 20 map-unit position effect. *Nature*, **196**, 65–70.

Gilbert, D.A. (1978) Why genes in pieces? *Nature*, **271**, 501.

Gilbert, D.A., Lehman, N., O'Brien, S.J. and Wayne, R.K. (1990) Genetic fingerprinting reflects population differentiation in the California Channel Island fox. *Nature*, **344**, 764–67.

Giles, N. (1978) The organization, function and evolution of gene clusters in eucaryotes. *Am. Nat.*, **112**, 641–57.

Gillespie, J.H. (1976) A general model to account for enzyme variation in natural populations. II. Characterization of the fitness functions. *Am. Nat.*, **110**, 800–21.

Gillespie, J.H. (1981a) The role of migration in the genetic structure of populations in temporally and spatially varying environments. III. Migration modification. *Am. Nat.*, **117**, 223–33.

Gillespie, J.H. (1981b) Mutation modification in a random environment. *Evolution*, **35**, 468–76.

Gillespie, J.H. (1986) Natural selection and the molecular clock. *Mol. Biol. Evol.*, **3**, 138–55.

Gillespie, J.H. and Turelli, M. (1989) Genotype–environment interactions and the maintenance of polygenic variation. *Genetics*, **121**, 129–38.

Gillies, C.B. (1981) Electron microscopy of spread maize pachytene synaptonemal complexes. *Chromosoma*, **83**, 575–91.

Gillies, C.B. (1985) An electron microscopic study of synaptonemal complex formation at zygotene in rye. *Chromosoma*, **92**, 165–75.

Gillois, M. (1991) Gene mapping today: applications to farm animals. *Genet. Sel. Evol.*, **23**, 19–48.

Ginsburg, E. Kh. (1984) *Formal Description of Quantitative Trait Inheritance*, Nauka, Novosibirsk.

Ginsburg, E.Kh., Fedotov, A.M., Chepkasov, I.L. (1986) *A Program Package for Testing Major-Gene Models of Quantitative Traits (MAN-1)*, ICG Press, Novosibirsk.

Ginsburg, E. Kh. and Mullokandov, M.P. (1987) *A Program Package for Linkage Analysis of Qualitative Traits (REK-1)*, ICG Press, Novosibirsk.

Ginsburg, E.Kh., Kolpakov, V.G., Fedotov, A.M. *et al.* (1987) On the possibility of major-gene control of scale *E* of Cattel's inventory. *Genetika* (USSR), **23**, 2157–68.

Ginzburg, L.R., Akcakaya, H.R. and Kim, J. (1988) The evolution of community structure: competition. *J. Theor. Biol.*, **133**, 513–23.

Giovannoni, J.J., Wing, R.A., Ganal, M.W., Tanksley, S.D. (1991) Isolation of molecular markers from specific chomosomal intervals using DNA pools from existing mapping populations. *Nucleic Acids Res.*, **19**, 6553–58.

Glaser, V.H.-J. (1985) Zur Bedeutung der sexuellen Fortflanzung in der Evolution. *Biol. Zbl.*, **104**, 385–402.

Glesener, R.R. (1979) Recombination in a simulated predator-prey interaction. *Am. Zool.*, **19**, 763–71.

Glotov, N.V. and Tarakanov, V.V. (1985) The reaction norm of genotype and genotype-environment interaction in a natural population. *J. Gen. Biol.* (USSR), **46**, 760–70.

Goddard, M.E. (1992) A mixed model for analyses of data on multiple genetic markers. *Theor. Appl. Genet.*, **83**, 878–86.

Gohil, R.N. and Kaul, R. (1980). Studies on male and female meiosis in Indian *Allium*. I. Four diploid species. *Chromosoma*, **77**, 123–27.

Goldar, D.E. and Fain, P.R. (1988) Models of multilocus recombination: nonrandomness in chiasma number and crossover positions. *Am.J. Hum. Genet.*, **43**, 38–45.

Goldar, D.E., Fain, P.R. and Kimberling, W.J. (1989) Chiasma-based models of multilocus recombination: Increased power for exclusion mapping and gene ordering. *Genomics*, **5**, 283–90.

Golden, J.W., Mulligan, M.E. and Haselkorn, R. (1987) Different recombination site specificity of two developmentally regulated genome rearrangements. *Nature*, **327**, 526–29.

Golding, G.B. and Strobeck, C. (1980) Linkage disequilibrium in a finite population that is partially selfing. *Genetics*, **94**, 777–89.

Goldman, S.L. and Coe, E.H. (1968) Recombination frequency and coincidence in maize. *Maize Genet. Coop. Newsl.*, **42**, 130–33.

Goldway, M., Sherman, A., Zenvirth, D., Arbel, T. and Simchen, G. (1993) A short chromosomal region with major roles in yeast chromosome *III* meiotic disjunction, recombination and double strand breaks. *Genetics*, **133**, 159–69.

Golenberg, E.M., Giannasi, D.E., Clegg, M.T. *et al.* (1990) Chloroplast DNA sequence from a Miocene *Magnolia* species. *Nature*, **344**, 656–58.

Golenberg, E.M. and Nevo, E. (1987) Multilocus differentiation and population structure in a selfer, wild emmer wheat, *Triticum dicoccoides*. *Heredity*, **58**, 451–56.

Golic, K.G. (1991) Site-specific recombination between homologous chromosomes in *Drosophila*. *Science*, **252**, 958–61.

Golubovsky, M.D. (1977) Genome organization of *D. melanogaster*, localization of mutations and their phenotypic diversity, in *Problems of Genetics in Studies on Drosophila* (eds V.V. Khvostova, L.I. Korochkin and M.D. Golubovsky), Nauka, Novosibirsk, pp. 152–203.

Golubovsky, M.D. (1985) Organization of genotype and types of genetic variation in eucaryotes., *Adv. Modern Biol.* (USSR), **3**, 323–39.

Goradia, T.M. and Lange, K. (1990) Multilocus ordering strategies based on sperm typing. *Am. J. Hum. Genet.*, **54**, 49–77.

Goradia, T.M., Lange, K., Miller, P.L. and Nadkarni, P.M. (1992) Fast computation of genetic likelihoods on human pedigree data. *Hum. Hered.*, **42**, 42–62.

Gorlov, I.P., Agulnik, A.I. and Agulnik, S.I. (1987) Localizing genes on the house mouse chromosomes by comparing the genetic map and chiasma distribution pattern. *Genetika* (USSR), **23**, 63–70.

Gorlov, I.P. and Borodin, P.M. (1986) The effect of emotional stress on the frequency of meiotic abnormalities in the mouse males. *Genetika* (USSR), **22**, 1019–24.

Gorlov, I.P., Schuler, L., Bunger, L. and Borodin, P.M. (1992) Chiasma frequency in strains of mice selected for litter size and for high body weight. *Theor. Appl. Genet.*, **84**, 640–42.

Gorodetsky, V., Zhuchenko, A.A. and Korol, A.B. (1990) Efficiency of feedback selection for recombination in *Drosophila*. *Genetika* (USSR), **26**, 1942–52.

Gowen, J.W. (1919) A biometrical study of crossing-over. On the mechanism of crossing-over in the third chromosome of *Drosophila melanogaster*. *Genetics*, **4**, 205–50.

Gowen, M.S. and Gowen, J.W. (1922) Complete linkage in *Drosophila melanogaster*. *Am. Nat.*, **56**, 286–88.

Grafen, A. (1988) A centrosomal theory of the short-term evolutionary maintenance of sexual reproduction. *J. Theor. Biol.*, **131**, 163–73.

Grandbastien, M.-A. (1992) Retroelements in higher plants. *Trends Genet.*, **8**, 103–08.

Graner, A:, Jahoor, A., Schondelmaier, J. *et al.* (1991) Construction of an RFLP map of barley. *Theor. Appl. Genet.*, **83**, 250–56.

Grant, V. (1958) The regulation of recombination in plants. *Cold Spring Harbor Symp. Quant. Biol.*, **23**, 337–63.

Grant, V. (1964) *The Architecture of the Germplasm*, Wiley, New York.

Grant, V. (1966) Block inheritance of viability genes in plant species. *Am. Nat.*, **100**, 591–601.

Grant, V. (1975) *Genetics of Flowering Plants*, Columbia University Press, New York.

Grant, V. (1981) *Plant Speciation*, 2nd edn, Columbia University Press, New York.

Grant, V. (1985) *The Evolutionary Process*, Columbia University Press, New York.

Green, M.M. (1975) Conversion as a possible mechanism of high coincidence values in the centromeric region of *Drosophila*. *Mol. Gen. Genet.*, **39**, 57–66.

Green, P., Lipman, D., Hillier, L.D. *et al.* (1993) Ancient conserved regions in new gene sequences and the protein databases. *Science*, **259**, 1711–16.

Greenblatt, I.M. (1981) Enhancement of crossing-over by the transposable element, Modulator, in maize. *Maydica*, **26**, 133–40.

Grell, R.F. (1978) A comparison of heat and interchromosomal effects on recombination and interference in *Drosophila melanogaster*. *Genetics*, **89**, 65–77.

Grell, R.F. and Day, J.W. (1974) Intergenic recombination, DNA replication and synaptonemal complex formation in the *Drosophila* oocyte, in *Mechanisms in Recombination* (ed. R.F. Grell), Plenum Press, New York, pp. 327–49.

Griffing, B. and Langridge, J. (1963) Factors affecting crossing-over in the tomato. *Aust. J. Biol. Sci.*, **16**, 826–37.

Griffing, B. and Scholl, R.L. (1991) Qualitative and quantitative genetics studies of *Arabidopsis thaliana*. *Genetics*, **129**, 605–09.

Griffiths, D.A. (1973) Maximum likelihood estimation for the beta-binomial distribution and an application to the household distribution of the total number of cases of a disease. *Biometrics*, **29**, 637–48.

Gusella, J.F., Tanzi, R.E., Anderson, M.A. *et al.* (1984) DNA markers for nervous diseases. *Science*, **225**, 1320–26.

Gustafson, J.P., Butler, E. and McIntyre, C.L. (1990) Physical mapping of a low-copy DNA sequence in rye (*Secale cereale* L.). *Proc. Natl Acad. Sci. USA*, **87**, 1899–902.

Haber, J.E. (1992) Mating-type gene switching in *Saccharomyces cerevisiae*. *Trends Genet.*, **8**, 446–51.

Hahn, C.S., Lustig, S. and Strauss, E.G. (1988) Western equine encephalitis virus is a recombinant virus. *Proc. Natl Acad. Sci. USA*, **85**, 5997–6001.

Haigh, J. (1978) The accumulation of deleterious genes in a population – Muller's ratchet. *Theor. Pop. Biol.*, **14**, 257–67.

Haldane, J.B.S. (1919) The combination of linkage values and the calculation of distances between the loci of linked factors. *J. Genet.*, **8**, 299–309.

Haldane, J.B.S. (1922) Sex ratio and unisexual sterility in hybrid animals. *J. Genet.*, **12**, 102–09.

Haldane, J.B.S. (1931) The cytological basis of genetical interference. *Cytologia*, **3**, 54–65.

Haldane, J.B.S. and Waddington, C.H. (1931) Inbreeding and linkage. *Genetics*, **16**, 357–74.

Haley, C.S. and Knott, S.A. (1992) A simple regression method for mapping quantitative trait loci in line crosses using flanking markers. *Heredity*, **69**, 315–24.

Hamilton, W.D. (1980) Sex versus non-sex versus parasite. *Oikos*, **35**, 282–90.

Hamilton, W.D. (1982) Pathogens as causes of genetic diversity in their host populations, in *Population Biology of Infectious Diseases* (eds R.M. Anderson and R.M. May), Springer, Berlin, pp. 269–96.

Hamilton, W.D., Axelrod, R. and Tanese, R. (1990) Sexual reproduction as an adaptation to resist parasites (a review). *Proc. Natl Acad. Sci. USA*, **87**, 3566–73.

Hamrick, J.L. and Holden, L.R. (1979) Influence of microhabitat heterogeneity on gene frequency distribution and gametic phase disequilibrium in *Avena barbata*. *Evolution*, **33**, 523–33.

Hanson, W.D. (1959) Theoretical distribution of lengths of parental gene blocks in the gametes of an F_1 individual. *Genetics*, **44**, 197–209.

Hardison, R. and Miller, W. (1993) Use of long sequence alignment to study the evolution and regulation of mammalian globin gene clusters. *Mol. Biol. Evol.*, **10**, 73–102.

Harinarayana, G. and Murty, B.R. (1971) Cytological regulation of recombination in *Pennisetum* and *Brassica*. *Cytologia*, **36**, 345–448.

Harland, S.C. (1936) The genetical conception of the species. *Biol. Rev.*, **11**, 83–112.

Hartley, D.A., Davies, K.E., Drayana, D. *et al.* (1984) A cytological map of the human *X* chromosome – evidence for non-random recombination. *Nucleic Acids Res.*, **12**, 5277–85.

Hartung, J. (1980) Genome parliaments and sex with the Red Queen, in *Natural Selection and Social Behaviour: Recent Research and Theory* (eds R.D. Alexander and D.W. Tinkle), Chiron Press, New York, pp. 382–402.

Haseman, J.K. and Kupper, L.L. (1979) Analysis of dichotomous response data from certain toxicological experiments. *Biometrics*, **35**, 281–93.

Hastings, A. (1972) Eliminating viability differences in computing recombination percentages. *J. Heredity*, **63**, 129–31.

Hastings, A. (1981) Stable cycling in discrete time genetic models. *Proc. Natl Acad. Sci. USA*, **78**, 7224–25.

Hastings, A. (1985) Four simultaneously stable polymorphic equilibria in two-locus, two-allele models. *Genetics*, **109**, 255–61.

Hastings, A. (1986) Limits to the relationship among recombination, disequilibrium and epistasis in two-locus models. *Genetics*, **113**, 177–85.

Hastings, A. and Hom, C.L. (1990) Multiple equilibria and maintenance of additive genetic variance in a model of pleiotropy. *Evolution*, **44**, 1153–63.

Hastings, I.M. (1991) Germline selection: population genetic aspects of the sexual/asexual life cycle. *Genetics*, **129**, 1167–76.

Hastings, P.J. (1988) Conversion events in fungi, in *Genetic Recombination* (eds R. Kucherlapati and G.R. Smith), American Society for Microbiology, Washington DC, pp. 397–428.

Hawley, R.S. and Marcus, C.H. (1989) Recombinational control of *r*DNA redundancy in *Drosophila*. *Annu. Rev. Genet.*, **23**, 87–120.

Hawley, R.S. (1980) Chromosomal sites necessary for normal levels of meiotic recombination in *Drosophila melanogaster*. I. Evidence for and mapping of the sites. *Genetics*, **94**, 625–46.

Hawley, R.S. (1988) Exchange and chromosome segregation in eucaryotes, in *Genetic Recombination* (eds R. Kucherlapati and G.R. Smith), American Society for Microbiology, Washington DC, pp. 497–527.

Hawley, R.S. and Tartof, K.D. (1983) The effect of *mei-41* on *r*DNA redundancy in *Drosophila melanogaster*. *Genetics*, **104**, 63–80.

Hayman, D.L., Moore, H.D.M. and Evans, E.P. (1988) Further evidence of novel sex differences in chiasma distribution in marsupials. *Heredity*, **61**, 455–58.

Healy, J., Corr, C., DeYoung, J. and Baker, B. (1993) Linked and unlinked transposition of a genetically marked *Dissociation* element in transgenic tomato. *Genetics*, **134**, 571–84.

Hedrick, P.W. (1980) Hitch-hiking: a comparison of linkage and partial selfing. *Genetics*, **94**, 791–808.

Hedrick, P.W. (1982) Genetic hitch-hiking: a new factor in evolution. *BioScience*, **32**, 845–53.

Hedrick, P.W. (1985) Coat variants in cats: gametic disequilibrium between unlinked loci. *J. Heredity*, **76**, 127–31.

Hedrick, P.W. (1986) Genetic polymorphism in heterogeneous environments: a decade later. *Annu. Rev. Ecol. Syst.*, **17**, 535–66.

Hedrick, P.W., Ginevan, M.E. and Ewing, E.P. (1976) Genetic polymorphism in heterogeneous environments. *Annu. Rev. Ecol. Syst.*, **7**, 1–32.

Hedrick, P.W. and Holden, L. (1979) Hitch-hiking: an alternative to coadaptation for the barley and slender wild oat examples. *Heredity*, **41**, 79–86.

Hedrick, P., Jain, S. and Holden, L. (1978) Multilocus systems in evolution. *Evol. Biol.*, **4**, 101–84.

Hedrick, P.W. and Thomson, G. (1988) Maternal–fetal interactions and the maintenance of *HLA* polymorphism. *Genetics*, **119**, 205–12.

Hedrick, P.W., Thomson, G. and Klitz, W. (1986) Evolutionary genetics: *HLA* as an exemplary system, in *Evolutionary Processes and Theory* (eds S. Karlin and E. Nevo), Academic Press, Orlando, FL, pp. 583–606.

Helentjaris, T., Slocum, M., Wright, S. *et al.* (1986) Construction of genetic linkage maps in maize and tomato using restriction fragment length polymorphisms. *Theor. Appl. Genet.*, **72**, 761–69.

Helentjaris, T., Weber, D. and Wright, S. (1988) Identification of the genomic location of duplicate nucleotide sequences in maize by analysis of restriction fragment length polymorphisms. *Genetics*, **118**, 353–63.

Henderson, S.A. (1963) Chiasma distribution at diplotene in a locust. *Heredity*, **18**, 173–190.

Henderson, S.A. (1988) Four effects of elevated temperature on chiasma formation in the locust *Schistocerca gregaria*. *Heredity*, **60**, 387–401.

Herbert, P.D.N. (1987) Genotypic characteristics of cyclic parthenogens and their obligately asexual derivatives, in *The Evolution of Sex and its Consequences* (ed. S.C. Stearns), Birkhauser, Basle, pp. 175–195.

Herzberg-Frankel, O. (1925) Facktorenkoppelungen bei Pflanzen. *Z. Vererbungsl.*, **38**, 324–48.

Hibner, B.L., Burke, W.D., Lecanidou, R. *et al.* (1988) Organization and expression of three genes from the silkmoth early chorion locus. *Dev. Biol.*, **125**, 423–31.

Hill, A.P. (1975) Quantitative linkage: a statistical procedure for its detection and estimation. *Ann. Hum. Genet.*, **38**, 439–48.

Hill, A., Allsop, C., Kwiatkowski, D. *et al.* (1991) Common West African *HLA* antigens are associated with protection from severe malaria. *Nature*, **352**, 595–600.

Hillel, J., Schaap, T., Haberfeld, A. *et al.* (1990) *DNA* fingerprints applied to gene introgression in breeding programs. *Genetics*, **124**, 783–89.

Hilliker, A.J., Appels, R. and Schalet, A. (1980) The genetic analysis of *D. melanogaster* heterochromatin. *Cell,* **21**, 607–19.

Hilliker, A.J. and Trusis-Coulter, S.N. (1987) Analysis of functional significance of linkage group conservation in *Drosophila. Genetics,* **117**, 233–44.

Hinton, C.W. (1966) Enhancement of recombination associated with the *c(3)G* mutant of *Drosophila melanogaster. Genetics,* **53**, 157–64.

Hoeijmakers, J.H.J. (1993a) Nucleotide excision repair I: from *E. coli* to yeast. *Trends Genet.,* **9**, 173–77.

Hoeijmakers, J.H.J. (1993b) Nucleotide excision repair II: from yeast to mammals. *Trends Genet.,* **9**, 211–17.

Hoekstra, R.F. (1982) On the asymmetry of sex: evolution of mating types in isogamous populations. *J. Theor. Biol.,* **98**, 427–51.

Hoekstra, R.F. (1987) The evolution of sexes, in *The Evolution of Sex and its Consequences* (ed. S.C. Stearns). Birkhauser, Basle, pp. 59–91.

Hoffmann, A.A. and Parsons, P.A. (1991) *Evolutionary Genetics and Environmental Stress,* Oxford University Press, Oxford.

Holden, L.R. (1979) New properties of the two-locus partial selfing model with selection. *Genetics,* **93**, 217–36.

Holliday, R. (1977) Recombination and meiosis. *Phil. Trans. R. Soc. London B,* **277**, 359–70.

Holliday, R. (1984) The biological significance of meiosis, in *Controlling Events in Meiosis* (eds C.E. Evans and H.C. Dickenson), S.E.B. Symposia 38, Cambridge University Press, Cambridge.

Holliday, R. (1987) The inheritance of epigenetic defects. *Science,* **238**, 163–70.

Holliday, R. (1988) A possible role for meiotic recombination in germ line reprogramming and maintenance, in *The Evolution of Sex: An Examination of Current Ideas* (eds R.E. Michod and B.R. Levin), Sinauer, Sunderland, MA, pp. 45–55.

Holmquist, G.P. (1992) Chromosome bands, their chromatin flavors, and their functional features. *Am. J. Hum. Genet.,* **51**, 17–37.

Holsinger, K.E. (1988) Inbreeding depression doesn't matter: the genetic basis of mating-system evolution. *Evolution,* **42**, 1235–44.

Holsinger, K.E. and Feldman, M.W. (1983a) Modifiers of mutation rate: Evolutionary optimum with complete selfing. *Proc. Natl. Acad. Sci. USA,* **80**, 6732–34.

Holsinger, K.E. and Feldman, M.W. (1983b) Linkage modification with mixed random mating and selfing: a numerical study. *Genetics,* **103**, 323–33.

Holsinger, K.E., Feldman, M.W. and Altenberg, L. (1986) Selection for increased mutation rates with fertility differences between matings. *Genetics,* **112**, 909–22.

Hormaza, J.I. and Herrero, M. (1992) Pollen selection. *Theor. Appl. Genet.,* **83**, 663–72.

Hotta, Y., Chandley, A.C. and Stern, H. (1977) Meiotic crossing-over in lily and mouse. *Nature,* **269**, 240–42.

Hotta, Y. and Stern, H. (1984) The organization of DNA segments undergoing repair synthesis during pachytene. *Chromosoma,* **89**, 127–37.

Hubby, J.L. and Lewontin, R.C. (1966) A molecular approach to the study of genic heterozygosity in natural populations. I. The number of alleles at different loci in *Drosophila pseudoobscura. Genetics,* **54**, 577–94.

Hudson, R.R., Kreitman, M. and Aguade, M. (1987) A test of natural molecular evolution based on nucleotide data. *Genetics,* **116**, 153–59.

Hughes, A.L., Hughes, M.K. and Watkins, D.L. (1993) Contrasting roles of interallelic recombination at the *HLA-A* and *HLA-B* loci. *Genetics*, **133**, 669–80.

Hughes, A.J. and Lambert, D.M. (1984) Functionalism, structuralism, and 'ways of seeing'. *J. Theor. Biol.*, **111**, 787–800.

Hunkapiller, T., Huang, H., Hood, L. and Campbell, J.H. (1982) The impact of modern genetics on evolutionary theory, in *Perspectives in Evolution* (ed. R. Milkman), Sinauer, Sunderland, MA, pp. 164–89.

Hurst, G.D.D., Hurst, L.D. and Johnstone, R.A. (1992) Intranuclear conflict and its role in evolution. *Trends Ecol. Evol.*, **7**, 373–78.

Huskel, G. (1954) Correlated responses to polygenic selection in animals and plants. *Am. Nat.*, **88**, 5–19.

Hutson, B.V. and Law, R. (1981) Evolution of recombination in populations experiencing frequency-dependent selection with time delay. *Proc. R. Soc. London B*, **213**, 345–59.

Huxley, J.S. (1942) *Evolution. The Modern Synthesis*, Allen & Unwin, London.

Ikeda, H. and Kobayashi, I. (1977) Involvement of DNA-dependent RNA polymerase in a *recA*-independent pathway of genetic recombination in *Escherichia coli*. *Proc. Natl Acad. Sci. USA*, **74**, 3932–36.

Imai, H.T., Maruyama, T., Gojobory, Y. *et al.* (1986) Theoretical basis for karyotype evolution. I. The minimum-interaction hypothesis. *Am. Nat.*, **128**, 900–20.

Immer, F.R. (1930) Formulae and tables for calculating linkage intensities. *Genetics*, **15**, 81–98.

Inge-Vechtomov, S.G., Repnevskaya, M.V. and Karpova, T.O. (1987) On the possible commonness of mutation and modification mechanisms, in *Ecological Genetics and Evolution* (ed. A.A. Zhuchenko), Shtiintsa, Kishinev, pp. 103–16.

Inge-Vechtomov, S.G. and Soydla, T.P. (1978) Evolutionary aspects of the dominance problem and molecular interaction. *Achievements of Science and Engineering*. Series of General Genetics, VINITI Press, Moscow, pp. 7–37.

Ishii, K., Matsuda, H., Iwasa, I. and Sasaki, A. (1989) Evolutionary stable mutation rate in a periodically changing environment. *Genetics*, **121**, 163–74.

Ito, M., Takegami, M.H. and Noda, S. (1983) Achiasmate meiosis in the *Fritillaria japonica* group. II. Formation of synaptonemal complexes in microsporocytes. *Jpn J. Genet.*, **58**, 377–81.

Iwasa, I. (1988) Free fitness that always increases in evolution. *J. Theor. Biol.*, **135**, 265–81.

Jaenike, J. (1978) An hypothesis to account for the maintenance of sex within populations. *Evol. Theory*, **3**, 191–94.

Jain, S.K. (1990) Variation and selection in plant populations, in *Population Biology* (eds K. Wohrmann and S.K. Jain), Springer, Berlin, pp. 199–230.

Jain, S.K. and Gupta, D. (1971) Observations on chiasma frequency in primitive and evolved varieties of maize and their hybrids. *Maize Genet. Coop. Newsl.*, **45**, 50–52.

Jayakar, S.D. (1970) On the detection and estimation of linkage between a locus influencing a quantitative character and a marker locus. *Biometrics*, **26**, 451–64.

Jeffreys, A.Y., Wilson, V., Thein, S.L. (1985) Hypervariable 'minisatellite' regions in human DNA. *Nature*, **314**, 67–73.

Jensen, J. (1989) Estimation of recombination parameters between a quantitative trait locus (QTL) and two marker gene loci. *Theor. Appl. Genet.*, **78**, 613–18.

Jentsch, S., McGrath, J.P., Varshavsky, A. (1987) The yeast DNA repair gene *RAD6* encodes a ubiquitin-conjugating enzyme. *Nature*, **329**, 131–35.

Jiang, J.C. and Gibson, J.B. (1992) The alcohol dehydrogenase polymorphism in natural populations of *Drosophila melanogaster*: Restriction map variation in the region of the *Adh* locus in populations from two hemispheres. *Heredity*, **68**, 1–14.

John, B. (1990) *Meiosis*, Cambridge University Press, Cambridge.

John, B. and King, M. (1985) The inter-relationship between heterochromatin distribution and chiasma distribution. *Genetica*, **66**, 183–94.

John, B. and Lewis, K.R. (1965) *The Meiotic System*, Springer, New York.

John, B. and Miklos, G.L. (1979) Functional aspects of satellite DNA and heterochromatin. *Int. Rev. Cytol.*, **58**, 1–114.

Johnston, B.H. and Rich, A. (1989) A recombination hotspot from the mouse *MHC* locus containing the sequence $(CAGG)_{17}$ forms Z-DNA despite the absence of strict purine-pyrimidine alternation. *J. Cell. Biochem.*, **13 D** (Suppl.), p. 128.

Jones, G.H. (1967) The control of chiasma distribution in rye. *Chromosoma*, **22**, 69–90.

Jones, G.H. (1987) Chiasmata, in *Meiosis* (ed. P.B. Moens), Academic Press, London, pp. 213–44.

Jones, G.H. and Tease, C. (1981) Meiotic exchange analysis by molecular labelling. *Chrom. Today*, **7**, 114–25.

Jones, J.S. and Harvey, P.H. (1987) It pays to be different. *Nature*, **328**, 575–76.

Juneja, R.K., Sandberg, K., Andersson, L., Gahne, B. (1982) Close linkage between albumin and vitamin *D* binding protein (*Gc*) loci in chicken: a 300 million year old linkage group. *Genet. Res.*, **40**, 95–98.

Justice, M.J., Jenkins, N.A., Copeland, N.G. (1992) Recombinant inbred mouse strains: models for disease study. *Trends Biotech.*, **10**, 120–26.

Kacser, H. (1987) Dominance not inevitable but very likely. *J. Theor. Biol.*, **126**, 505–06.

Kacser, H. and Burns, J.A. (1981) The molecular basis of dominance. *Genetics*, **97**, 639–66.

Kaidanov, L.Z., Bolshakov, V.N., Tsygvintsev, P.N., Gvozdev, V.A. (1991) The sources of genetic variability in highly inbred long-term selected strains of *Drosophila melanogaster*. *Genetica*, **85**, 73–78.

Kale, P.G. (1968) Spontaneous crossing over in the males of *Drosophila ananassae*. Two-way selection for recombination values. *Jpn J. Genet.*, **43**, 27–31.

Kammerer, C.M. and MacCluer, J.W. (1988) Empirical power of three preliminary methods for ordering loci. *Am. J. Hum. Genet.*, **43**, 964–70.

Kamshilov, M.M. (1939) Dominance and selection. *Dokl. Akad. Nauk SSSR*, **22**, 360–63.

Kandel, E.S. and Korol, A.B. (1991) Dynamics of the recombination system in ecologically subdivided populations, in *Ecological Genetics of Plants and Animals* (Abstracts of All-Union Conference) (ed. A.G. Jacote), Shtiintsa, Kishinev, pp. 464–65.

Kapitonov, V.V. and Kolchanov, N.A. (1988) Evolutionary significance of the presence in mobile genetic elements of regulatory sites responding to environment: regulatory site as a trigger, *Genetika* (USSR), **24**, 1696–703.

Kaplan, M.M. and Webster, R.G. (1977) The epidemiology of influenza. *Sci. Am.*, **237**, 88–106.

Karlin, S. (1969) *Equilibrium Behavior of Population Genetics Models with Non-random Mating*, Gordon & Breach, New York.

Karlin, S. (1975) General two-locus selection models: some objectives, results and interpretations. *Theor. Pop. Biol.*, **7**, 364–98.

Karlin, S. (1984) Theoretical aspects of genetic map functions in recombination processes, in *Human Population Genetics* (The Pittsburgh Symposium) (ed. A. Chakravarti), Van Nostrand Reinhold, New York, pp. 209–28.

Karlin, S. (1988) Non-Gaussian phenotypic models of quantitative traits, in *Quantitative Genetics* (eds B.S. Weir, E. Eisen, M. Goodman and G. Namkoong), Sinauer, Sunderland, MA, pp. 123–44.

Karlin, S. and Avivi, H. (1981) Analysis of central equilibria in multilocus systems: a generalized symmetric viability regime. *Theor. Pop. Biol.*, **20**, 241–80.

Karlin, S. and Brendel, V. (1993) Patchiness and correlations in DNA sequences. *Science*, **259**, 677–80.

Karlin, S. and Feldman, M.W. (1970) Linkage and selection: two locus symmetric viability model. *Theor. Pop. Biol.*, **1**, 39–71.

Karlin, S. and Feldman, M.W. (1978) Simultaneous stability of $D=0$ and $D \neq 0$ for multiplicative viabilities at two loci. *Genetics*, **90**, 813–25.

Karlin, S. and Liberman, U. (1979) A natural class of multilocus recombination and related measures of crossover interference. *Adv. Appl. Probab.*, **11**, 479–501.

Karlin, S. and McGregor, J. (1974) Towards a theory of the evolution of modifier genes. *Theor. Pop. Biol.*, **5**, 59–103.

Karlin, S., Williams, P.T. and Carmelli, D. (1981) Structured exploratory data analysis (SEDA) for determining mode of inheritance of quantitative traits. I. Simulation studies on the effect of background distributions. *Am. J. Hum.Genet.*, **33**, 262–81.

Karp, A. and Jones, R.N. (1982) Cytogenetics of *Lolium perenne*. *Theor. Appl. Genet.* **62**, 177–83.

Karp, M.L. (1940) Inzuht and heterosis. I. The influence of intrachromosome and interchromosome interaction on inzuht-depression and heterosis. *Izv. Akad. Nauk SSSR*, **2**, 219–50.

Kashi, Y., Tickochinsky, Y., Genislav, E. *et al.* (1990) Large restriction fragments containing poly-*TG* are highly polymorphic in a variety of vertebrates. *Nucleic Acids Res.*, **18**, 1129–32.

Keats, B.J.B., Morton, N.E. and Rao, D.C. (1981) Reduction of physical assignments to a standard lod table: chromosome 1. *Hum. Genet.*, **56**, 353–59.

Keim, P., Diers, B.W. and Shoemaker, R.C. (1990). Genetic analysis of soybean hard seededness with molecular markers. *Theor. Appl. Genet.*, **79**, 465–69.

Keim, P., Diers, B.W., Olson, T. and Shoemaker, R.C. (1990) RFLP mapping in soybean: association between marker loci and variation in quantitative traits. *Genetics*, **126**, 735–42.

Kelley, S.E., Antonovics, J. and Schmitt, J. (1988) A test of the short-term advantage of sexual reproduction. *Nature*, **331**, 714–16.

Kempthorne, O. (1988) An overview of the field of quantitative genetics, in *Quantitative Genetics* (eds B.S. Weir, E. Eisen, M. Goodman and G. Namkoong), Sinauer Associates, Sunderland, MA, pp. 47–55.

Kendall, M.G. and Stuart, A. (1963) *Distribution Theory*, Vol. 1, Charles Griffin, London.

Kendall, M.G. and Stuart, A. (1967) *Inference and Relationship*, Vol. 2, Charles Griffin, London.

Khesin, R.B. (1984) *Inconstancy of the Genome*, Nauka, Moscow.

Kidwell, M.G. (1972) Genetic change of recombination value in *Drosophila melanogas-*

ter. I. Artificial selection for high and low recombination and some properties of recombination modifying genes. *Genetics*, **70**, 419–32.

Kidwell, M.G., Kidwell, J.F. and Sved, J.A. (1977) Hybrid dysgenesis in *Drosophila melanogaster*: a syndrome of aberrant traits including mutation, sterility, and male recombination. *Genetics*, **86**, 813–33.

Kimura, M. (1967) On the evolutionary adjustment of spontaneous mutation rates. *Genet. Res.*, **9**, 23–34.

Kimura, M. (1983) *The Neutral Theory of Molecular Evolution*, Cambridge University Press, Cambridge.

Kimura, M. (1991) Recent development of the neutral theory viewed from the Wrightian tradition of theoretical population genetics. *Evolution*, **88**, 5969–73.

King, M. and Hayman, D. (1978) Seasonal variation of chiasma frequency in *Phyllodactylus marmoratus* (Gray) (*Gekkonidae-Reptilia*). *Chromosoma*, **69**, 132–54.

Kirkpatrick, M. (1986) The handicap mechanism of sexual selection does not work. *Am. Nat.*, **127**, 222–40.

Kirkpatrick, M. and Jenkins, C.D. (1989) Genetic segregation and the maintenance of sexual reproduction. *Nature*, **339**, 300–01.

Kirzhner, V.M. and Lyubich, Yu.I. (1974) An evolutionary equation and a limit theorem for general genetic systems without selection. *Dokl. Akad. Nauk SSSR*, **215**, 776–79.

Kirzhner, V.M., Korol, A.B. and Ronin, Y.I. (1993) The dynamics of linkage disequilibrium under temporal environmental fluctuations. Two-locus selection, in *Evolution 93*, Abstracts of the 4th Congress of the European Society for Evolutionary Biology, University of Montpellier, Montpellier, p. 219.

Kirzhner, V.M., Korol, A.B., Ronin, Y.I. and Nevo, E. (1993) Cyclical like behavior of genotype frequencies in a two-locus population under fluctuating haploid selection. *Proc. Natl Acad. Sci. USA* (submitted).

Klein, T.M., Harper, E.C., Svab, Z. *et al.* (1988) Stable genetic transformation of intact *Nicotiana* cells by the particle bombardment process. *Proc. Natl Acad. Sci. USA*, **85**, 8502–05.

Kliman, R.M. and Hey, J. (1993) DNA sequence variation at the *period* locus within and among species of the *Drosophila melanogaster* complex. *Genetics*, **133**, 375–87.

Klitz, W. and Thomson, G. (1987) Disequilibrium pattern analysis. II. Application to Danish *HLA-A* and *B* locus data. *Genetics*, **116**, 633–43.

Kmiec, E.B. and Holloman, W.K. (1986) Homologous pairing of DNA molecules by *Ustilago Recl* protein is promoted by sequences of Z-DNA. *Cell*, **44**, 545–54.

Knapp, S.J., Bridges, W.C. and Birkes, D. (1990) Mapping quantitative trait loci using molecular marker linkage maps. *Theor. Appl. Genet.*, **79**, 583–92.

Knibb, W.R. and Baker, J.S.F. (1988) Polymorphic inversion and Esterase complex loci on chromosome 2 of *Drosophila buzzatii*. II. Spatial variation. *Aust. J. Biol. Sci.*, **41**, 239–46.

Knott, S.A. and Haley, C.S. (1992) Aspects of maximum likelihood methods for mapping of quantitative trait loci in line crosses. *Genet. Res.*, **60**, 139–51.

Knott, S.A., Haley, C.S. and Thompson, R. (1992a) Methods of segregation analysis for animal breeding data: a comparison of power. *Heredity*, **68**, 299–312.

Knott, S.A., Haley, C.S. and Thompson, R. (1992b) Methods of segregation analysis for animal breeding data: parameter estimates. *Heredity*, **68**, 313–20.

Kobori, J.A., Strauss, E., Minard, K. and Hood, L. (1986) Molecular analysis of the hotspot of recombination in the murine major histocompatibility complex. *Science*, **234**, 173–79.

Koella, J.C. (1993) Ecological correlates of chiasma frequency and recombination index of plants. *Biol. J. Linnean Soc.*, **48**, 227–38.

Koltin, Y., Stamberg, J. and Ronen, R. (1975) Meiosis as a source of spontaneous mutations in *Schizophyllum commune*. *Mutat. Res.*, **27**, 319–25.

Kondrashov, A.S. (1982) Selection against harmful mutations in large sexual and asexual populations. *Genet. Res.*, **40**, 325–32.

Kondrashov, A.S. (1984) Deleterious mutations as an evolutionary factor. I. The advantage of recombination. *Genet. Res.*, **44**, 199–217.

Kondrashov, A.S. (1988a) Deleterious mutations and the evolution of sexual reproduction. *Nature*, **336**, 435–40.

Kondrashov, A.S. (1988b) Deleterious mutations as an evolutionary factor. III. Mating preference and some general remarks. *J. Theor. Biol.*, **131**, 487–96.

Kondrashov, A.S. and Crow, J.F. (1991) Haploidy or diploidy: which is better? *Nature*, **351**, 314–15.

Kopnin, B.P. and Gudkov, A.V. (1983) Amplification of genome sections in somatic cells of mammals resistant to colchicine. 4. Gene amplification induction in Djungarian hamster and mouse cells. *Genetika* (USSR), **19**, 872–80.

Kordyum, V.A. (1982) *Evolution and Biosphere*, Naukova Dumka, Kiev.

Korochkin, L.I. (1977) *Gene Interaction in Development*, Nauka, Moscow.

Korochkin, L.I. (1981) Physiological and biochemical phylogenetics, in *Physiological Genetics and Genetics of Behaviour*, Nauka, Moscow, pp. 95–155.

Korochkin, L.I. (1983) The evolutionary significance of mobile genetic elements: a hypothesis. *Tsytol. Genet.* (USSR), **4**, 67–75.

Korol, A.B. (1973) Analysis of genetic processes in tomato hybridization, in *Variability and Inheritance of Tomato Agronomic Characters* (ed. A.A. Zhuchenko), Cartya Moldovenyasca, Kishinev, pp. 362–414.

Korol, A.B. (1976) Analysis of Genetic Processes in Tomato Hybridization. Institute of General Genetics, Moscow. PhD Thesis.

Korol, A.B. (1987) Variability of Crossing Over in Higher Organisms. Leningrad State University, Leningrad, Dr Sci Thesis.

Korol, A.B. (1988) Increasing genetic variability available to selection, in *Ecological Genetic Basis of Tomato Breeding* (ed. V.N. Lysikov), Shtiintsa, Kishinev, pp. 213–46.

Korol, A.B. and Bocharnikova, N.I. (1988) Recombination and interspecies gene transfer, in *Interspecific Hybridization in Plants*. Proceedings of the International Symposium (eds J. Georghieva-Todorova and R. Pandeva), Bulgarian Institute of Genetics, Sofia, pp. 14–24.

Korol, A.B. and Iliadi, K.G. (1994) Recombination increase resulting from directional selection for geotaxis in *Drosophila*. *Heredity*, **72**, 64–8.

Korol, A.B., Kirzhner, V.M. and Ronin, Y.I. (1993) Polymorphism caused by stabilizing selection with cyclically changing optimum, in *Evolution 93*, Abstracts of the 4th Congress of the European Society for Evolutionary Biology, University of Montpellier II, Montpellier, p. 229.

Korol, A.B. and Mustyatsa, I.V. (1991) Density-dependent selection as a factor in the evolution of recombination, in *Ecological Genetics of Plants and Animals* (Abstracts of All-Union Conference) (ed. A.G. Jacote), Shtiintsa, Kishinev, pp. 467–68.

Korol, A.B. and Preygel, I.A. (1986) Selection for increased recombination under cyclical environmental changes. A multilocus model, in *Recombinogenesis: Its Significance for Evolution and Breeding* (Conference Proceedings) (ed. A.A. Zhuchenko), Shtiintsa, Kishinev, pp. 160–67.

Korol, A.B. and Preygel, I.A. (1988a) Using marker loci for analyzing quantitative trait complexes, in *Biometrical Genetics Methods in Plant Breeding* (ed. J.J. Vozhda), Mendelium Institute, Lednice (Brno), pp. 123–31.

Korol, A.B. and Preygel, S.I. (1988b) Host–parasite interaction as a factor of genetic recombination evolution. *Genetika* (USSR), **24**, 1113–20.

Korol, A.B. and Preygel, I.A. (1989) Increase in recombination in a multilocus system under environmental fluctuations. *Genetika* (USSR), **25**, 923–31.

Korol, A.B., Preygel, I.A. and Bocharnikova, N.I. (1987) Linkage between loci of quantitative traits and marker loci. 5. Simultaneous analysis of a set of marker and quantitative traits. *Genetika* (USSR), **23**, 1421–31.

Korol, A.B., Preygel, I.A. and Nyutin, Yu.I. (1989) Quasi-linkage in tomato: formal analysis. *Genetika* (USSR), **25**, 2225–33.

Korol, A.B., Preygel, I.A. and Preygel, S.I. (1990). *Variability of Crossing Over in Higher Organisms: Algorithms of Estimation and Population Genetic Models.* Shtiintsa Press, Kishinev, 407pp.

Korol, A.B., Preygel, S.I. and Preygel, I.A. (1986) Host–parasite interaction as a factor in evolution of genetic recombination, in '*Synergetics-86*' (Conference Proceedings) (eds V.A. Kovarsky *et al.*), Shtiintsa, Kishinev, pp. 108–10.

Korol, A.B., Ronin, Y.I. and Kirzhner, V.M. (1993) Linkage between loci of quantitative traits and marker loci. 6. Some ways to increase the resolution in single marker analysis (submitted).

Korol, A.B. and Zhuchenko, A.A. (1978) Non-additivity and interference in detecting factors of quantitative traits by the marker method. *Tsytol. Genet.* (USSR), **12**, 550–54.

Korol, A.B. and Zhuchenko, A.A. (1980) Changes in parameters of quantitative trait distributions in the progeny of heterozygotes upon mutagen treatment: formal analysis. *Genetika* (USSR), **16**, 1234–42.

Korol, A.B., Zhuchenko, A.A. and Preygel, I.A. (1983) Linkage between quantitative trait loci and marker loci. 4. The mean squares evaluation of parameters. *Genetika* (USSR), **19**, 594–601.

Korol, A.B., Zhuchenko, A.A. and Samovol, A.P. (1981) Linkage between quantitative trait loci and marker loci. 3. Biasing of estimates resulting from departures from initial assumptions. *Genetika* (USSR), **17**, 1234–47.

Korol, A.B., Kovtyukh, L.P., Preygel, I.A. *et al.* (1984) Genotypic differences in the rate and distribution of exchanges in *Drosophila*, in *Recombinogenesis: Its Significance for Evolution and Breeding* (Conference Proceedings) (ed. A.A. Zhuchenko), Shtiintsa, Kishinev, pp. 126–29.

Kosambi, D.D. (1944) The estimation of recombination values. *Ann. Eugen.*, **12**, 172–75.

Kozhevnikov, B. (1947) Experimental analysis of preferential segregation of non-homologous chromosomes with respect to the nature of forces causing chromosome segregation. *Dokl. Akad. Nauk SSSR*, **158**, 663–66.

Kramer, H.H. and Burnham, C.R. (1947) Methods of combining linkage intensity values from backcross, F_2 and F_3 genetic data. *Genetics*, **32**, 379–90.

Krasilov, V.A. (1979) The role of chance in evolution. *Proc. Inst. Biol. Soil* (Far-eastern Centre of the USSR Academy of Sciences), **52**, 7–19.

Krasilov, V.A. (1986) *Unsolved Problems of the Evolution Theory*, Far-eastern Centre of the USSR Academy of Sciences, Vladivostok.

Kruleva, M.M., Korol, A.B., Dankov, T.G., Scorpan, V. and Preygel, I.A. (1992) The effect of genotype × cytoplasm interaction on meiotic behaviour of maize chromosomes. *Genome*, **34**, 653–58.

Kubai, D.F. (1987) Nonrandom chromosome arrangements in germ line nuclei of *Sciara coprophila* males: the basis for nonrandom chromosome segregation on the meiosis *I* spindle. *J. Cell. Biol.*, **105**, 2433–46.

Kullback, S. (1959) *Information Theory and Statistics*. John Wiley, New York.

Kumar, A. and Gupta, J.P. (1988) Temporal variation and gametic disequilibrium in *Drosophila nasuta*. *Hereditas*, **109**, 113–17.

Kumar, D., Paria, P. and Basak, S.L. (1981) Genotypic control of chromosome behaviour in *Corchorus olitorius* L. *Cytologia*, **46**, 643–47.

Kunzel, G. (1982) Differences between genetic and physical centromere distances in the case of two genes for male sterility in barley. *Theor. Appl. Gènet.*, **64**, 25–29.

Kushev, V.V. (1974) *Mechanisms of Genetic Recombination*, Consultants Bureau, New York.

Ladygina, T.Yu., Gorlov, I.P. and Borodin, P.M. (1989) The influence of double insertions in homogeneously stainable regions in the first house mouse chromosome on recombination frequency. *Genetika* (USSR), **25**, 220–25.

Lai, E. and Birren, B.W. (eds) (1990) *Electrophoresis of Large DNA Molecules. Theory and Applications*, Cold Spring Harbor Laboratory Press, Cold Spring Harbor, NY.

Laing, E. (1972) Disturbed linkage. 1. Partial manifestation and differential viability at the same locus in backcross data. *Heredity*, **29**, 1–10.

Lake, S. (1986) Recombination frequencies and the coincidence in proximal *X*-chromosome regions including heterochromatin in *Drosophila melanogaster*. *Hereditas*, **105**, 263–68.

Lalley, P.A. and McKusick, V.A. (1985) Report of the committee on comparative mapping. *Cyt. Cell Genet.*, **40**, 536–66.

Lalouel, J.M., Rao, D.C., Morton, N.E. and Elston, R.C. (1983) A unified model for complex segregation analysis. *Am. J. Hum. Genet.*, **35**, 816–26.

Lamb, B.C. (1969) Evidence from *Sordaria* that recombination and conversion frequencies are partially determined before meiosis, and a general model of the control of recombination frequencies. *Genetics*, **63**, 807–20.

Lambert, C., Couto, L.B., Weiss, W.A. *et al.* (1988) A yeast DNA repair gene partially complements defective excision repair in mammalian cells. *EMBO J.*, **7**, 3245–53.

Lambie, E.J. and Roeder, G.S. (1988) A yeast centromere acts in *cis* to inhibit meiotic gene conversion of adjacent sequences. *Cell*, **52**, 863–73.

Lande, R. (1976) The maintenance of genetic variability by mutation in a polygenic character with linked loci. *Genet. Res.*, **26**, 221–35.

Lande, R. (1980) The genetic covariance between characters maintained by pleiotropic mutations. *Genetics*, **94**, 203–15.

Lande, R. and Arnold, S.J. (1983) The measurement of selection on correlated characters. *Evolution*, **37**, 1210–26.

Lande, R. and Schemske, D.W. (1985) The evolution of self-fertilization and inbreeding depression in plants. I. Genetic models. *Evolution*, **39**, 24–40.

Lande, R. and Thompson, R. (1990) Efficiency of marker assisted selection in the improvement of quantitative traits. *Genetics*, **124**, 743–56.

Lander, E.S. and Botstein, D. (1986) Strategies for studying heterogeneous genetic traits in humans by using a linkage map of restriction fragment length polymorphisms. *Proc. Natl Acad. Sci. USA*, **83**, 7353–57.

Lander, E.S. and Botstein, D. (1989) Mapping Mendelian factors underlying quantitative traits using RFLP linkage maps. *Genetics*, **121**, 185–99.

Lander, E.S. and Green, P. (1987) Construction of multilocus genetic linkage maps in humans. *Proc. Natl Acad. Sci. USA*, **84**, 2363–67.

Landner, L. (1970) Variation of recombination frequency in *Neurospora crassa* following meiosis and evidence for a premeiotic sensitive stage. *Mol. Gen. Genet.*, **27**, 385–92.

Landner, L. (1971) Genetic control of recombination in *Neurospora crassa*: correlated regulation in unlinked chromosome intervals. *Heredity*, **27**, 385–92.

Landry, B.S., Kesseli, R.V., Farrara, B. and Michelmore, R.W. (1987) A genetic map of lettuce (*Lactuca sativa L.*) with restriction fragment length polymorphism, isozyme, disease resistance and morphological markers. *Genetics*, **116**, 331–37.

Lange, K. and Boehnke, M. (1992) Bayesian methods and optimal experimental design for gene mapping by radiation hybrids. *Am. J. Hum. Genet.*, **56**, 119–44.

Lange, K., Spence, M.A. and Frank, M.B. (1976) Application of the lod method to the detection of linkage between a quantitative trait and a qualitative marker: a simulation experiment. *Am. J. Hum. Genet.*, **28**, 167–73.

Langley, C.H., Ito, K. and Voelker, R.A. (1977) Linkage disequilibrium in natural populations of *Drosophila melanogaster*. Seasonal variation. *Genetics*, **86**, 447–54.

Langley, C.H., Smith, D.B. and Johnson, F.M. (1978) Analysis of linkage disequilibrium between allozyme loci in natural populations of *Drosophila melanogaster*. *Genet. Res.*, **32**, 215–30.

Langley, C.H., Tobari, Y.N. and Kojima, K.J. (1974). Linkage disequilibrium in natural populations of *Drosophila melanogaster*. *Genetics*, **78**, 921–36.

Lathrop, G.M., Lalouel, J.M., Julier, C. and Ott, J. (1984) Strategies for multilocus linkage analysis in humans. *Proc. Natl Acad. Sci. USA*, **81**, 3443–46.

Laurie, C.C., Bridgham, J.T. and Choudhary, M. (1991) Associations between DNA sequence variation and variation in expression of the *Adh* gene in natural populations of *Drosophila melanogaster*. *Genetics*, **129**, 489–99.

Laurie, D.A., Palmer, R.W. and Hulten, M.A. (1982) Chiasma derived genetic length and recombination fractions. *Ann. Hum. Genet.*, **46**, 233–44.

Laurie-Ahlberg, G.C. and Weir, B.S. (1979) Allozymic variation and linkage disequilibrium in some laboratory populations of *Drosophila melanogaster*. *Genetics*, **92**, 1295–314.

Lavie, B. and Nevo, E. (1988) Multilocus genetic resistance and susceptibility to mercury and cadmium pollution in the marine gastropod, *Cerithium scabridum*. *Aqua. Tox.*, **13**, 291–96.

Law, C.N. (1966) The location of genetic factors affecting a quantitative character in wheat. *Genetics*, **53**, 487–98.

Law, C.N. (1967) The location of genetic factors controlling a number of quantitative characters in wheat. *Genetics*, **56**, 445–61.

Law, C.N. (1981) Intraspecific chromosome manipulation. *Phil. Trans. Roy. Soc. Lond. B*, **292**, 509–18.

Law, C.N., Snape, J.W. and Worland, A.J. (1983) Chromosome manipulation and its exploitation in the genetics and breeding of wheat. *Stadler Genet. Symp.*, **15**, 5–23.

Lawrence, M.J. (1958) Genotypic control of crossing-over in the first chromosome of *Drosophila melanogaster. Nature*, **182**, 889–90.

Lawrence, M.J. (1963) The control of crossing-over in the X-chromosome of *Drosophila melanogaster. Heredity*, **18**, 27–46.

Lawrence, M.J., Cornish, M.A. and Hyward, M.D. (1979) Simultaneous estimation of male and female linkage from a single intercross family. *Heredity*, **43**, 107–21.

Lebel, E.G., Masson, J., Bogucki, A. and Paszkowski, J. (1993) Stress-induced intrachromosomal recombination in plant somatic cells. *Proc. Natl Acad. Sci. USA*, **90**, 422–26.

Lebo, R.V., Chakravarti, A., Buetow, K.H. *et al.* (1983) Recombination within and between the human insulin and β-globin gene loci. *Proc. Natl Acad. Sci. USA*, **80**, 4808–12.

Lebowitz, B.J., Soller, M. and Beckmann, J.S. (1987) Trait-based analyses for the detection of linkage between marker loci and quantitative trait loci in crosses between inbred lines. *Theor.Appl. Genet.*, **73**, 556–62.

Leigh, E.G. Jr (1970) Natural selection and mutability. *Amer. Natur.*, **104**, 301–05.

Leigh, E.G. Jr (1973) The evolution of mutation rate. *Genetics*, **73**, Suppl., 1–18.

Lein, V. and Lelley, T. (1987) A separate control for frequency and within-bivalent distribution of chiasmata in rye (*Secale cereale L.). Genome*, **29**, 419–24.

Lenski, R.E. and Mittler, J.E. (1993) The directed mutation controversy and neo-Darwinism. *Science*, **259**, 188–94.

Le Roy, P. and Elsen, J.M. (1992) Simple test statistics for major gene detection: a numerical comparison. *Theor. Appl. Genet.*, **83**, 635–44.

Lessard, S. (1985) The role of recombination in the modifier theory of autosomal segregation distortion. *Theor. Pop. Biol.*, **28**, 133–49.

Lessard, S. (1987) The role of recombination and selection in the modifier theory of sex ratio distortion. *Theor. Pop. Biol.*, **31**, 339–58.

Levin, B.R., Allison, A.C., Bremermann, H.J. *et al.* (1982) Evolution of parasites and hosts, in *Population Biology of Infectious Diseases* (eds R.M. Anderson and R.M. May), Springer, Berlin, pp. 213–243.

Levin, D.A. (1975) Pest pressure and recombination systems in plants. *Am. Nat.*, **109**, 437–51.

Levin, S. and Castillo-Chavez, C. (1990) Topics in evolutionary ecology, in *Mathematical and Statistical Developments of Evolutionary Theory* (ed. S. Lessard), Kluwer, Dordrecht, pp. 327–58.

Levitan, M. and Scheffer, S.J. (1993) Studies of linkage in populations. X. Altitude and autosomal gene arrangements in *Drosophila robusta. Genet. Res.*, **61**, 9–20.

Lewis, W.M. Jr (1983) Interruption of synthesis as a cost of sex in small organisms. *Am. Nat.*, **121**, 825–34.

Lewis, W.M. Jr (1987) The costs of sex, in *The Evolution of Sex and its Consequences* (ed. S.C. Stearns), Birkhauser, Basle, pp. 33–57.

Lewontin, R.C. (1964) The interaction of selection and linkage. I. General considerations; heterotic models. *Genetics*, **49**, 49–67.

Lewontin, R.C. (1971) The effect of genetic linkage on the mean fitness of a population. *Proc. Natl Acad. Sci. USA*, **68**, 984–86.

Lewontin, R.C. (1974) *The Genetic Basis of Evolutionary Change*, Columbia University Press, New York.

Lewontin, R.C. (1977) The relevance of molecular biology to plant and animal breeding in *Proceedings of the International Conference on Quantitative Genetics* (eds E. Pollak, O. Kempthorne and T.B. Bailey Jr), Iowa University Press, Ames,IA, pp. 55–62.

Lewontin, R.C. and Feldman, M.W. (1988) A general asymptotic property of two-locus selection models. *Theor. Pop. Biol.*, **94**, 177–93.

Li, H., Cui, X. and Arnheim, N. (1990) Direct electrophoretic detection of the allelic state of single DNA molecules in human sperm by using the polymerase chain reaction. *Proc. Natl Acad. Sci. USA*, **87**, 4580–84.

Li, Z. and Wu, Q. (1984) A hypothesis on the action of chromosomal genes. *J. Theor. Biol.*, **108**, 645–61.

Liberman, U. and Feldman, M.W. (1986) Modifiers of mutation rate: a general reduction principle. *Theor. Pop. Biol.*, **30**, 125–42.

Liberman, U. and Karlin, S. (1984) Theoretical models of genetic map function. *Theor. Pop. Biol.*, **25**, 331–46.

Lima-de-Faria, A. (1983) *Molecular Evolution and Organization of the Chromosome*, Elsevier Science, Amsterdam.

Limborskaya, C.A. (1989) The human genome, in *Genome Organization* (eds Yu.F. Bogdanov and A.A. Prozorov), Nauka, Moscow, pp. 198–217.

Linde-Laursen, L. (1982) Linkage map of the long arm barley chromosome 3 using C-bands and marker genes. *Heredity*, **49**, 27–35.

Lindgren, D. (1975) Sensitivity of premeiotic and meiotic stages to spontaneous and induced mutation in barley and maize. *Hereditas*, **79**, 227–38.

Lindsley, D.L. and Sandler L. (1977) The genetic structure of meiosis in female *Drosophila melanogaster*. *Phil. Trans. R. Soc. London B*, **277**, 295–312.

Lippert, L.F., Smith, P.G. and Bergh, B.O. (1966) Cytogenetics of the vegetable crops. Garden pepper, *Capsicum* sp. *Bot. Rev.*, **32**, 24–55.

Lloyd, D.G. (1980) Benefits and handicaps of sexual reproduction. *Evol. Biol*, **13**, 69–108.

Lloyd, D.G. (1984) Gene selection of Mendel's rules. *Heredity*, **53**, 613–24.

Lloyd, D.G. (1988) Benefits and costs of biparental and uniparental reproduction in plants, in *The Evolution of Sex: An Examination of Current Ideas* (eds R.E. Michod and B.R. Levin), Sinauer, Sunderland, MA, pp. 233–52.

Lobashev, M.E., Ponomarenko, V.V., Polyanskaya, G.G. and Tsapygina, R.I. (1973) On the role of nervous system in regulation of various genetic and cytological processes. *J. Evol. Biochem.* (USSR), **9**, 398–405.

Loidl, J. (1987) Heterochromatin and differential chromosome staining in meiosis. *Biol. Zbl.*, **106**, 641–62.

Love, J.M., Knight, A.M., McAleer, M.A. and Todd, J.A. (1990) Towards construction of a high resolution map of the mouse genome using PCR-analysed microsatellites. *Nucleic Acids Res.*, **18**, 4123–30.

Lucchesi, J.C. and Suzuki, D.T. (1968) The interchromosomal control of recombination. *Annu. Rev. Genet.*, **2**, 53–86.

Lukaszewski, A.J. (1992) A comparison of physical distribution of recombination in chromosome *1R* in diploid rye and in hexaploid triticale. *Theor. Appl. Genet.*, **83**, 1048–53.

Lukaszewski, A.J. and Curtis, C.A. (1993) Physical distribution of recombination in B-genome chromosomes of tetraploid wheat. *Theor. Appl. Genet.*, **86**, 121–27.

Luo, Z.W. and Kearsey, M.J. (1992) Interval mapping of quantitative trait loci in an F_2 population. *Heredity*, **69**, 236–42.

Luo, Z.W. and Woolliams, J.A. (1993) Estimation of genetic parameters using linkage between a marker gene and a locus underlying a quantitative character in F_2 populations. *Heredity*, **70**, 245–53.

Lynch, M. and Gabriel, W. (1983) Phenotypic evolution and parthenogenesis. *Am. Nat.*, **122**, 745–64.

Lyon, M.F. (1976) Distribution of crossing-over in mouse chromosomes. *Genet. Res.*, **28**, 291–99.

Lyttle, T.W. (1991) Segregation distorters. *Annu. Rev. Genet.*, **25**, 511–57.

Lyttle, T.W. (1993) Cheaters sometimes prosper: distortion of mendelian segregation by meiotic drive. *Trends Genet.*, **9**, 205–10.

Lyubich, Yu.I. (1971) Basic notions and theorems of the evolutionary genetics of free populations. *Uspekhi Matem. Nauk* (USSR), **26**, 51–116.

Lyubich, Yu.I. (1992) *Mathematical Structures in Population Genetics*, Springer, Berlin, 373 pp.

Lyubischev, A.A. (1982) *The Problems of Form, Systematics and Evolution of Organisms*, Nauka, Moscow.

Mackay, T.F.G. (1981) Genetic variation in varying environments. *Genet. Res.*, **37**, 79–93.

Mackay, T.F.G. (1984) Jumping genes meet abdominal bristles: hybrid dysgenesis-induced quantitative variation in *Drosophila melanogaster. Genet. Res.*, **44**, 231–37.

Mackay, T.F.G. (1985) Transposable element-induced response to artificial selection in *Drosophila melanogaster. Genetics*, **111**, 351–74.

Mackay, T.F.G. (1987) Transposable element-induced polygenic mutations in *Drosophila melanogaster. Genet. Res.*, **49**, 225–33.

Maddox, J. (1993) The dark side of molecular biology. *Nature*, **363**, 13.

Magni, G.E. (1964) Origin and nature of spontaneous mutation in meiotic organisms. *J. Cell. Comp. Physiol.*, **1** (Suppl.), 165–71.

Maguire, M.P. (1980) Adaptive advantage for chiasma interference: a novel suggestion. *Heredity*, **45**, 127–31.

Maguire, M.P. (1992) The evolution of meiosis. *J. Theor. Biol.*, **154**, 43–55.

Mahajan, S.K. (1988) Pathways of homologous recombination in *Esherichia coli*, in *Genetic Recombination* (eds R. Kucherlapati and G.R. Smith), American Society for Microbiology, Washington DC, pp. 87–140.

Maletsky, S.I. (1976) Limiting reactions and inheritance of quantitative traits, in *Genetics and Selection of Quantitative Traits*, Naukova Dumka, Kiev, pp. 25–30.

Malinowsky, A.A. (1941) Rules of inheritance viewed in the light of selection theory. *Adv. Modern Biol.* (USSR), **14**, 171–76.

Malinowsky, E. (1927) The hypothesis of chromosome affinity and the phenomenon of suppression of characters on crossing. *J. Genet.*, **18**, 223–31.

Manga, V. and Pantulu, J.V. (1974) Chiasma frequency in homozygotes and heterozygotes of pearl millet. *Proc. Indian Acad. Sci.*, **80**, 105–20.

Mangelsdorf, P.C. (1958) The mutagenic effect of hybridizing maize and teosinte. *Cold Spring Harbor Symp. Quant. Biol.*, **23**, 409–21.

Mangelsdorf, P.C. and Galinat, W.C. (1964) The tunicate locus in maize dissected and reconstituted. *Proc. Natl Acad. Sci. USA*, **51**, 147–52.

Mani, G.S. (1985) Evolution of resistance in the presence of two insecticides. *Genetics*, **109**, 761–83.

Manly, B.F.J. (1985) *The Statistics of Natural Selection*, Chapman & Hall, London.

Manning, J.T. (1982) Hitch-hiking and a short-term advantage for sex. *J. Theor. Biol.*, **98**, 703–05.

Manning, J.T. (1984) Males and the advantage for sex. *J. Theor. Biol.*, **108**, 215–20.

Manning, J.T. and Dickson, D.P.E. (1986) Environmental change, mutational load and the advantage of sexual reproduction. *Acta Biotheor.*, **35**, 149–62.

Marani, A. (1975) Simulation of response of small self-fertilizing populations to selection for quantitative traits: I. Effect of number of loci, selection intensity and initial heritability under conditions of no dominance. *Theor. Appl. Genet.*, **46**, 221–31.

Marshall, D.L. and Ellstrand, N.C. (1986) Sexual selection in *Rafanus sativus*. Experimental data on non-random fertilization, maternal choice and consequences of multiple paternity. *American Naturalist*, **127**, 446–61.

Martin, B., Nienhius, J., King, G. and Schaefer, A. (1989) Restriction fragment length polymorphisms associated with water use efficiency in tomato. *Science*, **243**, 1725–28.

Martin, G.B., Williams, J.G.K. and Tanksley, S.D. (1991) Rapid identification of markers linked to a *Pseudomonas* resistance gene in tomato by using random primers and near-isogenic lines. *Proc. Natl Acad. Sci. USA*, **88**, 2336–40.

Martin, M.J., Perez-Tome, J.M. and Toro, M.A. (1988) Competition and genotypic variability in *Drosophila melanogaster*. *Heredity*, **60**, 119–23.

Martinez, O. and Curnow, R.N. (1992) Estimating the locations and the sizes of the effects of quantitative trait loci using flanking markers. *Theor. Appl. Genet.*, **85**, 480–88.

Mather, K. (1936a) The determination of position in crossing-over. I. *Drosophila melanogaster*. *J. Genet.*, **33**, 207–35.

Mather, K. (1936b) Competition between bivalents during chiasma formation. *Proc. R. Soc. London B*, **120**, 208–27.

Mather, K. (1943) Polygenic inheritance and natural selection. *Biol. Rev.*, **18**, 32–64.

Mather, K. (1953) The genetical structure of populations. *Symp. Soc. Exp. Biol.*, **7**, 66–95.

Mather, K. and Harrison, B.J. (1949) The manifold effects of selection. *Heredity*, **3**, 1–52, 131–62.

Mather, K. and Jinks, J.L. (1971) *Biometrical Genetics*, 2nd edn, Chapman & Hall, London.

Mather, K. and Lawrence, M.J. (1979) An intercross family with an unusual property. *Heredity*, **43**, 123–28.

Matsuda, M. and Tobari, Y.N. (1987) A new enhancer locus, *En(2)-hn*, and a new allele of the enhancer, *En(2)-cc*, of male crossing-over in *Drosophila ananassae*. *Jpn J. Genet.*, **62**, 217–24.

May, R.M. (1974) Biological populations with non-overlapping generations: Stable points, stable cycles and chaos. *Science*, **186**, pp. 645–47.

May, R.M. (1987) Chaos and the dynamics of biological populations. *Proc. R. Soc. London A*, **413**, 27–44.

May, R.M. and Anderson, R.M. (1983) Epidemiology and genetics in the coevolution of parasites and hosts. *Proc. R. Soc. London B*, **219**, 281–313.

Maynard Smith, J. (1977) Why the genome does not congeal. *Nature*, **268**, 693–96.

Maynard Smith, J. (1978) *The Evolution of Sex*, Cambridge University Press, Cambridge.

Maynard Smith, J. (1979) The effects of normalizing and disruptive selection on genes for recombination. *Genet. Res.*, **33**, 121–28.

Maynard Smith, J. (1980) Selection for recombination in a polygenic model. *Genet. Res.*, **35**, 269–77.

Maynard Smith, J. (1982) Overview – unsolved evolutionary problems, in *Genome Evolution* (eds G.A. Dover and R.B.L. Flavell), Academic Press, San Diego, pp. 375–382.

Maynard Smith, J. (1988a) The evolution of recombination, in *The Evolution of Sex: An Examination of Current Ideas* (eds R.E. Michod and B.R. Levin), Sinauer, Sunderland, MA, pp. 106–125.

Maynard Smith, J. (1988b) Selection for recombination in a polygenic model: the mechanism. *Genet. Res.*, **51**, 59–63.

Maynard Smith, J., Dowson, C.G. and Spratt, B.G. (1991). Localized sex in bacteria. *Nature*, **349**, 29–31.

Mayo, O. (1966) On the evolution of dominance. *Heredity*, **21**, 499–511.

Mayr, E. (1963) *Animal Species and Evolution*, Harvard University Press, Cambridge, MA.

Mayr, E. (1970) *Populations, Species and Evolution*, Harvard University Press, Cambridge, MA.

Mayr, E. (1980) Some thoughts on the history of the evolutionary synthesis, in *The Evolutionary Synthesis* (eds E. Mayr and W.B. Provine), Harvard University Press, Cambridge, MA, pp. 1–48.

Mayr, E. (1982) Speciation and macroevolution. *Evolution*, **36**, 1119–32.

McCartney, P.R., Renwick, J.H. and Munday, M.R. (1977) The pattern of loci on *Drosophila* chromosomes. *Heredity*, **38**, 37–45.

McClintock, B. (1965) The control of gene action in maize. *Brookhaven Symp. Biol.*, **18**, 162–84.

McClintock, B. (1984) The significance of responses of the genome to challenge. *Science*, **226**, 792–801.

McCouch, S.R., Kochert, G., Yu, Z.H. *et al.* (1988) Molecular mapping of rice. *Theor. Appl. Genet.*, **76**, 815–29.

McDonald, J.F. (1983) The molecular basis of adaptation: a critical review of relevant ideas and observations. *Annu. Rev. Ecol. Syst.*, **14**, 77–102.

McGrath, J. M. and Quiros, C.F. (1991) Inheritance of isozyme and RFLP markers in *Brassica campestris* and comparison with *B. oleracea*. *Theor. Appl. Genet.*, **82**, 668–73.

McLachlan, G.J. and Basford, K.E. (1988) *Mixture Models*. Marcel Dekker, New York.

McMillan, I. and Robertson, A. (1974) The power of methods for the detection of major genes affecting quantitative characters. *Heredity*, **32**, 349–56.

McNelly-Ingle, C.A., Lamb, B.C. and Frost, L.C. (1966) The effect of temperature on recombination frequency in *Neurospora crassa*. *Genet. Res.*, **7**, 169–83.

Meagher, R.B., McLean, M.D. and Arnold, J. (1988) Recombination within a subclass of restriction fragment length polymorphisms may help link classical and molecular genetics. *Genetics*, **120**, 809–18.

Mednikov, B.M. (1982) Regularities of genome evolution, in *Molecular Mechanisms of Genetic Processes* (ed. N.P. Dubinin), Nauka, Moscow, pp. 76–84.

Meijen, S.V. (1984) Principles of historic reconstructions in biology, in *Systematics and Evolution*, Nauka, Moscow, pp. 7–31.
Meijen, S.V. (1987) The geography of macroevolution in higher plants. *J. Gen. Biol.* (USSR), **48**, 291–309.
Melzer, A.L. and Koeslag, J.H. (1991) Mutations do not accumulate in asexual isolates capable of growth and extinction–Muller's ratchet re-examined. *Evolution*, **45**, 649–55.
Messmer, M.M., Melchinger, A.E., Lee, M. *et al.* (1991) Genetic diversity among progenitors and elite lines from the Iowa Stiff Stalk Synthetic (BSSS) maize population: comparison of allozyme and RFLP data. *Theor. Appl. Genet.*, **83**, 97–107.
Meyerowitz, E.M. (1987) *Arabidopsis thaliana. Annu. Rev. Genet.*, **21**, 93–111.
Michie, D. (1953) Affinity: a new genetic phenomenon in the house mouse. *Nature*, **171**, 26–27.
Michod, R.E. and Levin, B.R. (eds) (1988) *The Evolution of Sex: An Examination of Current Ideas*, Sinauer, Sunderland, MA.
Mike, V. (1977) Theories of quasi-linkage and 'affinity': one implication for population structure. *Proc. Natl Acad. Sci. USA*, **74**, 3513–17.
Mikhailov, M.E. (1991) Analysis of quasi-linkage between loci of chromosomes 2, 6 and 9 in some hybrids of maize, in *Ecological Genetics of Plants and Animals* (Abstracts of All-Union Conference) (ed. A.G. Jacote), Shtiintsa, Kishinev, p. 46.
Miklos, G.L.G. (1982) Sequencing and manipulating highly repeated DNA, in *Genome Evolution* (eds G.A. Dover and R.B.L. Flavell), Academic Press, San Diego, pp. 41–68.
Miller, J.C. and Tanksley, S.D. (1990) RFLP analysis of phylogenetic relationship and genetic variation in the genus *Lycopersicon. Theor. Appl. Genet.*, **80**, 37–48.
Mitchell-Olds, T. (1992) Does environmental variation maintain genetic variation? A question of scale. *Trends Ecol. Evol.*, 7, 397–98.
Mitchell-Olds, T. and Shaw, R.G. (1987) Regression analysis of natural selection: statistical inference and biological interpretation. *Evolution*, **41**, 1149–61.
Mitrofanov, V.G. (1977) Physiological basis and evolution of dominance, in *Problems of Experimental Biology*, Nauka, Moscow, pp. 21–31.
Mitton, J.B. and Grant, M.C. (1984) Associations among protein heterozygosity, growth rate, and developmental homeostasis. *Annu. Rev. Ecol. Syst.*, **15**, 479–99.
Mochi, M., Guerra, D. and Palenzona, D. (1978) Selection and recombination frequency: interaction with X-irradiation. *Genetica*, **48**, 55–59.
Mock, J.J. (1973) Manipulation of crossing-over with intrinsic and extrinsic factors. *Egypt. J. Genet. Cytol.* **2**, 158–75.
Modrich, P. (1991) Mechanisms and biological effects of mismatch repair. *Annu. Rev. Genet.*, **25**, 229–53.
Moens, P.B. (1969) Genetic and cytological effects of three desynaptic genes in the tomato. *Can. J. Genet. Cytol.*, **11**, 857–59.
Mont, S., Moya, A. and Marinkovic, D. (1988) A test to detect association between autosomal and sex-linked loci in random mating populations: an example with *Drosophila melanogaster. Genet. Sel. Evol.*, **20**, 417–23.
Montelone, B.A., Hoekstra, M.F. and Malone, R.E. (1988) Spontaneous mitotic recombination in yeast: the hyper-recombinational *rem1* mutations are alleles of the *RAD3* gene. *Genetics*, **119**, 289–301.
Montelone, B.A., Prakash, S. and Prakash, L. (1981) Recombination and mutagenesis in

RAD6 mutants of *Saccharomyces cerevisiae*: Evidence for multiple functions of the *RAD6* gene. *Mol. Gen. Genet.*, **184**, 410–15.

Montgomery, E., Charlesworth, B. and Langley, C.H. (1987) A test for the role of natural selection in the stabilization of transposable element copy number. *Genet. Res.*, **49**, 31–41.

Moore, C.W. and Sherman, F. (1975) Role of DNA sequences in genetic recombination in the iso-1-cytochrome gene of yeast. I. Discrepancies between physical distances and genetic distances determined by five mapping procedures. *Genetics*, **79**, 397–418.

Moore, W.S. and Hines, W.G.S. (1981) Sex in random environments. *J. Theor. Biol.*, **92**, 301–16.

Morell, V. (1993) The puzzle of the triple repeats. *Science*, **260**, 1422–23.

Moreno-Gonzalez, J. (1992) Genetic models to estimate additive and non-additive effects of marker-assisted QTL using multiple regression techniques. *Theor. Appl. Genet.*, **85**, 435–44.

Morgan, T.H., Bridges, C.B. and Sturtevant, A.H. (1925) The genetics of *Drosophila*. *Bibliogr. Genet.*, **2**, 1–262.

Moriwaki, D., Tobari, J.N. and Matsuda, M. (1979) Role of *Y*-chromosome in male crossing-over in *Drosophila ananassae*. *Jpn J. Genet.*, **54**, 295–302.

Morizot, D.C. (1990) Use of fish gene maps to predict ancestral vertebrate genome organization, in: *Isozymes: Structure, Function and Use in Biology and Medicine*. Wiley-Liss, New York, pp. 207–34.

Morton, N.E. (1955) Sequential tests for the detection of linkage. *Am. J. Hum. Genet.*, **7**, 277–318.

Morton, N.E. (1988) Multipoint mapping and the emperor's clothes. *Ann. Hum. Genet.*, **52**, 309–18.

Morton, N.E. and MacLean, C.J. (1974) Analysis of family resemblance. III. Complex segregation of quantitative traits. *Am. J. Hum. Genet.*, **54**, 295–302.

Morton, N.E. and MacLean, C.J. (1984) Multilocus recombination frequencies. *Genet. Res.*, **44**, 99–108.

Morton, N.E., MacLean, C.J. and Lew, R. (1985) Tests of hypotheses on recombination frequencies. *Genet. Res.*, **45**, 279–86.

Motro, U. and Soller, M. (1993) Sequential sampling in determining linkage between marker loci and quantitative trait loci. *Theor. Appl. Genet.*, **85**, 658–64.

Mukai, T. (1988) Genotype-environment interaction in relation to the maintenance of genetic variability in populations of *Drosophila melanogaster*, in *Quantitative Genetics* (eds B.S. Weir *et al.*), Sinauer, Sunderland, MA, pp. 21–33.

Mukai, T., Mettler, L.E. and Chigusa, S.I. (1971) Linkage disequilibrium in a local population of *Drosophila melanogaster*. *Proc. Natl Acad. Sci. USA*, **68**, 1065–69.

Mukai, T., Watanabe, T.K. and Yamaguchi, O. (1974) The genetic structure of natural populations of *Drosophila melanogaster*. XII. Linkage disequilibrium in a large local population. *Genetics*, **77**, 771–93.

Mukherjee, A.S. (1961) Effect of selection on crossing-over in the males of *Drosophila ananassae*. *Am. Nat.*, **95**, 57–59.

Mulcahy, D.L. and Mulcahy, G.B. (1987) The effects of pollen competition. *Am. Sci.*, **75**, 44–50.

Muller, A.J., Mendel, R.R., Schiemann, J. *et al.* (1987) High meiotic stability of a foreign gene introduced into tobacco by *Agrobacterium*-mediated transformation. *Mol. Gen. Genet.*, **207**, 171–75.

Muller, H.J. (1916) The mechanism of crossing-over. *Am. Nat.*, **45**, 65–78.

Muller, H.J. (1932) Some genetic aspects of sex. *Am. Nat.*, **66**, 118–38.

Muller, H.J. (1964) The relation of recombination to mutational advance. *Mutat. Res.*, **1**, 2–9.

Nadeau, J.H. (1989) Maps of linkage and synteny homologies between mouse and man. *Trends Genet.*, **5**,82–86.

Nadeau, J.H. and Taylor, B.A. (1984) Lengths of chromosomal segments conserved since divergence of man and mouse. *Proc. Natl Acad. Sci. USA*, **81**, 814–18.

Nagylaki, T. (1989) The maintenance of genetic variability in two-locus models of stabilizing selection. *Genetics*, **122**, 235–48.

Nagylaki, T. (1992) *Theoretical Population Genetics*, Springer, Berlin.

Nakamura, Y., Leppert, M. and O'Connell, P. (1987) Variable number of tandem repeat (VNTR) markers for human gene mapping. *Science*, **235**, 1616–22.

Nakamura, Y., Lathrop, M., O'Connell, P. *et al.* (1989) Frequent recombination is observed in the distal end of the long arm of chromosome 14. *Genomics*, **4**, 76–81.

Naora, H. (1987) Gene ecology. An attempt to find a universal principle. *Biol. Forum*, **80**, 257–59.

Nassif, N. and Engels, W. (1993) DNA homology requirements for mitotic gap repair in *Drosophila. Proc. Natl Acad. Sci. USA*, **90**, 1262–66.

Naveira, H.F. (1992) Location of *X*-linked polygenic effects causing sterility in male hybrids of *Drosophila simulans* and *D. mauritiana. Heredity*, **68**, 211–17.

Navidi, W., Arnheim, N., Waterman, M.S. (1992) A multiple-tubes approach for accurate genotyping of very small DNA samples by using PCR: statistical considerations. *Am. J. Hum. Genet.*, **50**, 347–59.

Nee, S. (1988) Deleterious mutation and genetic recombination. *Nature*, **331**, 308.

Nee, S. (1989) On the evolution of sex in RNA viruses. *J. Theor. Biol.*, **138**, 407–12.

Neel, J.V. (1941) A relation between larval nutrition and the frequency of crossing-over in the third chromosome of *Drosophila melanogaster. Genetics*, **26**, 506–16.

Nei, M. (1967) Modification of linkage intensity by natural selection. *Genetics*, **57**, 625–41.

Nei, M. (1969) Linkage modification and sex difference in recombination. *Genetics*, **63**, 681–99.

Nei, M. (1975) *Molecular Population Genetics and Evolution*, North-Holland, Amsterdam.

Nevo, E. (1978) Genetic variation in natural populations: patterns and theory. *Theor. Pop. Biol.*, **13**, 121–77.

Nevo, E. (1987) Plant genetic resources: prediction by isozyme markers and ecology. *Isozymes: Curr. Top. Biol. Med. Res.*, **16**, 247–67.

Nevo, E. (1988a) Genetic diversity in nature: patterns and theory. *Evol. Biol.*, **23**, 217–46.

Nevo, E. (1988b) Genetic differentiation in evolution. *ISI Atlas of Science: Animal and Plant Sciences*, Institute for Scientific Information, Philadelphia, pp. 195–202.

Nevo, E. (1991) Evolutionary theory and processes of active speciation and adaptive radiation in subterranean mole rats, *Spalax ehrenbergii* superspecies in Israel. *Evol. Biol.*, **25**, 1–125.

Nevo, E. (1992) Origin, evolution, population genetics and resources for breeding of wild barley, *Hordeum spontaneum*, in the fertile crescent, in *Barley: Genetics, Biochemistry, Molecular Biology and Biotechnology* (ed. P.R. Shewry), C.A.B. International, pp. 19–43.

Nevo, E. and Beiles, A. (1988) Genetic parallelism of protein polymorphism in nature: ecological test of the neutral theory of molecular evolution. *Biol. J. Linn. Soc.*, **35**, 229–45.

Nevo, E. and Beiles, A. (1989) Genetic diversity of wild emmer wheat in Israel and Turkey: structure, evolution, and application in breeding. *Theor. Appl. Genet.*, **77**, 421–55.

Nevo, E. and Beiles, A. (1992) Selection for class II *Mhc* heterozygosity by parasites in subterranean mole rats. *Experientia*, **48**, 512–15.

Nevo, E., Beiles, A. and Ben-Shlomo, R. (1984) The evolutionary significance of genetic diversity: Ecological demographic and life history correlates, in *Evolutionary Dynamics of Genetic Diversity* (ed. G.S. Mani). pp. 13–213.

Nevo, E., Joh, K., Hori, K. and Beiles, A. (1990) Aldolase DNA polymorphism in subterranean mole-rats: genetic differentiation and environmental correlates. *Heredity*, **65**, 307–20.

Nevo, E., Nishikawa, K., Furuta, Y., Gonokami, Y. and Beiles, A. (1993) Genetic polymorphism of α- and β-amylase isozymes in wild emmer wheat, *Triticum dicoccoides*, in Israel. *Theor. Appl. Genet.*, **85**, 1029–42.

Newman, S.M., Harris, E.H., Johnson, A.M., Boynton, J.E., Gillham, N.W. (1992) Non-random distribution of chloroplast recombination events in *Chlaminomodas reinhardtii*: evidence for a hotspot and an adjacent cold region. *Genetics*, **132**, 413–39.

Newmark, P. (1984) Molecular diagnostic medicine. *Nature*, **307**, 11–12.

Nicolas, A. (1979) Variation of gene conversion and intragenic recombination frequencies in the genome of *Ascobolus immersus*. *Mol. Gen. Genet.*, **176**, 129–138.

Nillson-Ehle, H. (1927) Das Verhalten partieller Speltoidmutationen bei Kreuzung untereinander (Untersuchungen uber Speltoidmutationen Beim Weizen IY). *Hereditas*, **9**, 369–79.

Nodari, R.O., Tsai, S.M., Gilbertson, R.L., Gepts, P. (1993) Towards an integrated linkage map of common bean. 2. Development of an RFLP-based linkage map. *Theor. Appl. Genet.*, **85**, 513–20.

Nothel, H. (1987) Adaptation of *Drosophila melanogaster* populations to high mutation pressure: evolutionary adjustment of mutation rates. *Proc. Natl Acad. Sci. USA*, **84**, 1045–49.

Nunney, L. (1989) The maintenance of sex by group selection. *Evolution*, **43**, 245–57.

Nunney, L. and Campbell, K.A. (1993) Assessing minimum viable population size: demography meets population genetics. *Trends Ecol. Evol.*, **8**, 234–39.

O'Brien, S.J. (ed.) (1990) *Genetic Maps*, Cold Spring Harbor Laboratory Press, Cold Spring Harbor, NY.

O'Connell, P., Lathrop, C.M., Nakamura, Y. *et al.* (1989) Twenty-eight loci form a continuous linkage map of markers for human chromosome 1. *Genomics*, **4**, 12–20.

Ogawa, T., Yu, X., Shinohara, A. and Egelman, E.H. (1993) Similarity of the yeast *RAD51* filament to the bacterial *RecA* filament. *Science*, **259**, 1896–99.

Ohno, S. (1970) *Evolution by Gene Duplication*, Springer, Berlin.

Ohno, S. (1974) Conservation of ancient linkage groups in evolution and some insight into genetic regulatory mechanism of X-inactivation. *Cold Spring Harbor Symp. Quant. Biol.*, **38**, 155–64.

Ohta, T. (1986) Population genetics theory of multigene families with emphasis on genetic variation contained in the family, in *Evolutionary Processes and Theory* (eds S. Karlin and E. Nevo), Academic Press, Orlando, FL, pp. 239–53.

Olson, M.V. (1993) The human genome project. *Proc. Natl Acad. Sci. USA*, **90**, 4338–44.

Osborne, B.I., Corr, C.A., Prince, J.P. *et al.* (1991) *Ac* transposition from a *T*-DNA can generate linked and unlinked clusters of insertions in the tomato genome. *Genetics*, **129**, 833–44.

Osman, F., Tomsett, B. and Strike, P. (1993) The isolation of mutagene-sensitive *nuv* mutants of *Aspergillus nidulans* and their effects on mitotic recombination. *Genetics*, **134**, 445–54.

Ostergard, H. and Hovmoller, M.S. (1991) Gametic disequilibria between virulence genes in barley powdery mildew populations in relation to selection and recombination I. Models. *Plant. Pathol. (Oxford)*, **40**, 166–77.

Ott, J. (1985) *Analysis of Human Genetic Linkage*, Johns Hopkins University Press, Baltimore.

Otto, S.P. (1991) On evolution under sexual and viability selection: a two-locus diploid model. *Evolution*, **45**, 1443–57.

Otto, S.P. and Goldstein, D.B. (1992) Recombination and the evolution of diploidy. *Genetics*, **131**, 745–51.

Owen, A.R.G. (1950) The theory of genetical recombination. *Adv. Genet.*, **3**, 117–57.

Owen, A.R.G. (1951) An extension of Kosambi's formula. *Nature*, **168**, 208–9.

Ozias-Akins, P., Lubbers, E.L., Hanna, W.W. and McNay, J.W. (1993) Transmission of the apomictic mode of reproduction in Pennisetum: co-inheritance of the trait and molecular markers. *Theor. Appl. Genet.*, **85**, 632–38.

Padmore, R., Cao, L. and Kleckner, N. (1991) Temporal comparison of recombination and synaptonemal complex formation during meiosis in *S. cerevisiae*. *Cell*, **66**, 1239–56.

Palenzona, D.L. and Zattoni, A. (1970) Selection and recombination in *Drosophila melanogaster*. *Theor. Appl. Genet.*, **40**, 345–46.

Palladino, F. and Klein, H.L. (1992) Analysis of mitotic and meiotic defects in *Saccharomyces cerevisiae SRS2* DNA helicase mutants. *Genetics*, **132**, 23–37.

Pamilo, P. (1988) Genetic variation in heterogeneous environments. *Ann. Zool. Fenn.*, **25**, 99–106.

Pamilo, P., Nei, M., Li, W.H. (1987) Accumulation of mutations in sexual and asexual populations. *Genet. Res.*, **49**, 135–46.

Paquin, C.E. and Williamson, W.M. (1984) Temperature effect on the rate of *Ty* transposition. *Science*, **226**, 53–55.

Parker, J.S. (1975) Chromosome-specific control of chiasma formation. *Chromosoma*, **49**, 391–406.

Parker, M.A. (1988a) Disequilibrium between disease-resistance variants and allozyme loci in an annual legume. *Evolution*, **42**, 239–47.

Parker, M.A. (1988b) Genetic uniformity and disease resistance in a clonal plant. *Am. Nat.*, **132**, 538–42.

Parker, M.A. (1991) Nonadaptive evolution of disease resistance in an annual legume. *Evolution*, **45**, 1209–17.

Parsons, P.A. (1958) Selection for increased recombination in *Drosophila melanogaster*. *Am. Nat.*, **92**, 255–56.

Parsons, P.A. (1987) Evolutionary rates under environmental stress. *Evol. Biol.*, **21**, 311–47.

Parsons, P.A. (1988) Evolutionary rates: effects of stress upon recombination. *Biol. J. Linn. Soc.*, **35**, 49–68.

Passananti, C., Davies, B., Ford, M. and Fried, M. (1987) Structure of an inverted duplication formed as a first step in a gene amplification event: implications for a model of a gene amplification. *EMBO J.*, **6**, 1697–703.

Paterson, A.H., Lander, E.S., Hewitt, J.D. *et al.* (1988) Resolution of quantitative traits into Mendelian factors by using a complete RFLP linkage map. *Nature*, **335**, 721–26.

Paterson, A.H., Deverna, J.W., Lanini, B., Tanksley, S.D. (1990) Fine mapping of quantitative trait loci using overlapping recombinant chromosomes, in an interspecies cross of tomato. *Genetics*, **124**, 735–42.

Paterson, A.H., Damon, H.S., Hewitt, J.D. *et al.* (1991) Mendelian factors underlying quantitative traits in tomato: comparison across species, generations, and environments. *Genetics*, **127**, 181–97.

Paul, S.R., Liang, K.V., Self, S.G. (1989) On testing departure from the binomial and multinomial assumptions. *Biometrics*, **45**, 231–36.

Payne, F. (1918) An experiment to test the nature of the variation on which selection acts. *Indiana Univ. Studies*, **5**, 1–45.

Pearlman, R.E, Tsao, N., Moens, P.B. (1992) Synaptonemal complexes from DNase-treated rat pachytene chromosomes contain $(GT)_n$ and *LINE/SINE* sequences. *Genetics*, **130**, 865–72.

Peixoto, A.A. and Klaczko, L.B. (1991) Linkage disequilibrium analysis of chromosomal inversion polymorphisms of *Drosophila*. *Genetics*, **129**, 773–77.

Perkins, D.D. and Barry, E.G. (1977) The cytogenetics of *Neurospora*. *Adv. Genet.*, **19**, 133–85.

Perrot, V., Richerd, S. and Valero, M. (1991) Transition from haploidy to diploidy. *Nature*, **351**, 315–17.

Perry, P.E. and Jones, G.H. (1974) Male and female meiosis in grasshoppers. I. *Stethophyma grossum*. *Chromosoma*, **47**, 227–36.

Petes, T.D. and Hill, C.W. (1988) Recombination between repeated genes in microorganisms. *Annu. Rev. Genet.*, **22**, 147–68.

Pfeiffer, T.W. (1993) Recombination rates of soybean varieties from different periods of introduction and release. *Theor. Appl. Genet.*, **86**, 557–61.

Pillen, K., Steinrucken, G., Wricke, G. *et al.* (1992) A linkage map of sugar beet (*Beta vulgaris* L.). *Theor. Appl. Genet.*, **84**, 129–35.

Plaetke, R., Lumme, J. and Koehler, W. (1988) Selection on recombination in subdivided populations with stabilizing selection. *Hereditas*, **109**, 61–67.

Plasterk, R.H.A. (1992) Genetic switches: mechanism and function. *Trends Genet.*, **8**, 403–6.

Plough, H.H. (1917) The effect of temperature on crossing-over in *Drosophila*. *J. Exp. Zool.*, **24**, 147–209.

Pollard, J.W. (1987) The moveable genome: Weismann's doctrine and new models for speciation. *Biol. Forum*, **80**, 11–54.

Prakash, S. and Lewontin, R.C. (1968) A molecular approach to the study of genic heterozygosity in natural populations. III. Direct evidence of coadaptation in inversions of *Drosophila*. *Proc. Natl Acad. Sci. USA*, **59**, 398–405.

Preisler, H.K. (1988) Maximum likelihood estimates for binary data with random effects. *Biometr. J.*, **30**, 339–50.

Preygel, I.A. (1986) *Genetical-Statistical Analysis of the Level and Spectrum of Recombination Variability*. Institute of General Genetics, Moscow. Ph. D. Thesis.

Preygel, I.A. and Korol, A.B. (1989) Marker analysis of quantitative traits. *Adv. in Modern Genet.* (USSR), **16**, 123–38.

Preygel, I.A., Garbuz, L.I. and Korol, A.B. (1986) Evaluation of the variability spectrum in segregating populations using trait complexes, in *Genetic Methods for Accelerating the Breeding Process* (ed. A.A. Zhuchenko), Shtiintsa, Kishinev, 115–26.

Preygel, I.A., Preygel, S.I. and Korol, A.B. (1991) A computer model of artificial selection in introgression and transgression breeding schemes, in *Plant Breeding: New Genetic Approaches and Solutions* (ed. N.N. Balashova), Shtiintsa, Kishinev, pp. 22–31.

Preygel, S.I. and Korol, A.B. (1990) Evolution of recombination in systems of 'host–parasite' type. Multilocus models. *Genetika* (USSR), **26**, 349–58.

Price, D.J. and McBride, J. (1981) Chiasma frequency variation in *Cepaea* populations from a localised area of South-West England. *Caryologia*, **34**, 363–75.

Prokofjeva-Belgovskaya, A.A. (1986) *Heterochromatic Regions in Chromosomes*, Nauka Press, Moscow.

Purugganan, M.D. (1993) Transposable elements as introns: evolutionary connections. *Trends Ecol. Evol.*, **8**, 239–43.

Rao, D.C., Morton, N.E., Lindstrem, J. *et al.* (1977) A mapping function for man. *Hum. Heredity*, **39**, 255–66.

Rao, D.C., Keats, B.J.B., Morton, N.E. *et al.* (1978) Variability of human linkage data. *Am. J. Hum. Genet.*, **30**, 516–29.

Rasmussen, S.W. and Holm, P.B. (1984) The synaptonemal complex. Recombination nodules and chiasmata in human spermatocytes, in *Controlling Events in Meiosis* (eds C.W. Evans and H.G. Dickinson), Society of Experimental Biology, Cambridge, pp. 271–92.

Rasmusson, J. (1927) Genetically changed linkage values in *Pisum*. *Hereditas*, **10**, 1–152.

Rasooly, R.S. and Robbins, L.G. (1991) *Rex* and a suppressor of *Rex* are repeated neomorphic loci in the *Drosophila melanogaster* ribosomal DNA. *Genetics*, **129**, 119–32.

Ratner, V.A. (1990) Towards a unified theory of molecular evolution (TME). *Theor. Pop. Biol.*, **38**, 233–61.

Ratner, V.A., Zharkikh, A.A., Kolchanov, N.A. *et al.* (1985) *Problems of the Molecular Evolution Theory*, Nauka, Novosibirsk.

Rattner, J.B. (1992) Integrating chromosome structure with function. *Chromosoma*, **101**, 259–64.

Rattray, B. and Rose, A.M. (1988) Increased intragenic recombination and nondisjunction in the *Rec-1* strain of *Caenorhabditis elegans*. *Genet. Res.*, **51**, 89–93.

Read, A.F. (1987) Comparative evidence supports the Hamilton and Zuk hypothesis on parasites and sexual selection. *Nature*, **328**, 68–70.

Reanney, D.C. (1984) RNA splicing as an error-screening mechanism. *J. Theor. Biol.*, **110**, 315–21.

Redfield, R.J. (1988) Evolution of bacterial transformation: is sex with dead cells ever better than no sex at all? *Genetics*, **119**, 213–21.

Redfield, R.J. (1993) Evolution of natural transformation: testing the DNA repair hypothesis in *Bacillus subtilis* and *Haemophilus influenzae*. *Genetics*, **133**, 755–61.

Rees, H. and Durrant, A. (1986) Recombination and genome size. *Theor. Appl. Genet.*, **73**, 72–76.

Rees, H., Jenkins, G., Seal, A.G. and Hutchinson, J. (1982) Assays of the phenotypic effects of changes in DNA amounts, in *Genome Evolution* (eds G.A. Dover and R.B.L. Flavell), Academic Press, San Diego, pp. 287–97.

Rees, H. and Jones, R.N. (1977) *Chromosome Genetics*, Edward Arnold, London.

Rees, H. and Thompson, J.B. (1956) Genotypic control of chromosome behavior in rye. III. Chiasma frequency in homozygotes and heterozygotes. *Heredity*, **10**, 409–24.

Reiter, R.S., Coors, J.G., Sussman, M.R., Gabelman, W.H. (1991) Genetic analysis of tolerance to low-phosphorus stress in maize using restriction fragment length polymorphisms. *Theor. Appl. Genet.*, **82**, 561–68.

Reiter, R.S., Williams, J.G.K., Feldmann, K.A. *et al.* (1992) Global and local genome mapping in *Arabidopsis thaliana* by using recombinant inbred lines and random ampliphied polymorphic DNAs. *Proc. Natl Acad. Sci. USA*, **89**, 1477–81.

Rendel, J.M. (1957) Relationship between coincidence and crossing-over in *Drosophila*. *J. Genet.*, **55**, 95–99.

Renwick, J.H. (1971) The mapping of human chromosomes. *Annu. Rev. Genet.*, **5**, 81–120.

Rhoades, M.M. (1978) Genetic effects of heterochromatin in maize, in *Maize Breeding and Genetics* (ed. D.B. Walden), John Wiley, New York, pp. 641–71.

Rhyne, C.L. (1962) Enhancing linkage-block breakup following interspecific hybridization and backcross transference of genes in *Gossypium hirsutum* L. *Genetics*, **47**, 61–69.

Rice, W.R. (1983a) Parent–offspring pathogen transmission – a selective agent promoting sexual reproduction. *Am. Nat.*, **121**, 187–203.

Rice, W.R. (1983b) Sexual reproduction: an adaptation reducing parent–offspring contagion. *Evolution*, **37**, 1317–20.

Rice, W.R. (1987) The accumulation of sexually antagonistic genes as a selective agent promoting the evolution of reduced recombination between primitive sex chromosomes. *Evolution*, **41**, 911–14.

Rich, A., Nordhein, A., Wand, A.H.-J. (1984) The chemistry and biology of left-handed Z-DNA. *Annu. Rev. Biochem.*, **53**, 791–846.

Rick, C.M. (1969) Controlled introgression of chromosomes of *Solanum pennellii* into *Lycopersicon esculentum*: segregation and recombination. *Genetics*, **62**, 753–68.

Rick, C.M. (1971) Some cytogenetic features of the genome in diploid plant species. *Stadler Genet. Symp.*, **2**, 153–74.

Rick, C.M. (1972) Further studies on segregation and recombination in backcross derivatives of a tomato species hybrid. *Biol. Zbl.*, **91**, 209–20.

Rick, C.M. and Butler, L. (1956) Cytogenetics of the tomato. *Adv. Genet.*, **6**, 267–382.

Risch, N. (1988) A new statistical test for linkage heterogeneity. *Am. J. Hum. Genet.*, **42**, 353–64.

Risch, N. and Lange, K. (1979) An alternative model of recombination and interference. *Ann. Hum. Genet.*, **48**, 61–70.

Robbins, L.G. and Swanson, E.E. (1988) *Rex*-induced recombination implies bipolar organization of the ribosomal RNA genes of *Drosophila melanogaster*. *Genetics*, **120**, 1053–59.

Roberts, P.A. (1965) Difference in the behavior of eu- and heterochromatin: crossing-over. *Nature*, **205**, 725–26.

Roberts, R.G., Montadon, A.J., Bodrow, M., Bentley, D.R. (1989) Detection of novel genetic markers by mismatch analysis. *Nucl Acids Res.*, **17**, 5961–71.

Robertson, B.D. and Meyer, T.F. (1992) Genetic variation in pathogenic bacteria. *Trends Genet.*, **8**, 422–6.

Robertson, D.S. (1984) Different frequency in the recovery of crossover products from male and female gametes of plants hypoploid for *B-A* translocations in maize. *Genetics*, **107**, 117–130.

Robertson, D.S. (1985) A possible technique for isolating genic DNA for quantitative traits in plants. *J. Theor. Biol.*, **117**, 1–10.

Robertson, H.M. (1993) The *mariner* transposable element is widespread in insects. *Nature*, **362**, 241–45.

Robertson, M. (1982) The evolutionary past of the major histocompatibility complex and the future of cellular immunology. *Nature*, **297**, 629–32.

Robinson, W.P., Asmussen, M.A. and Thomson, G. (1991) Three-locus systems impose additional constraints on pairwise disequilibria. *Genetics*, **129**, 925–30.

Robinson, W.P., Cambon-Thomson, A., Borot, N. *et al.* (1991) Selection, hitchhiking and disequilibrium analysis at three linked loci with application to *HLA* data. *Genetics*, **129**, 931–48.

Roderick, T.H., Lally, P.A., Davisson, M.T. *et al.* (1984) Report of the committee on comparative mapping. *Cyt. Cell Genet.*, **37**, 312–39.

Rodin, S.N. and Rzhetsky, A.Yu. (1987) *The Problems of Somatic Hypermutagenesis of Immunoglobulin Genes: A Coevolutionary Aspect*, ICG Press, Novosibirsk.

Roff, D.A. (1986) Evolution of wing polymorphism and its impact on life cycle adaptation in insects, in the *Evolution of Insect Life Cycles* (eds F. Taylor and R. Kabran), Springer, Berlin, pp. 204–21.

Rogowsky, P.M., Gidet, F.L.Y., Langridge, P. *et al.* (1991). Isolation and characterization of wheat-rye recombinants involving chromosome arm *1DS* of wheat. *Theor. Appl. Genet.*, **82**, 537–44.

Ronin, Y. I., Kirzhner, V.M. and Korol, A.B. (1993) Linkage between loci of quantitative traits and marker loci. 7. Multi-trait analysis with a single marker (submitted).

Ronin, Y.I., Korol, A.B. and Kirzhner, V.M. (1993) The dynamics of the *rec*-system in variable environment. I. Haploid selection in cyclical two-state environment, in *Evolution 93*, Abstracts of the 4th Congress of the European Society for Evolutionary Biology, University of Montpellier II, Montpellier, p. 380.

Rose, A.M. and Baillie, D.L. (1979) A mutation in *Caenorhabditis elegans* that increases recombination frequency more than three-fold. *Nature*, **281**, 599–600.

Roughgarden, J. (1971) Density-dependent natural selection. *Ecology*, **52**, 453–68.

Ruiz, A., Fontdevila, A., Santos, M. *et al.* (1986) The evolutionary history of *Drosophila buzzatii*. VIII. Evidence for endocyclic selection acting on the inversion polymorphism in a natural population. *Evolution*, **40**, 740–55.

Ruvinsky, A.O. (1991) Sex, meiosis and progressive evolution, in *Problems of Genetics and Theory of Evolution* (eds V.K. Shumny and A.O. Ruvinsky), Nauka, Novosibirsk, pp. 214–28.

Sagan, D. and Margulis, Ł. (1987) Cannibal's relief: the origins of sex. *New Sci.*, **115**, 36–40.

Sall, T. and Bengtsson, B.O. (1989) Apparent negative interference due to variation in recombination frequencies. *Genetics*, **122**, 935–42.

Salmon, R.K., Berg, R.T., Yeh, F.C. and Hodgetts, R.B. (1988) Identification of a variant growth hormone haplotype in mice selected for high body weight. *Genet. Res.*, **52**, 7–15.

Sandler, L. and Szauter, P. (1978) The effect of recombination-defective meiotic mutants on fourth-chromosome crossing-over in *Drosophila melanogaster. Genetics*, **90**, 699–712.

Sano, Y., Chu, Y. and Oka, H. (1980) Genetic studies of speciation in cultivated rice. 2. Character variations in backcross derivatives between *Oryza sativa* and *O. graberrima*: *M-V* linkage and key characters. *Jpn. J. Genet.*, **55**, 19–39.

Santos, J.L. (1993) The effect of supernumerary segments on recombination in *Euchorthippus pulvinatus* (grasshopper): a comparative study between sexes and populations. *Heredity*, **70**, 130–34.

Sapre, A.B. and Deshpande, D.S. (1987) Spontaneous emergence of parents from the F_1 interspecific hybrids of *Coix* L. *J. Heredity*, **78**, 357–60.

Sarfatti, M., Abu-Abied, M., Katan, J. and Zamir, D. (1991) RFLP mapping of *I1*, a new locus in tomato conferring resistance against *Fusarium oxysporum* f. sp. *lycopersici* race 1. *Theor. Appl. Genet.*, **82**, 22–26.

Sasaki, A. and Iwasa, Y. (1987) Optimal recombination rate in fluctuating environments. *Genetics*, **115**, 377–88.

Sax, K. (1923) The association of size differences with seed coat pattern and pigmentation in *Phaseolus vulgaris. Genetics*, **8**, 552–60.

Saunders, J.H. (1969) Changes in recombination and genetic disturbance on transferring the H_2Lc_2 linkage group from *Gossypium tomentosum* to *G. barbadense* and *G. hirsutum. Genet. Res.*, **13**, 1–15.

Schaap, T. and Simchen, G. (1971) Genetic control of recombination affecting mating factors in a population of *Schizophyllum* and its relation to inbreeding. *Genetics*, **68**, 67–75.

Schatz, D.G. and Baltimore, D. (1988) Stable expression of immunoglobulin gene *V(D)J* recombinase activity by gene transfer into *3T3* fibroblasts. *Cell*, **53**, 107–15.

Scherer, G., Bausch, E., Gaa, A. and Deimling, O. (1989) Gene mapping on mouse chromosome 8 by interspecific crosses: new data on a linkage group conserved on human chromosome *16q*. *Genomics*, **5**, 275–82.

Scherly, D., Nouspikel, T., Corlet, J. *et al.* (1993) Complementation of the DNA repair defect in xeroderma pigmentosum group G cells by a human cDNA related to yeast *RAD2*. *Nature*, **363**, 182–85.

Schimke, R.T. (1984) Gene amplification in cultured animal cells. *Cell*, **37**, 705–13.

Schluter, D. (1988) Estimating the form of natural selection on a quantitative trait. *Evolution*, **42**, 849–61.

Schmalhausen, I.I. (1942) *Organism as a Whole Entity in the Individual and Historic Development*, USSR Academy of Sciences, Moscow.

Schmalhausen, I.I. (1949) *Factors of Evolution: The Theory of Stabilizing Selection*, Blakiston, Philadelphia.

Schmalhausen, I.I. (1968) *Cybernetic Problems of Biology*, Nauka, Novosibirsk.

Schnee, F.B. and Thompson, J.N. Jr (1984) Conditional polygenic effects in the sternopleural bristle system of *Drosophila melanogaster. Genetics*, **108**, 409–24.

Schultes, N.P. and Szostak, J.W. (1991) A poly(*dA dT*) tract is a component of the recombination initiation site at the *ARG4* locus in *Saccharomyces cerevisiae*. *Mol. Cell. Biol.*, **11**, 322–28.

Schultz, R.J. (1973) Origin and synthesis of a unisexual fish, in *Genetics and Mutagenes. Fish*, Berlin, pp. 207–11.

Schwartz-Sommer, Z. and Saedler, H. (1987) Can plant transposable elements generate novel regulatory systems? *Mol. Gen. Genet.*, **209**, 207–9.

Scrinber, K.T. and Avise, J.C. (1993) Cytonuclear genetic architecture in mosquitofish populations and the possible roles of introgressive hybridization. *Mol. Ecol.*, **2**, 139–49.

Seager, R.D., Ayala, F.J. and Marks, R.W. (1982) Chromosome interactions in *Drosophila melanogaster*. II. Total fitness. *Genetics*, **102**, 485–502.

Searle, A.G., Peters, J., Lyon, M.F. *et al.* (1987). Chromosome maps of man and mouse, III. Genomics, **1**, 3–18.

Seger, J. and Hamilton, W.D. (1988) Parasites and sex, in *The Evolution of Sex: An Examination of Current Ideas* (eds R.E. Michod and B.R. Levin), Sinauer, Sunderland, MA, pp. 176–93.

Seldin, M.F., Howard, T.A. and D'Eustachio, P. (1989) Comparison of linkage maps of mouse chromosome 12 derived from laboratory strain intraspecific backcrosses. *Genomics*, **5**, 24–28.

Seldin, M.F. and Kruh, G.D. (1989) Mapping of *Abll* within a conserved linkage group on distal mouse chromosome 1 syntenic with human chromosome 1 using an interspecific cross. *Genomics*, **4**, 221–23.

Semenov, E.P. and Smirnov, A.F. (1981) Somatic chromosome pairing in *Drosophila melanogaster. Genetika* (USSR), **17**, 77–84.

Serebrovsky, A.S. (1926) An effect of *purple* gene on crossing over between *black* and *cinnabar* in *Drosophila melanogaster. J. Exp. Biol. Ser. A* (USSR), **2**, 55–100.

Serebrovsky, A.S. (1936) Experience of applying a new method of genetic analysis of quantitative traits. *Dokl. Akad. Nauk SSSR*, **2**, 45–47.

Serebrovsky, A.S. (1938) Genes *scute* and *achaeta* in *Drosophila melanogaster* and a hypothesis of their divergence. *Dokl. Akad. Nauk SSSR*, **19**, 77–81.

Serebrovsky, A.S. (1970) *Genetic Analysis*, Nauka, Moscow.

Serebrovsky, A.S. (1973) *Some Problems of Organic Evolution*, Nauka, Moscow.

Serov, O.L., Gradov, A.A., Rubtsov, N.B. *et al.* (1987) A genetic map of the American mink: gene conservation and organization of chromosomes. *Isozymes: Curr. Top. Biol. Med. Res.*, **15**, 179–215.

Severtsov, A.S. (1981) *Introduction to the Evolution Theory*, Moscow State University Press, Moscow.

Shapiro, N.I. (1938) The mutation process as an adaptive character of the species. *Zool. J.* (USSR), **17**, 592–601.

Shapiro, N.I. and Ignatjev, M.V. (1945) The evolution of mutability. *Adv. in Modern Biol.* (USSR), **20**, 325–44.

Sharma, A.K. (1978) Change in chromosome concept. *Proc. Indian Acad. Sci.*, **87**, 161–90.

Sharp, P.J. and Hayman, D.L. (1988) An examination of the role of chiasma frequency in the genetic system of marsupials. *Heredity*, **60**, 77–85.

Shattuck-Eidens, D.M., Bell, R.N., Neuhausen, S.L., Helentjaris, T. (1990) DNA sequence variation within maize and melon: observations from polymerase chain reaction amplification and direct sequencing. *Genetics*, **126**, 207–17.

Shaw, D.D. (1972) Genetic and environmental components of chiasma control. II. The response to selection in *Schistocerca. Chromosoma*, **37**, 297–308.

Shaw, D.D. (1974) Genetic and environmental components of chiasma control. III. Genetic analysis of chiasma frequency variation in two selected lines of *Schistocerca gregaria* Forsk. *Chromosoma*, **46**, 365–74.

Shaw, D.D., Wilkinson, P. and Coates, D.J. (1982) The chromosomal component of reproductive isolation in the grasshopper *Caledia captiva. Chromosoma*, **86**, 533–49.

Sheppard, P.M. and Ford, E.B. (1966) Natural selection and the evolution of dominance. *Heredity*, **21**, 139–47.

Shields, W.M. (1982) *Philopatry, Inbreeding and the Evolution of Sex*, State University of New York Press, Albany, NY.

Shields, W.M. (1988) Sex and adaptation, in *The Evolution of Sex: An Examination of Current Ideas* (eds R.E. Michod and B.R. Levin) Sinauer, Sunderland, MA, pp. 253–69.

Shimmin, L.C., Chang, B.H.-J., Li, W.-H. (1993) Male-driven evolution of DNA sequences. *Nature*, **362**, 745–47.

Shinohara, A., Ogawa, H., Ogawa, T. (1992) *Rad51* protein involved in repair and recombination in *S. cerevisiae* is a *RecA*-like protein. *Cell*, **69**, 457–70.

Shnoll, S.E. (1979) *Physico-Chemical Factors of Biological Evolution*, Nauka, Moscow.

Shrimpton, A.E. and Robertson, A. (1988a) The isolation of polygenic factors controlling bristle score in *Drosophila melanogaster*. I. Allocation of the third chromosome sternopleural bristle effects to chromosome sections. *Genetics*, **118**, 437–43.

Shrimpton, A.E. and Robertson, A. (1988b) The isolation of polygenic factors controlling bristle score in *Drosophila melanogaster*. II. Distribution of the third chromosome bristle effects within chromosome sections. *Genetics*, **118**, 445–59.

Shwarz, S.S. (1980) *Ecological Regularities of Evolution*, Nauka, Moscow.

Sih, A., Growley, P., McPeek, M. *et al.* (1985) Predation, competition, and prey communities: a review of field experiments. *Annu. Rev. Ecol. Syst.*, **16**, 269–311.

Silver, J. (1985) Confidence limits for estimates of gene linkage based on analysis of recombinant inbred strains. *J. Heredity*, **76**, 436–40.

Silver, L.M. (1993) The peculiar journey of a selfish chromosome: mouse *t* haplotypes and meiotic drive. *Trends Genet.*, **9**, 250–54.

Simchen, G. (1967) Genetic control of recombination in *Schizophyllum commune. Genet. Res.*, **9**, 195–210.

Simchen, G. and Stamberg, J. (1969) Fine and coarse controls of genetic recombination. *Nature*, **222**, 329–32.

Simchen, G., Kassir, Y., Horesh-Cabilly, O. and Friedmann, A. (1981) Elevated recombination and pairing structures during meiotic arrest in yeast of the nuclear division mutant *cdc*5. *Mol. Gen. Genet.*, **184**, 46–51.

Simms, E.L. and Rausher, M.D. (1987) Costs and benefits of plant resistance to herbivore. *Am. Nat.*, **130**, 570–81.

Simpson, P. (1993) Flipping fruit-flies: a powerful new technique for generating *Drosophila* mosaics. *Trends Genet.*, **9**, 227–28.

Sinclair, D.A. (1975) Crossing-over between closely linked markers spanning the centromere of chromosome 3 in *Drosophila melanogaster. Genet. Res.*, **11**, 173–85.

Singer, M. and Parmesan, C. (1993) Sources of variations in patterns of plant-insect associations. *Nature*, **361**, 251–53.

Singh, R.S. (1990) Patterns of species divergence and genetic theories of speciation, in *Population Biology*, Springer, Berlin, pp. 231–266.

Siracusa, L.D., Buchberg, A.M., Copeland, N.G. and Jenkins, N.A. (1989) Recombinant inbred strain and interspecific backcross analysis of molecular markers flanking the murine *agouti* coat color locus. *Genetics*, **122**, 669–79.

Siracusa, L.D., Silan, C.M., Justice, M.J. *et al.* (1990) A molecular genetic linkage map of mouse chromosome 2. *Genomics*, **6**, 491–504.

Slatkin, M. (1972) On treating the chromosome as the unit of selection. *Genetics*, **72**, 157–68.

Slatkin, M. (1975) Gene flow and selection in a two-locus system. *Genetics*, **81**, 787–802.

Slatkin, M. (1986) Interchromosomal biased gene conversion, mutation and selection in a multigene family. *Genetics*, **112**, 681–98.

Slatkin, M. and Lande, R. (1976) Niche width in a fluctuating environment – density independent selection. *Am. Nat.*, **110**, 31–55.

Slatko, B.E. (1978) Parameters of male and female recombination influenced by the *T-007* second chromosome in *Drosophila melanogaster. Genetics*, **90**, 257–76.

Slocum, M.K., Figdore, S.S., Kennard, W.C. *et al.* (1990) Linkage arrangement of restriction fragment length polymorphism loci in *Brassica oleracea. Theor. Appl. Genet.*, **80**, 57–67.

Smirnov, A.F. and Smaragdov, M.G. (1979) Characterization and molecular organization of *Drosophila* heterochromatin. *Genet. Studies* (Leningrad State University), **8**, 30–39.

Smirnov, V.G. (1989) Analysis of genetic basis of the parallelism in hereditary variation, in *Vavilov's Heritage in Modern Biology* (ed. V.K. Shumny), Nauka, Moscow, pp. 15–26.

Smirnov, V.G. (1991) *Cytogenetics*. Visshaya Shkola, Moscow, 247pp.

Smith, C.A.B. (1975) A non-parametric test for linkage with a quantitative character. *Ann. Hum. Genet.*, **38**, 451–60.

Smith, C.A.B. (1983) Heterogeneity in human linkage analysis, in *Contributions to Statistics: Essays in Honour of Norman L. Johnson* (ed. P.K. Sen), North-Holland, Amsterdam, pp. 379–90.

Smith, C.A.B. (1986) The development of human linkage analysis. *Ann. Hum. Genet.*, **50**, 293–311.

Smith, G.R. (1987) Mechanism and control of homologous recombination in *Escherichia coli. Annu. Rev. Genet.*, **21**, 179–201.

Smit-McBride, Z., Moya, A. and Ayala, F.J. (1988) Linkage disequilibrium in natural and experimental populations of *Drosophila melanogaster. Genetics*, **120**, 1043–51.

Smouse, P.E. (1988) Combining genetic information from two-point testcrosses and recombinant inbred lines. *J. Heredity*, **79**, 478–79.

Snape, J.W. (1988) The detection and estimation of linkage using doubled haploid or single seed descent populations. *Theor. Appl. Genet.* **76**, 125–28.

Snow, A.A. and Spira, T.P. (1991) Pollen vigour and the potential for sexual selection in plants. *Nature*, **352**, 796–97.

Sogaard, B. (1974). The localization of *eceriferum* loci in barley. III. Three point tests of genes on chromosome 1 in barley. *Hereditas*, **76**, 41–48.

Soller, M. (1990) Genetic mapping of the bovine genome using DNA-level markers with particular attention to loci affecting quantitative traits of economic importance. *J. Dairy Sci.*, **73**, 2628–46.

Soller, M. and Beckmann, J.S. (1983) Genetic polymorphism in varietal identification and genetic improvement. *Theor. Appl. Genet.*, **67**, 25–33.

Soller, M. and Beckmann, J.S. (1987) Cloning quantitative trait loci by insertional mutagenesis. *Theor. Appl. Genet.*, **74**, 369–78.

Soller, M. and Beckmann, J.S. (1988) Genomic genetics and the utilization for breeding purposes of genetic variation between populations, in *Quantitative Genetics* (eds B.S. Weir *et al.*), Sinauer, Sunderland, MA, pp. 161–188.

Soller, M. and Beckmann, J.S. (1990) Marker-based mapping of quantitative trait loci using replicated progeny. *Theor. Appl. Genet.*, **80**, 205–8.

Soller, M., Brody, T. and Genizi, A. (1976) On the power of experimental designs for the detection of linkage between marker loci and quantitative loci in crosses between inbred lines. *Theor. Appl. Genet.*, **47**, 35–39.

Soller, M., Brody, T. and Genizi, A. (1979) The expected distribution of marker-linked quantitative effects in crosses between inbred lines. *Heredity*, **43**, 179–90.

Soller, M. and Genizi, A. (1978) The efficiency of experimental designs for the detection of linkage between a marker locus and a locus affecting a quantitative trait in segregating populations. *Biometrics*, **34**, 47–55.

Soller, M. and Plotkin–Hazan, J. (1977) The use of marker alleles for the introgression of linked quantitative alleles. *Theor. Appl. Genet.*, **51**, 133–37.

Song, K.M., Susuki, J.Y., Slocum, M.K. *et al.* (1991) A linkage map of *Brassica rapa* (syn. *campestris*) based on restriction fragment length polymorphism loci. *Theor. Appl. Genet.*, **82**, 296–304.

Spickett, S.G. and Thoday, J.H. (1966) Regular responses to selection. 3. Interaction between located polygenes. *Genet. Res.*, **7**, 96–121.

Spurr, N.K., Gough, A., Goodfellow, P.J. *et al.* (1988) Evolutionary conservation of the human homologue of the yeast cell cycle gene *cdc2* and assignment of *Cd2* to chromosome 10. *Hum. Genet.*, **78**, 333–37.

Srivastava, H.K. (1980) Heterosis for chiasma frequency and quantitative traits in common beans (*Phaseolous vulgaris* L.). *Theor. Appl. Genet.*, **56**, 25–29.

Stack, S.M. and Soulliere, D.L. (1984) The relation between synapsis and chiasma formation in *Rhoeo spathacea. Chromosoma*, **90**, 72–83.

Stadler, D.R. and Towe, A.M. (1962) Genetic factors influencing crossing-over frequency in *Neurospora. Genetics*, **47**, 839–46.

Stahl, F.W., Chattoraj, D., Crasemann, J.M. *et al.* (1980) What accounts for the orientation dependence and directionality of *Chi*, in *Mechanistic Studies of DNA Replication and Genetic Recombination* (ed. B. Alberts), Academic Press, New York, pp. 919–26.

Stallings, R.L. and Siciliano, M.J. (1983) Evolutionary conservatism in the arrangement of autosomal gene loci in mammals. *Isozymes: Curr. Top. Biol. Med. Res.*, **10**, 313–21.

Stallings, R.L., Munk, A.C., Longmire, J.L. *et al.* (1985) Oncogenes and linkage groups: conservation during mammalian chromosome evolution. *Chromosoma*, **92**, 156–63.

Stam, P. (1979) Interference in genetic crossing-over and chromosome mapping. *Genetics*, **92**, 573–94.

Stanley, S.M. (1978) Chronospecies' longevities, the origin of genera and the punctuational model of evolution. *Paleobiology*, **4**, 26–40.

Stearns, S.C. (ed.) (1987a) *The Evolution of Sex and its Consequences*, Birkhauser, Basle.

Stearns, S.C. (1987b) Why sex evolved and the differences it makes, in *The Evolution of Sex and its Consequences* (ed. S.C. Stearns), Birkhauser, Basle, pp. 15–31.

Stebbins, G.L. (1945) The cytological analysis of species hybrids. *Bot. Rev.*, **11**, 463–86.

Stebbins, G.L. (1950) *Variation and Evolution in Plants*, Columbia University Press, New York.

Stebbins, G.L. (1957) Self fertilization and population variability in higher plants. *Am. Nat.*, **91**, 337–54.

Stebbins, G.L. (1982) Modal themes: a new framework for evolutionary syntheses, in *Perspectives on Evolution* (ed. R. Milkman), Sinauer, Sunderland, MA, pp. 1–14.

Stegniy, V.N. (1991) Systemic rearrangements of genome in speciation, in *Problems of Genetics and Theory of Evolution* (eds V.K. Shumny and A.O. Ruvinsky), Nauka, Novosibirsk, pp. 242–52.

Steinhauer, D.A. and Holland, J.J. (1987) Rapid evolution of RNA viruses. *Annu. Rev. Microbiol.*, **41**, 409–33.

Stenseth, N.C. and Maynard Smith, J. (1984) Coevolution in ecosystems: Red Queen evolution or stasis. *Evolution*, **38**, 870–80.

Stephan, W. (1986) Recombination and the evolution of satellite DNA. *Genet. Res.*, **47**, 167–74.

Stephan, W. (1987) Quantitative variation and chromosomal location of satellite DNAs. *Genet. Res.*, **50**, 41–52.

Stephens, S.G. (1950) The internal mechanism of speciation in *Gossypium*. *Bot. Rev.*, **16**, 115–49.

Stern, H. (1986) Meiosis: some considerations. *J. Cell Sci.*, **4**, Suppl., 29–43.

Strain, L., Curtis, A., Mennie, M. *et al.* (1988) Use of linkage disequilibrium data in prenatal diagnosis of cystic fibrosis. *Hum. Genet.*, **80**, 77–78.

Strauss, J.H. and Strauss, E.G. (1988) Evolution of RNA viruses. *Annu. Rev. Microbiol.*, **42**, 657–83.

Strobeck, C. (1976) The three locus model with multiplicative fitness values: the crystallization of the genome, in *Population Genetics and Ecology* (eds S. Karlin and E. Nevo), Academic Press, New York, pp. 781–90.

Strobeck, C., Maynard Smith, J. and Charlesworth, B. (1976) The effects of hitch-hiking on a gene for recombination. *Genetics*, **82**, 547–58.

Stuber, C.W., Edwards, M.D. and Wendell, J.E. (1987) Molecular marker facilitated investigations of quantitative trait loci in maize. II. Factors influencing yield and its component traits. *Crop. Sci.*, **27**, 639–48.

Stubblefield, E.A. (1986) A theory for developmental control by a program encoded in the genome. *J. Theor. Biol.*, **118**, 129–43.

Sturt, E. (1975) The use of lod scores for the determination of the order of loci on a chromosome. *Ann. Hum. Genet.*, **39**, 255–60.

Sturt, E. (1976) A mapping function for human chromosomes. *Ann. Hum. Genet.*, **40**, 147–63.

Sturtevant, A.H. (1921) A case of rearrangement of genes in *Drosophila*. *Proc. Natl Acad. Sci. USA*, **7**, 235–37.

Sturtevant, A.H. (1937) Essays in evolution. I. On the effects of selection on mutation rate. *Q. Rev. Biol.*, **12**, 464–67.

Sturtevant, A.H. and Mather, K. (1938) The interrelations of inversions, heterosis and recombinations. *Am. Nat.*, **72**, 447–52.

Sturtevant, A.H. and Novitski, E. (1941) The homologies of the chromosome elements in the genus Drosophila. *Genetics*, **26**, 517–41.

Subramani, S. and Rubnitz, J. (1985) Recombination events after transient infection and stable integration of DNA into cells. *Cell. Biol.*, **5**, 659–66.

Sun, S. and Rees, H. (1964) Genotypic control of chromosome behavior in rye. 7. Unadaptive heterozygotes. *Heredity*, **19**, 357–67.

Surzycki, S. and Paszewski, A. (1964) Non-random segregation of chromosomes in *Ascobolus immersus*. *Genet. Res.*, **5**, 20–26.

Sved, J.A. (1966) Telomere attachment of chromosomes. Some genetical consequences. *Genetics*, **53**, 747–56.

Sved, J.A. (1968) The stability of linked systems of loci with a small population size. *Genetics*, **59**, 543–63.

Sved, J.A. (1976) Hybrid dysgenesis in *Drosophila melanogaster*: A possible explanation in terms of spatial organization of chromosomes. *Aust. J. Biol. Sci.*, **29**, 375–88.

Sved, J.A., Blackman, L.M., Gilchrist, A.S. and Engels, W.R. (1991) High level of recombination induced by homologous *P* elements in *Drosophila melanogaster*. *Mol. Gen. Genet.*, **225**, 443–47.

Sybenga, J. (1975) *Meiotic Configurations*, Springer, Berlin.

Syvanen, M. (1985) Cross-species gene transfer: implications for a new theory of evolution. *J. Theor. Biol.*, **112**, 333–43.

Szathmary, E. and Kover, S. (1991) A theoretical test of the DNA repair hypothesis for the maintenance of sex in eukaryotes. *Genet. Res.*, **58**, 157–65.

Szauter, P. (1984) An analysis of regional constraints on exchange in *Drosophila melanogaster* using recombination-defective meiotic mutants. *Genetics*, **106**, 45–71.

Taccioli, G.E., Rathbun, G., Ottz, E., Stamato, T., Jeggo, P.A. and Alt, F.W. (1993) Impairment of V(D)J recombination in double-strand break repair mutants. *Science*, **260**, 207–10.

Takahata, N. (1993) Allelic genealogy and human evolution. *Molec. Biol. Evolution*, **10**, 2–22.

Tanksley, S.D. (1983) Molecular markers in plant breeding. *Plant Mol. Biol. Rep.*, **1**, 3–8.

Tanksley, S.D. (1987) Organization of the nuclear genome in tomato and related diploid species. *Am. Nat.*, **130** (Suppl.), 46–56.

Tanksley, S.D. and Hewitt, J. (1988) Use of molecular markers in breeding for soluble solids content in tomato – a re-examination. *Theor. Appl. Genet.*, **75**, 811–23.

Tanksley, S.D., Medina–Filho, H. and Rick, C.M. (1981) The effect of isozyme selection on metric characters in an interspecific backcross of tomato – basis of an early screening procedure. *Theor. Appl. Genet.*, **60**, 291–96.

Tanksley, S.D. and Rick, C.M. (1980) Isozyme linkage map of the tomato: applications in genetics and breeding. *Theor. Appl. Genet.*, **57**, 161–70.

Tanksley, S.D., Bernatzky, R., Lapitan, N.L. and Prince, J.P. (1988) Conservation of gene repertoire but not gene order in pepper and tomato. *Proc. Natl Acad. Sci. USA*, **85**, 6419–23.

Tanksley, S.D., Young, N.D., Paterson, A.H. and Bonierbale, M.W. (1989) RFLP mapping in plant breeding: new tools for an old science. *Biotechnology*, **7**, 257–64.

Tanksley, S.D., Ganal, M.W., Prince, J.P. *et al.* (1992) High density molecular linkage maps of the tomato and potato genomes. *Genetics*, **132**, 1141–60.

Tanzi, R.E., Watkins, P.C., Stewart, G.D. *et al.* (1992) A genetic linkage map of human chromosome 21: analysis of recombination as a function of sex and age. *Am. J. Hum. Genet.*, **50**, 551–58.

Tchaikovsky, Yu.V. (1976) The problem of inheritance and genetic search (description of the problem and simplest searching method), in *Theoretical and Experimental Biophysics*, Kaliningrad Press, Kaliningrad, pp. 147–165.

Tease, C. and Jones, G.H. (1976) Chromosome-specific control of chiasma formation in *Crepis capillaris. Chromosoma*, **57**, 33–49.

Templeton, A.R. (1979) The unit of selection and the origin of coadapted genomes in parthenogenetic strains. *Genetics*, **92**, 1265–82.

Templeton, A.R. (1981) Mechanisms of speciation – a population genetic approach, *Annu. Rev. Ecol. Syst.*, **12**, 23–48.

Templeton, A.R., Sing, C.F., Brokaw, B. (1976) The unit of selection in *Drosophila mercatorum*. 1. The interaction of selection and meiosis in parthenogenetic strains. *Genetics*, **82**, 349–76.

Terzi, M. (1974) *Genetics and the Animal Cell*, Wiley, London.

Thoday, J.M. (1961) Location of polygenes. *Nature*, **191**, 368–70.

Thoday, J.M. (1967) Genes in the study of continuous variation. *Ciencia e cultura*, **19**, 54–63.

Thoday, J.M. and Boam, T.B. (1956) A possible effect of the cytoplasm on recombination in *Drosophila melanogaster. J. Genet.*, **53**, 456–61.

Thompson, E.A. (1984) Information gain in joint linkage analysis. *IMA J. Math. Appl. Med. Biol.*, **1**, 31–49.

Thompson, J.N. (1975) Quantitative variation and gene number. *Nature*, **258**, 665–68.

Thompson, J.N. and Burdon, J.J. (1992) Gene-for-gene coevolution between plants and parasites. *Nature*, **360**, 121–25.

Thompson, J.N. and Thoday, J.M. (1979) Synthesis: polygenic variation in perspective, in *Quantitative Genetic Variation* (eds J.N. Thompson and J.M. Thoday), Academic Press, London, pp. 295–301.

Thompson, R.E. (1962) Asynapsis and mutability in *Drosophila melanogaster. Genetics*, **47**, 337–49.

Thompson, V. (1986) Half-chromosome viability and synthetic lethality in *Drosophila melanogaster. J. Heredity*, **77**, 385–88.

Thomson, G. (1977) The effect of a selected locus on neutral loci. *Genetics*, **85**, 753–88.

Thomson, G. (1988) HLA disease associations: models for insulin dependent diabetes mellitus and the study of complex human genetic disorders. *Annu. Rev. Genet.*, **22**, 31–50.

Thomson, G.J. and Feldman, M.W. (1974) Population genetics of modifiers of meiotic drive. II. Linkage modification in the segregation distortion system. *Theor. Pop. Biol.*, **5**, 155–62.

Thomson, G. and Klitz, W. (1987) Disequilibrium pattern analysis. I. Theory. *Genetics*, **116**, 623–32.

Thomson, R.L., Westerman, M. Murray, N.D. (1984). *B* chromosomes in *Rattus fuscipes*. I. Mitotic and meiotic chromosomes and the effects of *B* chromosomes on chiasma frequency. *Heredity*, **52**, 355–62.

Thuriaux, P. (1977) Is recombination confined to structural genes of the eukaryotic chromosome? *Nature*, **268**, 460–62.

Timofeeff-Ressovsky, N.V. (1984) Genetics, evolution and theoretical biology, in

Biology and Information (ed. V.D. Pekelis), Nauka, Moscow, pp. 18–24.

Timofeeff-Ressovsky, N.V., Yablokov, A.V. and Glotov, N.V. (1973) *An Essay on Population Theory*, Nauka, Moscow.

Titterington, D.M., Smith, A.F.M. and Makov, U.E. (1985) *Statistical Analysis of Finite Mixture Distributions*. John Wiley, Chichester.

Tiwari, J.L. and Terasaki, P.I. (1985) *HLA and Disease Associations*, Springer, New York.

Tomekpe, K. and Lumaret, R. (1991) Association between quantitative traits and allozyme heterozygosity in a tetrasomic species: *Dactylis geomerata*. *Evolution*, **45**, 359–70.

Tonegawa, S. (1983) Somatic generation of antibody diversity. *Nature*, **302**, 575–581.

Tonegawa, S., Maxam, A.M., Tizard, R. *et al.* (1978) Sequence of a mouse germline gene for a variable region of an immunoglobulin light chain. *Proc. Natl Acad. Sci. USA*, **75**, 1485–89.

Tooby, J. (1982) Pathogens, polymorphisms, and evolution of sex. *J. Theor. Biol.*, **97**, 557–76.

Torkamanzehi, A., Moran, C. and Nicholas, F.W. (1992) *P*-element transposition contributes substantial new variation for a quantitative trait in *Drosophila melanogaster*. *Genetics*, **131**, 73–78.

Treger, J.M., Heichman, K.A. and McEntee, K. (1988) Expression of the yeast *UB14* gene increases in response to DNA-damaging agents and in meiosis. *Mol. Cell. Biol.*, **8**, 1123–36.

Trankle, E. (1975) Digital computer analysis of electrophoresis and ultracentrifugation patterns by a sum of gaussian-like distributions. *Z. Naturforsch.*, **30**, 311–17.

Treco, D. and Arnheim, N. (1986) The evolutionary conserved repetitive sequence $d(TG \cdot AC)_n$ promotes reciprocal exchange and generates unusual recombinant tetrads during yeast meiosis. *Mol. Cell. Biol.*, **6**, 3934–47.

Trifonov, E.N. (1990) Making sense of the human genome, in *Human Genome Initiative and DNA Recombination*, Vol. 1, *Structure and Methods* (eds R.H. Sharma and M.H. Sharma), Adenine Press, New York, pp. 69–77.

Trivers, R. (1988) Sex differences in rates of recombination and sexual selection, in *The Evolution of Sex: An Examination of Current Ideas* (eds R.E. Michod and B.R. Levin), Sinauer, Sunderland, MA, pp. 270–86.

Tsapygina, R.I., Dayev, E.V. and Nikonov, S.N. (1981) The action of exogenous metabolites of house mouse males on the cell division process in the generative tissue of young animals upon single and multiple exposures. *Genet. Studies* (Leningrad University Press), **9**, 17–23.

Tucic, N., Ayala, F.J. and Marinkovic, D. (1981) Correlation between recombination frequency and fitness in *Drosophila melanogaster*. *Genetica*, **56**, 61–69.

Turelli, M. (1984) Heritable genetic variation via mutation–selection balance: Lerch's zeta meets the abdominal bristle. *Theor. Pop. Biol.*, **25**, 138–93.

Turelli, M. (1988) Phenotypic evolution, constant covariances and the maintenance of additive variance. *Evolution*, **42**, 1342–47.

Turelli, M. and Barton, N.H. (1990) Dynamics of polygenic characters under selection. *Theor. Pop. Biol.*, **38**, 1–57.

Turner, J.R.G. (1967) Why does the genotype not congeal? *Evolution*, **21**, 645–56.

Turner, J.R.G. (1979) Genetic control of recombination in the silkworm. I. Multigenic control of chromosome 2. *Heredity*, **43**, 273–93.

Twomey, M.J. and Feldman, M.W. (1990) Mutation-modification with multiplicative fertility selection. *Theor. Pop. Biol.*, **37**, 320–42.

Ukai, Y. (1987) Fixation of genes and breakup of linkage blocks in selfed generations. *Jpn. J. Breed.*, **37**, 40–53.

Ukai, Y. (1988) Fitting of a binomial distribution to the number of chiasmata per bivalent. *Jpn. J. Genet.*, **63**, 259–65.

Ushakov, B.P. (1978) Statistical treatment of experimental data and their interpretation in terms of population biology. *J. Gen. Biol.* (USSR), **39**, 602–11.

Uyenoyama, M.K. (1986) Inbreeding and the 'cost of meiosis': the evolution of selfing in populations practising biparental inbreeding. *Evolutiŏn*, **40**, 388–404.

Uyenoyama, M.K. (1988) On the evolution of genetic incompatibility systems. II. Initial increase of strong gametophytic self-incompatibility under partial selfing and half-sib mating. *Am. Natur.*, **131**, 700–22.

Uyenoyama, M.K. and Bengtsson, B.O. (1989) On the origin of meiotic reproduction: a genetic modifier model. *Genetics*, **123**, 873–85.

Valentin, J. (1973) Heritability of recombination frequency: observations and report on small-scale tests in *Drosophila*. *Hereditas*, **75**, 1–4.

Van der Plank, J.E. (1978) *Genetic and Molecular Basis of Plant Pathogenesis*, Springer, Berlin.

van Noordwijk, A.J. (1990) The methods of genetical ecology applied to the study of evolutionary change, in *Population Biology* (eds K. Wohrmann and S.K. Jain), Springer, Berlin, pp. 291–319.

Vasiljeva, L.A., Zabanov, S.A., Ratner, V.A. *et al.* (1987) Expression of the *radius incompletus* quantitative trait, temperature effects and localization of mobile elements in *Drosophila*. II. Mobile genetic element *Dm-412*. *Genetika* (USSR), **23**, 81–92.

Verdonck, M.V. (1987) Adaptation to environmental heterogeneity in populations of *Drosophila melanogaster*. *Genet. Res.*, **49**, 1–10.

Vereschagina, N.M. and Korol, A.B. (1991) The effect of hybrid dysgenesis factors of the *T-007* chromosome on the recombination system of *Drosophila melanogaster*, in *Intrapopulation Variability and Direct Selection for Recombination* (12th European Drosophila Research Conference, Late Abstract Book), Mainz.

Via, S. and Lande, R. (1987) Evolution of genetic variability in spatially heterogeneous environment: effects of genotype–environment interaction. *Genet. Res.*, **49**, 147–56.

Vizir, I.Yu. and Korol, A.B. (1990) Sex difference in recombination frequency in *Arabidopsis*. *Heredity*, **65**, 379–83.

Voelkel-Meiman, K., Keil, R.L. and Roeder, G.S. (1987) Recombination stimulating sequences in yeast ribosomal DNA correspond to sequences regulating transcription by RNA polymerase I. *Cell*, **48**, 1071–79.

Voelkel-Meiman, K. and Roeder, G.S. (1991) Gene conversion tracts stimulated by *HOT1*-promoted transcription are long and continuous. *Genetics*, **126**, 851–67.

Vogt, P. (1990) Potential genetic functions of tandem repeated DNA sequence blocks in the human genome are based on highly conserved 'chromatin-folding-code'. *Hum. Genet.*, **84**, 301–36.

Vol, I.A. and Koval, G.M. (1985) Study of mechanisms of maintenance of variability in a population under stabilizing selection. *Genetika* (USSR), **21**, 265–73.

von Wettstein, D., Rasmussen, S.W. and Holm, P.B. (1984) The synaptonemal complex in genetic segregation. *Ann. Rev. Genet.*, **18**, 331–413.

Vrijenhoek, R.C. (1990) Genetic diveristy and the ecology of asexual populations, in *Population Biology. Ecological and Evolutionary Viewpoints* (eds K. Wohrmann and S.K. Jain), Springer, Berlin, pp. 175–97.

Vysotskaya, L.V., Bugrov, A.G. and Stebaev, I.V. (1983) Chiasma frequency as a cytogenetic criterion of evolutionary relationships in the family *Acrididae. J. General Biol.* (USSR), **44**, 480–90.

Wagner, G.P. (1981) Feedback selection and the evolution of modifiers. *Acta Biotheor.*, **30**, 99–102.

Wagner, G.P. (1989) Multivariate mutation–selection balance with constrained pleiotropic effects. *Genetics*, **122**, 223–34.

Wagner, G.P. and Burger, R. (1985) On the evolution of dominance modifiers. II. A nonequilibrium approach to the evolution of genetic systems. *J. Theor. Biol.*, **113**, 475–500.

Wagner, G.P. and Gabriel, W. (1990) Quantitative variation in finite parthenogenetic populations: what stops Muller's ratchet in the absence of recombination? *Evolution*, **44**, 715–31.

Walbot, V. and Cullis, C.A. (1985) Rapid genomic change in higher plants. *Annu. Rev. Plant Physiol.*, **36**, 367–96.

Walker, I. (1978) The evolution of sexual reproduction as a repair mechanism. I. A model for self-repair and its biological implications. *Acta Biotheor.*, **27**, 133–58.

Wallace, B., King, J.C., Madden, C.V. *et al.* (1953) An analysis of variability arising through recombination. *Genetics*, **38**, 272–307.

Wallace, H. (1974) Chiasmata have no effect on fertility. *Heredity*, **33**, 423–29.

Wallace, M.E. (1953) Affinity: a new genetic phenomenon in the house mouse. Evidence from within laboratory stocks. *Nature*, **171**, 27–28.

Wallace, M.E. (1957) The use of affinity in chromosome mapping. *Biometrics*, **13**, 98–110.

Wallace, M.E. (1958) New linkage and independence data for *ruby* and *jerker* in the mouse. *Heredity*, **12**, 453–62.

Wallace, M.E. (1959) An experimental test of the hypothesis of affinity. *Genetica*, **29**, 243–55.

Wallace, M.E. (1960a) Possible cases of affinity in cotton. *Heredity*, **14**, 263–74.

Wallace, M.E. (1960b) Possible cases of affinity in tomatoes. *Heredity*, **14**, 275–83.

Wallis, J.W., Chrebet, G., Brodsky, G. *et al.* (1989) A hyperrecombination mutation in *S. cerevisiae* identifies a novel eukaryotic topoisomerase. *Cell*, **58**, 409–19.

Walsh, J.B. (1987) Sequence-dependent gene conversion: can duplicated genes diverge fast enough to escape conversion? *Genetics*, **117**, 543–57.

Wang, S.-S. and Zakian, V.A. (1990) Telomere–telomere recombination provides an express pathway for telomere acquisition. *Nature*, **345**, 456–60.

Wang, Z.Y., Second, G. and Tanksley, S.D. (1992) Polymorphism and phylogenetic relationship among species in the genus *Oryza* as determined by analysis of nuclear RFLP. *Theor. Appl. Genet.*, **83**, 565–81.

Weaver, R., Helms, C., Mishra, S.K. and Donis-Keller, H. (1992) Software for analysis and manipulation of genetic linkage data. *Am. J. Hum. Genet.*, **50**, 1267–74.

Weber, D. and Helentjaris, T. (1989) Mapping RFLP loci in maize using *B-A* translocations. *Genetics.*, **121**, 583–90.

Weber, C., Salazar, E., Stewart, S. and Thompson, L. (1990) *ERCC2*: cDNA cloning and molecular characterization of a human nucleotide excision repair gene with high homology to yeast *RAD3*. *EMBO J.*, **9**, 1437–47.

Weeden, N.F. and Muehlbauer, F.J. (1992) Extensive conservation of linkage relationship between pea and lentil genetic maps. *J. Heredity*, **83**, 123–29.

Wehrhahn, C.F. and Allard, R.W. (1965) The detection and measurement of the effects of individual genes involved in the inheritance of a quantitative character in wheat. *Genetics*, **51**, 109–19.

Weining, S. and Langridge, P. (1991) Identification and mapping of polymorphisms in cereals based on the polymerase chain reaction. *Theor. Appl. Genet.*, **82**, 209–16.

Weinshall, D. and Eshel, I. (1987) On the evolution of an optimal rate of sexual reproduction. *Am. Nat.*, **130**, 758–74.

Weir, B.S. and Cockerham, C.C. (1973) Mixed self and random mating at two loci. *Genet. Res.*, **21**, 247–62.

Weissenbach, J., Gyapay, G., Dib, C. *et al.* (1992) A second-generation linkage map of the human genome. *Nature*, **359**, 794–801.

Weller, J.I. (1986) Maximum likelihood techniques for the mapping and analysis of quantitative trait loci with the aid of genetic markers. *Biometrics*, **42**, 627–40.

Weller, J.I. (1987) Mapping and analysis of quantitative trait loci in *Lycopersicon* (tomato) with the aid of genetic markers using approximate maximum likelihood methods. *Heredity*, **59**, 413–21.

Weller, J.I., Kashi, Y. and Soller, M. (1990) Estimation of sample size necessary for genetic mapping of quantitative traits in dairy cattle using genetic markers. *J. Dairy Sci.*, **73**, 2525–37.

Weller, J.I. and Wyler, A. (1992) Power of different sampling strategies to detect quantitative trait loci variance effect. *Theor. Appl. Genet.*, **83**, 582–88.

Welsh, J., Honeycutt, R.J., McClelland, M. and Sobral, B.W.S. (1991) Parentage determination in maize hybrids using the arbitrary primed polymerase chain reaction (AP-PCR). *Theor. Appl. Genet.*, **82**, 473–76.

Welsh, J. and McClelland, M. (1990) Fingerprinting genomes using PCR with arbitrary primers. *Nucleic Acids Res.*, **18**, 7213–18.

White, J.H., DiMartino, J.F., Anderson, R.W. *et al.* (1988) A DNA sequence conferring high postmeiotic segregation frequency to heterozygous deletions in *Saccharomyces cerevisiae* is related to sequences associated with eucaryotic recombination hotspots. *Mol. Cell. Biol.*, **8**, 1253–58.

White, M.J.D. (1954) An extreme form of chiasma localization in a species of *Bryodema (Orthoptera, Acrididae)*. *Evolution*, **8**, 350–58.

White, R. and Lalouel, J.M. (1988) Sets of linked genetic markers for human chromosomes. *Annu. Rev. Genet.*, **22**, 259–79.

Whitehouse, C., Edgar, L.A., Jones, G.H. and Parker, J.S. (1981) The population cytogenetics of *Crepis capillaris* I. Chiasma variation. *Heredity*, **47**, 95–103.

Whitehouse, H.L.K. (1966) An operator model of crossing over. *Nature*, **211**, 708–13.

Whitehouse, H.L.K. (1982) *Genetic Recombination. Understanding the Mechanisms*, John Wiley, New York.

Wiener, P. and Feldman, M.W. (1991) The evolution of dispersal in a model of mixed selfing and random mating. *Evolution*, **45**, 1717–26.

Williams, D.A. (1975) The analysis of binary responses from toxicological experiments involving reproduction and teratogenicity. *Biometrics*, **31**, 949–52.

Williams, G.C. (1975) Sex and Evolution, Princeton University Press, Princeton.

Williams, G.C. (1988) Retrospect on sex and kindred topics, in *The Evolution of Sex: An Examination of Current Ideas* (eds R.E. Michod and B.R. Levin), Sinauer, Sunderland, MA, pp. 287–98.

Williams, M.N.V., Pande, N., Nair, S. *et al.* (1991) Restriction fragment length polymorphism analysis of polymerase chain reaction products amplified from mapped loci of rice (*Oryza sativa L.*) genomic DNA. *Theor. Appl. Genet.*, **82**, 489–98.

Williams, S.M., Robbins, L.G., Cluster, P.D. *et al.* (1990) Superstructure of the *Drosophila* ribosomal gene family. *Proc. Natl Acad. Sci. USA*, **87**, 3156–60.

Williams, W.R., Morton, N.E., Lew, R. and Yee, S. (1979) The likely region of overlap (LRO) method for physical assignment of loci. *Hum. Genet.*, **47**, 297–304.

Wills, C. (1984) The possibility of stress-triggered evolution, in *Evolutionary Dynamics of Genetic Diversity* (ed. G.S. Mani). *Lect. Notes Biomath.*, **53**, 299–312.

Willson, M.E., Thomas, P.A., Hoppes, W.G. *et al.* (1987) Sibling competition in plants: an experimental study. *Am. Natur.*, **129**, 304–11.

Wohrmann, K. and Jain, S.K. (eds) (1990) *Population Biology. Ecological and Evolutionary Viewpoints*, Springer, Berlin.

Wolf, H.G., Wohrmann, K. and Tomiuk, J. (1987) Experimental evidence for the adaptive value of sexual reproduction. *Genetica*, **72**, 151–59.

Wright, S. (1965) Factor interaction and linkage in evolution. *Proc. R. Soc. London B*, **162**, 80–104.

Wright, S. (1968) *Evolution and the Genetics of Populations. Vol. 1. Genetic and Biometric Foundations*, University of Chicago Press, Chicago.

Yablokov, A.V. (1987) *Population Biology*, Vysshaya Shkola, Moscow.

Yamaizumi, M., Inaoka, T., Uchida, T., Ohtsuka, E. (1988) Microinjection of *T4* endonuclease *V* produced by a synthetic *denV* gene stimulates unscheduled DNA synthesis in both *xeroderma pigmentosum* and normal cells. *Mutat. Res. DNA Rep.*, **317**, 135–40.

Yamamoto, M. (1979a) Cytological studies of heterochromatin function in the *Drosophila melanogaster* male: autosomal meiotic pairing. *Chromosoma*, **72**, 293–328.

Yamamoto, M. (1979b) Interchromosomal effects of heterochromatic deletions on recombination in *Drosophila melanogaster*. *Genetics*, **93**, 437–48.

Yamamoto, M. and Miklos, G.L.G. (1978) Genetic studies on heterochromatin in *Drosophila melanogaster* and their implications for the functions of satellite DNA. *Chromosama*, **66**, 71–98.

Yamamoto, T. (1961) Progenies of sex-reversal females mated with sex-reversal males in the Medeka, *Oryzias latipes*. *J. Exp. Zool.*, **146**, 163–79.

Yeom, Y.I., Kuniya, A. and Artzt, K. (1992) Evolution of the mouse *H-2K* region: a hot spot of mutation associated with genes transcribed in embryos and/or germ cells. *Genetics*, **130**, 629–38.

Young, J.P.W. (1981) Sib competition can favor sex in two ways. *J. Theor. Biol.*, **88**, 755–56.

Young, N.D. and Tanksley, S.D. (1989) Restriction fragment length polymorphism maps and the concept of graphical genotypes. *Theor. Appl. Genet.*, **77**, 95–101.

Young, N.D., Zamir, D., Ganal, M. and Tanksley, S.D. (1988) Use of isogenic lines and simultaneous probing to identify DNA markers tightly linked to the *Tm-2a* gene in tomato. *Genetics*, **120**, 579–85.

Yurchenko, N.N., Zakharov, I.K. and Golubovsky, M.D. (1985) Recombination mediated by a mobile element in *Drosophila. Genetika* (USSR). **21**, 1125–34.

Zakharov, I.A. (1986) Some regularities of gene ordering in eukaryotic chromosomes. *Genetika* (USSR), **22**, 2620–24.

Zakharov, I.A. and Valejev, A.K. (1988) Quantitative analysis of the mammalian genome evolution based on genetic map comparisons. *Dokl. Akad. Nauk SSSR*, **301**, 1213–18.

Zamir, D. and Tanksley, S.D. (1988) Tomato genome is comprised largely of fast evolving low-copy-number sequences. *Mol. Gen. Genet.*, **213**, 254–61.

Zavadsky, K.M. and Kolchinsky, E.I. (1977) *The Evolution of Evolution*, Nauka, Leningrad.

Zelikman, A.L. (1967) Natural selection – the leading force of organic evolution in *Modern Problems of Evolution* (eds V.I. Polyansky and Yu.I. Polyansky), Nauka, Leningrad, pp. 98–114.

Zetka, M.C., Rose, A.M. and Baillie, D.L. (1987) A test of the Muller Ratchet Hypothesis using the *REC-1* strain of *Caenorhabditis elegans. Genetics*, **116**, (Suppl.), p. 30.

Zhang, Q., Maroof, M.A.S., Lu, T.Y. and Shen, B.Z. (1992) Genetic diversity and differentiation of *indica* and *japonica* rice detected by RFLP analysis. *Theor. Appl. Genet.*, **83**, 495–99.

Zhang, W. and Smith C. (1992) Computer simulation of marker-assisted selection utilizing linkage disequilibrium. *Theor. Appl. Genet.*, **83**, 813–30.

Zhivotovsky, L.A. (1984) *Integration of Polygenic Systems in Populations*, Nauka Press, Moscow.

Zhivotovsky, L.A. and Feldman, M.W. (1992) On models of quantitative genetic variability: a stabilizing selection–balance model. *Genetics*, **130**, 947–55.

Zhivotovsky, L.A. and Gavrilets, S. (1990) Stabilizing selection and linkage disequilibrium (English translation). *Sov. Genet.*, **26**, 222–31.

Zhuchenko, A.A. (1973) *Tomato Genetics*, Shtiintsa, Kishinev.

Zhuchenko, A.A. (1980) *Ecological Genetics of Cultivated Plants*, Shtiintsa, Kishinev.

Zhuchenko, A.A. (1988) *Adaptive Potential of Cultivated Plants*, Shtiintsa, Kishinev.

Zhuchenko, A.A. and Korol, A.B. (1982) The problem of recombination in ecological genetics. *Izv. Akad. Nauk SSSR*, **2**, 165–79.

Zhuchenko, A.A. and Korol, A.B. (1983) Ecological aspects of the recombination problem. *Theor. Appl. Genet.*, **64**, 177–85.

Zhuchenko, A.A. and Korol, A.B. (1985) *Recombination in Evolution and Breeding*, Nauka, Moscow.

Zhuchenko, A.A. and Korol, A.B. (1987) Possible ways of restructuring the recombination system in higher plants, in *Ecological Genetics of Plants and Animals* (Conference Proceedings) (ed. A.A. Zhuchenko), Shtiintsa, Kishinev, pp. 63–65.

Zhuchenko, A.A., Korol, A.B. and Andryuschenko, V.K. (1978) Linkage between loci of quantitative traits and marker loci. 1. The model. *Genetika* (USSR), **14**, 771–78.

Zhuchenko, A.A., Korol, A.B. and Gorodetsky, V.P. (1988) Experimental modelling of recombination microevolution. *Dokl. Akad. Nauk SSSR*, **308**, 452–56.

Zhuchenko, A.A., Korol, A.B. and Kovtyukh, L.P. (1985) Change of crossing over frequency in *Drosophila* during selection for resistance to temperature fluctuations. *Genetica*, **67**, 73–78.

Zhuchenko, A.A., Preygel, S.I. and Korol, A.B. (1986) The evolution of dominance of *rec* loci: a mathematical model. *Genetika* (USSR), **22**, 609–15.

Zhuchenko, A.A., Vereschagina, N.M. and Korol, A.B. (1988) Effect of the male recombination factor on the level of intrapopulation variability for recombination frequency in *Drosophila*. *Dokl. Akad. Nauk SSSR*, **298**, 477–79.

Zhuchenko, A.A., Andryuschenko, V.K., Korol, A.B. and Korochkina, S.K. (1975) Mapping genetic factors of quantitative characters in tomato. *Cytol. Genet.* (USSR), **9**, 101–5.

Zhuchenko, A.A., Andryuschenko, V.K., Nyutin, Y.I. Korol, A.B. and Vyrodov, D.A. (1977) Genetic effects of mutagenic treatment of tomato hybrids. III. Induced changes in recombination frequency between unlinked markers. *Genetika* (USSR), **13**, 1922–32.

Zhuchenko, A.A., Samovol, A.P., Korol, A.B. and Andryuschenko, V.K. (1979) Linkage between quantitative trait loci and marker loci. 2. Effects of three tomato chromosomes on the variation for five quantitative characters in backcross generations. *Genetika* (USSR), **15**, 672–83.

Zhuchenko, A.A., Korol, A.B., Gavrilenko, T.A. and Kibenko, T.Y. (1986) Relationship between the stability of the genotype and change in its recombination characteristics upon temperature treatments. *Genetika* (USSR), **22**, 966–74.

Zhuchenko, A.A., Korol, A.B., Kovtyukh, L.P. and Tabak, E.P. (1983) Increased crossing over frequency in a population undergoing selection for resistance to temperature fluctuations. *Dokl. Akad. Nauk SSSR*. **273**, 721–25.

Zhuchenko, A.A., Korol, A.B., Preygel, I.A. and Bronstein, S.I. (1985) The evolutionary role of the dependence of recombination on environment. *Theor. Appl. Genet.*, **69**, 617–24.

Zhuchenko, A.A., Preygel. I.A., Korol, A.B. and Kazantsev, E.F. (1982) On identifying 'rare' recombinants in segregating populations. *Tsytol. Genet.* (USSR), **16**, 41–44.

Zickler, D., Moreau, P., Huynh, A.D. and Slezec, A.-M. (1992) Correlation between pairing intitation sites, recombination nodules and meiotic recombination in *Sordaria macrospora*. *Genetics*, **132**, 135–48.

Zimmer, A. and Gruss, P. (1989) Production of chimaeric mice containing embryonic stem (ES) cells carrying a homoeobox *Hox 1.1* allele mutated by homologous recombination. *Nature*, **338**, 150–53.

Zivy, M., Devaux, P., Blaisonneau, J. *et al.* (1992) Segregation distortion and linkage studies in microspore-derived double haploid lines of *Hordeum vulgare L. Theor. Appl. Genet.*, **83**, 919–24.

Zujtin, A.I. (1938) The effect of changes in temperature regime on the frequency of lethal mutations in *Drosophila melanogaster*. *Dokl. Akad. Nauk SSSR*, **21**, 54–55.

Zujtin, A.I. (1941) The effect of replacing natural development conditions by laboratory ones on mutational variability in *Drosophila melanogaster*. *Dokl. Akad. Nauk SSSR*, **30**, 61–63.

Zuk, M. (1988) Parasite load, body size, and age of wild-caught male field crickets (*Orthoptera: Gryllidae*) effects on sexual selection. *Evolution*, **42**, 969–76.

Index